Abstract Algebra
with a Concrete Introduction

Abstract Algebra
with a Concrete Introduction

JOHN A. BEACHY

WILLIAM D. BLAIR

Northern Illinois University

Prentice Hall, Englewood Cliffs, New Jersey 07632

Library of Congress Cataloging-in-Publication Data

Beachy, John A.
  Abstract algebra with a concrete introduction / John A. Beachy,
William D. Blair.
    p. cm.
  Includes bibliographical references.
  ISBN 0-13-004425-3
  1. Algebra, Abstract. I. Blair, William D. II. Title.
QA162.B4 1990
512′.02—dc20

Editorial/production supervision and
  interior design: Maria McColligan
Cover design: Edsal Enterprises
Manufacturing buyer: Paula Massenaro

Printed in the United States of America
10 9 8 7 6 5 4 3 2 1

ISBN 0-13-004425-3

Prentice-Hall International (UK) Limited, *London*
Prentice-Hall of Australia Pty. Limited, *Sydney*
Prentice-Hall Canada Inc., *Toronto*
Prentice-Hall Hispanoamericana, S.A., *Mexico*
Prentice-Hall of India Private Limited, *New Delhi*
Prentice-Hall of Japan, Inc., *Tokyo*
Simon & Schuster Asia Pte. Ltd., *Singapore*
Editora Prentice-Hall do Brasil, Ltda., *Rio de Janeiro*

Contents

Preface

An abstract algebra course at the junior/senior level, whether for one or two semesters, has been a well-established part of the curriculum for mathematics majors for over a generation. Our book is intended for this course, and has grown directly out of our experience in teaching the course at Northern Illinois University.

As a prerequisite to the abstract algebra course, our students are required to have taken a sophomore level course in linear algebra that is largely computational, although they have been introduced to proofs to some extent. Our classes include students preparing to teach high school, but almost no computer science or engineering students. We certainly do not assume that all of our students will go on to graduate school in pure mathematics.

In searching for appropriate text books, we have found several texts that start at about the same level as we do, but most of these stay at that level, and they do not teach nearly as much mathematics as we desire. On the other hand, there are several fine books that start and finish at the level of our Chapter 6, but these books tend to begin immediately with the abstract notion of group (or ring), and then leave the average student at the starting gate. We have in the past used such books, supplemented by our Chapter 1.

Historically the subject of abstract algebra arose from concrete problems, and it is our feeling that by beginning with such concrete problems we will be able to generate the student's interest in the subject and at the same time build on the foundation with which the student feels comfortable.

Although the book starts in a very concrete fashion, we increase the level of sophistication as the book progresses, and, by the end of Chapter 6, all of the topics taught in our course have been covered. It is our conviction that the level of sophistication should increase, slowly at first, as the students become familiar with the subject. We think our ordering of the topics speaks directly to this assertion.

Recently there has been a tendency to yield to demands of "relevancy," and to include "applications" in this course. It is our feeling that such inclusions often tend to be superficial. In order to make room for the inclusion of applications, some important mathematical concepts have to be sacrificed. It is clear that one must have substantial experience with abstract algebra before any genuine applications can be treated. For this reason we feel that the most honest introduction concentrates on the algebra. One of the reasons frequently given for treating applications is that they motivate the student. We prefer to motivate the subject with concrete problems from areas that the students have previously encountered, namely, the integers and polynomials over the real numbers.

One problem with most treatments of abstract algebra, whether they begin with group theory or ring theory, is that the students simultaneously encounter for the first time both abstract mathematics and the requirement that they produce proofs of their own devising. By taking a more concrete approach than is usual, we hope to separate these two initiations.

In three of the first four chapters of our book we discuss familiar concrete mathematics: number theory, functions and permutations, and polynomials. Although the objects of study are concrete, and most are familiar, we cover quite a few nontrivial ideas and at the same time introduce the student to the subtle ideas of mathematical proof. (At Northern Illinois University, this course and Advanced Calculus are the traditional places for students to learn how to write proofs.) After studying Chapters 1 and 2, the students have at their disposal some of the most important examples of groups—permutation groups, the group of integers modulo n, and certain matrix groups. In Chapter 3 the abstract definition of a group is introduced, and the students encounter the notion of a group armed with a variety of concrete examples.

Probably the most difficult notion in elementary group theory is that of a factor group. Again this is a case where the difficulty arises because there are, in fact, two new ideas encountered together. We have tried to separate these by treating the notions of equivalence relation and partition in Chapter 2 in the context of sets and functions. We consider there the concept of factoring a function into "simpler" functions, and show how the notion of a partition arises in this context. These ideas are related to the integers modulo n, studied in Chapter 1. When factor groups are introduced in Chapter 3, we have partitions and equivalence relations at our disposal, and we are able to concentrate on the group structure introduced on the equivalence classes.

In Chapter 4 we return to a more concrete subject when we derive some important properties of polynomials. Here we draw heavily on the students' familiarity with polynomials from high school algebra and on the parallel between the properties of the integers studied in Chapter 1 and the polynomials. Chapter 5 then introduces the abstract definition of a ring after we have already encountered several important examples of rings: the integers, the integers modulo n, and the ring of polynomials with coefficients in any field.

From this point on our book looks more like a traditional abstract algebra textbook. After rings we consider fields, and we include a discussion of root adjunction as well as the three problems from antiquity: squaring the circle, duplicating the cube, and trisecting an angle. We also discuss splitting fields and finite fields here. We feel that the first six chapters represent the most that students at institutions such as ours can reasonably absorb in a year.

Chapter 7 returns to group theory to consider several more sophisticated ideas including those needed for Galois Theory, which is the subject matter of Chapter 8. These last two chapters will make the book suitable for an honors course or for classes of especially talented or well-prepared students. In these chapters the writing style is rather terse and demanding. Proofs are included for the Sylow theorems, the structure of finite abelian groups, the fundamental theorem of Galois theory, and the insolvability of the general quintic.

The only prerequisite for our text is a sophomore level course in linear algebra. We do not assume that the student has been required to write, or even read, proofs before taking our course. We do use examples from matrix algebra in our discussion of group theory, and we draw on the computational techniques learned in the linear algebra course—see, for example, our treatment of the Euclidean algorithm in Chapter 1.

We have included a number of appendices to which the student may be referred for background material. The appendices on induction and on the complex numbers might be appropriate to cover in class, and so they include some exercises.

In our classes we usually intend to cover Chapters 1, 2, and 3 in the first semester, and most of Chapters 4, 5, and 6 in the second semester. In practice, we usually begin the second semester with group homomorphisms and factor groups, and end with geometric constructions. We have rarely had time to cover splitting fields and finite fields. For students with better preparation, Chapters 1 and 2 could be covered more quickly. The development is arranged so that Chapter 7 on the structure of groups can be covered immediately after Chapter 3. On the other hand, the material from Chapter 7 is not really needed until section 8.4, at which point we need results on solvable groups.

ACKNOWLEDGMENTS

To list all of the many sources from which we have learned is almost impossible. Perhaps because we are ring theorists ourselves, we have been attracted to and influenced by the work of two ring theorists—I. N. Herstein in *Topics in Algebra* and N. Jacobson in *Basic Algebra I, II*. In most cases our conventions, notation, and symbols are consistent with those used by Jacobson. We certainly need to mention the legacy of E. Noether, which we have met via the classic text *Algebra* by B. L. van der Waerden. In many ways our approach to abstract concepts via concrete examples is similar in flavor to that of Birkhoff and MacLane in *A Survey of Modern Algebra*, although we have chosen to take a naive approach to the development of the number systems and have omitted any discussion of ordered fields. We have also been influenced by the historical approaches and choice of material in *Abstract Algebra: A First Course* by L. Goldstein and *Introduction to Abstract Algebra* by L. Shapiro.

A number of colleagues have taught from preliminary versions of parts of this book. We would like to thank several of them for their comments: Harvey Blau, Tac Kambayashi, Henry Leonard, John Lindsey, Martin Lorenz, Robert McFadden, Gunnar Sigurdsson, and Doug Weakley. Numerous students have offered comments and pointed out errors. We are particularly indebted to Penny Fuller, Susan Janes, and Michelle Mace for giving us lists of misprints. We would like to thank all of the reviewers of our manuscript including: Victor Camillo, The University of Iowa; John C. Higgins, Brigham Young University; I. Martin Issacs, University of Wisconsin; Paul G. Kumpel, State University of New York; and Mark L. Teply, The University of Wisconsin. We would also like to thank our editors at Prentice Hall: Bob Sickles, Priscilla McGeehon, Maria McColligan, and Dave Ostrow, together with the production staff.

This seems to be an appropriate place to record our thanks to Goro Azumaya and Lance Small (respectively) for their inspiration, influence, and contributions to our mathematical development. Finally, we would like to thank our families: Marcia, Gwen, Elizabeth, and Hannah Beachy and Kathy, Carla, and Stephanie Blair.

John A. Beachy
William D. Blair

To the Student

This book has grown out of our experiences in teaching abstract algebra over a considerable period of time. Our students have generally had three semesters of calculus, followed by a semester of linear algebra, and so we assume only this much background. This has meant that our students have had some familiarity with the abstract concepts of vector spaces and linear transformations. They have even had to write out a few short proofs in previous courses. But they have not usually been prepared for the depth in our course, where we require that almost everything be proved quite carefully. Learning to write proofs has always been a major stumbling block. The best advice we can give in this regard is to urge you to talk to your teacher. Each ten minutes of help in the early going will save hours later.

Don't be discouraged if you can't solve all of the exercises. Do the ones you are assigned, try lots of others, and come back to the ones you can't do on the first try. From time to time there will be "misplaced" exercises. By this we mean exercises for which you have sufficient tools to solve the problem as it appears, but which have easier solutions after better techniques have been introduced at a later stage. Simply attempting one of these problems (even if the attempt ends in failure) can help you understand certain ideas when they are introduced later. For this reason we urge you to keep coming back to exercises that you cannot solve on the first try.

We urge any student who feels in need of a pep talk to reread this part of the preface. The same general comment applies to the introductions to each chapter. When first read these introductory comments are meant to motivate the material that follows by indicating why it is interesting or important and at the same time relating this new material to things from the student's background. They are also intended to tie together various concepts to be introduced in the chapter, and so some parts will make more sense after the relevant part of the chapter has been covered in detail. Not only will the introductions themselves make more sense on rereading, but the way in which they tie the subject matter of the chapter into the broader picture should be easier to understand.

We often hear comments or questions similar to the ones we have listed below. We hope that our responses will be helpful to you as you begin studying our book.

"I have to read the text several times before I begin to understand it."

Yes, you should probably expect to have to do this. In fact, you might benefit from a "slow reading" rather than a "speed reading" course. There aren't many pages in a section, so you can afford to read them line by line. You should make sure that you can supply any reasons that may have been left out. We have written the book with the intention of gradually raising the level as it progresses. That simply means that we take more for granted, that we leave out more details. We hope to force you to become more sophisticated as you go along, so that you can supply more and more of the details on your own.

"I understand the definitions and theorems, but I can't do the problems."

Please forgive us for being skeptical of this statement. Often it just simply isn't the case. How do you really understand a definition or a theorem? Being able to write down a definition constitutes the first step. But to put it into context you need a good variety of examples, which should allow you to relate a new definition to facts you already know. With each definition you should associate several examples, simple enough to understand thoroughly, but complex enough to illustrate the properties inherent in the definition. In writing the book we have tried to provide good examples for each definition, but you may need to come up with your own examples based on your particular background and interests.

Understanding a proof is similar. If you can follow every step in the proof, and even write it out by yourself, then you have one degree of understanding. A complementary aspect of really understanding the proof is to be able to show exactly what it means for some simple examples that you can easily grasp. Sometimes it is helpful to take an example you understand well and follow through each step of the proof, applying it to the example you have in mind.

Trying to use "lateral thinking" is often important in solving a problem. It is easy to get stuck in one approach to a problem. You need to keep asking yourself if there are other ways to view the problem. Time to reflect is important. You need to do the groundwork in trying to understand the question and in reviewing relevant definitions and theorems from the text. Then you may benefit from simply taking a break and allowing your subconscious mind to sort some things out and make some connections. If you do the preparation well, you will find that a solution or method of attack may occur to you at quite unlikely times, when you are completely relaxed or even absorbed in something unrelated.

After emphasizing the use of examples, we need to discuss the next complaint, which is a standard one.

"I need more examples."

This is probably true, since even though we have tried to supply a good variety of examples, we may not have included the ones that best tie into your previous experience. We can't overemphasize the importance of examples in providing motivation, as well as in understanding definitions and theorems.

For real understanding you must learn to construct your own examples. A good example should be simple enough for you to grasp, but not so simple that it doesn't illuminate the relevant points. We hope that you will learn to construct good examples for yourself while reading our book. We have drawn our examples from areas that we hope are familiar. We use ordinary addition and multiplication of various sets of numbers, composition of functions, and multiplication of matrices to illustrate the basic algebraic concepts that we want to study.

"When I try to do a proof, I don't know where to start."

Partly, this is just a matter of experience. Just as in any area, it takes some practice before you will feel able to use the various ideas with some ease. It is also probably a matter of some "math anxiety," because actually doing a proof can seem a little mysterious. How is it possible to come up with all the right ideas, in just the right order?

It is true that there are certain approaches that an experienced mathematician would know to try first while attempting to solve a problem. In the text we will try to alert you to these. In fact, sometimes we have suggested a few techniques to keep in mind while attacking the assigned problems. If you get stuck, see what happens in some simple examples. If all else fails, make a list of all of the results in the text that have the hypothesis of your proof as their hypothesis. Also make a list of all those results which have the conclusion of your proof as their conclusion. Then you can use these results to help you narrow the gap between the hypothesis and conclusion of the proof you are working on.

"I'm no good at writing proofs."

There are really several parts to proving something: understanding the problem, finding a solution, and writing it down in a logical fashion.

What is involved in writing a proof? Isn't it just an explanation? Of course, it has to include enough detail to be convincing, but it shouldn't include unnecessary details which might only obscure the real reasons why things work as they do. One way to test this is to see if your proof will convince another student in the class. You should even ask yourself whether or not it will convince you when you read it while studying for the final exam.

Constructing a proof is like building a bridge. Construction begins at both ends and continues until it is possible to put in the final span that links both sides. In the same way, in actually constructing a proof, it is often necessary to simplify or rewrite or expand both the hypothesis and the conclusion. You need to try to make the gap between the two as small as possible, so that you can finally see the steps that link them.

The bridge is designed to be used by people who simply start at one side and move across to the other. In writing down a proof you should have the same goal, so that a reader can start at the hypothesis and move straight ahead to the conclusion. Writing a clear proof is like any writing—it will probably take several revisions, even after all of the key ideas are in place. (We want to make sure that you don't suffer from writer's block because you believe that a proof should appear on your paper, line after line, in perfect order.)

Of course, we can't avoid the real problem. Sometimes the proofs are quite difficult and require a genuine idea. In doing your calculus homework, you may have followed the time-honored technique used by most students. If you couldn't do a problem, you would look for an example of exactly the same type, reading the text only until you found one. That technique often is good enough to solve routine computational problems, but in a course such as this you should not expect to find models for all of the problems that you are asked to solve as exercises. These problems may very well be unique. The only way to prepare to do them is to read the text in detail.

"I keep trying, but I don't seem to be making any progress."

We can only encourage you to keep trying. Sometimes it seems a bit like learning to ride a bicycle. There is a lot of struggling and effort, trial and error, and it can be really discouraging to see your friends all of a sudden riding pretty well, while you keep falling over. Then one day it just seems to happen—you can do it, and you never really forget how.

"I should have taken a course in logic before I started to learn abstract algebra."

That isn't a bad idea. However, we think that your common sense plus a few comments from us and from your teacher will get you started on the right track.

Logic is the glue that holds together the proofs that you will be writing. Logical connectives such as "and," "or," "if . . . then . . .," and "not" are used to build compound statements out of simpler ones. We assume that you are more or less familiar with these terms, but we need to make a few comments because they are used in mathematics in a precise fashion.

The word "or" can be ambiguous in ordinary English usage. It may mean "$\mathscr{P}$ or $\mathscr{Q}$, but not both", which we call the *exclusive* "or", or it may mean "$\mathscr{P}$ or $\mathscr{Q}$, or possibly both", which we call the *inclusive* "or." In mathematics, it is generally agreed to use "or" only in the inclusive sense. That is, the compound statement "$\mathscr{P}$ or $\mathscr{Q}$" is true precisely when one of the following occurs: (i) $\mathscr{P}$ is true and $\mathscr{Q}$ is false; (ii) $\mathscr{Q}$ is true and $\mathscr{P}$ is false; (iii) $\mathscr{P}$ is true and $\mathscr{Q}$ is true.

The expression "if $\mathcal{P}$ then $\mathcal{Q}$" is called a *conditional expression*, and is the single most important form that we will use. Here $\mathcal{P}$ is the *hypothesis* and $\mathcal{Q}$ is the *conclusion*. This expression is true in all cases except when $\mathcal{P}$ is true and $\mathcal{Q}$ is false. If you are having a hard time agreeing that this expression is true in those cases in which $\mathcal{P}$ is false, try reading the following argument carefully. It should be obvious that "if $\mathcal{P}$ and $\mathcal{Q}$ are true, then $\mathcal{P}$ is true" is valid for all choices of $\mathcal{P}$ and $\mathcal{Q}$. In particular, it must be true when we replace $\mathcal{Q}$ by "not $\mathcal{P}$." Now the statement reads "if $\mathcal{P}$ and not $\mathcal{P}$ are true, then $\mathcal{P}$ is true." But "$\mathcal{P}$ and not $\mathcal{P}$" is always false, and so a false statement in the hypothesis always makes a conditional expression valid.

There are several equivalent ways to say "if $\mathcal{P}$ then $\mathcal{Q}$." We can say "$\mathcal{P}$ implies $\mathcal{Q}$," or "$\mathcal{P}$ is sufficient for $\mathcal{Q}$," or "$\mathcal{Q}$ if $\mathcal{P}$," or "$\mathcal{Q}$ necessarily follows from $\mathcal{P}$."

Two expressions related to "if $\mathcal{P}$ then $\mathcal{Q}$" are its *contrapositive* "if not $\mathcal{Q}$ then not $\mathcal{P}$" and its *converse* "if $\mathcal{Q}$ then $\mathcal{P}$." The expression "$\mathcal{P}$ implies $\mathcal{Q}$" is logically equivalent to its contrapositive "not $\mathcal{Q}$ implies not $\mathcal{P}$," but is not logically equivalent to its converse "$\mathcal{Q}$ implies $\mathcal{P}$."

The *biconditional* is the statement "$\mathcal{P}$ if and only if $\mathcal{Q}$." We can also say "$\mathcal{P}$ is equivalent to $\mathcal{Q}$," or "$\mathcal{P}$ is necessary and sufficient for $\mathcal{Q}$."

The precision of our mathematical language is abused at one point. Definitions are usually stated in a form such as "a number is said to be even if it is divisible by 2." It must be understood that the biconditional is being used, since the statement is clearly labeled as a definition, and so the meaning of the definition is "a number is said to be even if and only if it is divisible by 2."

Historical Background

The word "algebra" entered the mathematical vocabulary from Arabic over one thousand years ago, and for almost all of that time it has meant the study of equations. The "algebra" of equations is at a higher level of abstraction than arithmetic, in that symbols may be used to represent unknown numbers. "Modern algebra" or "abstract algebra" dates from the nineteenth century, when problems from number theory and the theory of equations led to the study of abstract mathematical models. In these models, symbols might represent numbers, polynomials, permutations, or elements of various other sets in which arithmetic operations could be defined. Mathematicians attempted to identify the relevant underlying principles, and to determine their logical consequences in very general settings.

One of the problems that has motivated a great deal of work in algebra has been the problem of solving equations by radicals. We begin our discussion with the familiar quadratic formula

$$x = \frac{-b \pm \sqrt{b^2 - 4ac}}{2a}.$$

This formula gives a solution of the equation

$$ax^2 + bx + c = 0,$$

expressed in terms of its coefficients, and using a square root. More generally, we say that an equation

$$a_n x^n + \ldots + a_1 x + a_0 = 0$$

is solvable by radicals if the solutions can be given in a form that involves sums, differences, products, or quotients of the coefficients $a_n, \ldots, a_1, a_0$, together with square roots, cube roots, etc., of such combinations of the coefficients.

Quadratic and even cubic equations were studied as early as Babylonian times. In the second half of the eleventh century, Omar Khayyam wrote a book on algebra, which contained a systematic study of cubic equations. His approach was mainly geometric, and he found the roots of the equations as intersections of conic sections. A general method for solving cubic equations numerically eluded the Greeks and later oriental mathematicians. The solution of the cubic equation represented for the Western world the first advance beyond classical mathematics.

General cubic equations were reduced to the form $x^3 + px + q = 0$. In the early sixteenth century, a mathematician by the name of Scipione del Ferro (1465–1526) solved one particular case of the cubic. He did not publish his solution, but word of the discovery became known, and several others were also successful in solving the equations. The solutions were published in a textbook by Gerolamo Cardano (1501–1576) in 1545. This caused a bitter dispute with another mathematician, who had independently discovered the formulas, and claimed to have given them to Cardano under a pledge of secrecy. (For additional details see the introduction to Chapter 4.) The solution of the equation $x^3 + px = q$ was given by Cardano in the form

$$x = \sqrt[3]{\frac{q}{2} + \sqrt{\frac{p^3}{27} + \frac{q^2}{4}}} - \sqrt[3]{-\frac{q}{2} + \sqrt{\frac{p^3}{27} + \frac{q^2}{4}}}.$$

A solution to the general quartic equation was also given, in which the solution could be expressed in terms of radicals involving the coefficients.

Subsequently, attempts were made to find similar solutions to the general quintic equation, but without success. The development of calculus led to methods for approximating roots, and the theory of equations became analytic. One result of this approach was the discovery by D'Alembert (1717–1783) in 1746 that every algebraic equation of degree n has n roots in the set of complex numbers. This changed the emphasis of the question from the existence of roots to whether equations of degree 5 or greater could be solved by radicals.

In 1798, Ruffini (1765–1822) published a proof claiming to show that the quintic could not be solved by radicals. The proof was not complete, although the general idea was correct. A full proof was finally given by Abel (1802–1829) in 1826. A complete answer to the question of which equations are solvable by radicals was found by Galois in 1832 (Galois lived from 1811 to 1832, and was killed in a duel). Galois considered certain permutations of the roots of a polynomial—those that leave the coefficients fixed—and showed that the polynomial is solvable by radicals if and only if the associated group of permutations (see the introduction to Chapter 3) has certain properties. This theory, named after Galois, contains deep and very beautiful results, and is the subject of Chapter 8. Although it is not always possible to cover that material in a beginning course in abstract algebra, it is toward this goal that many of the results in this book were originally directed.

The mathematics necessary to answer the question of solvability by radicals includes the development of a good deal of group theory and field theory, which have subsequently been applied in many other areas, including physics and computer science. In studying these areas we have used a modern, axiomatic approach rather than an historical one.

INDEX OF SYMBOLS

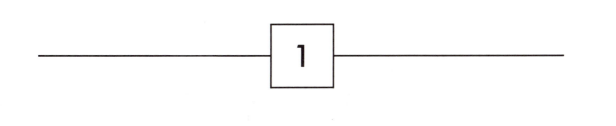

Integers

In this chapter we will develop some of the properties of the set of integers

$$\mathbf{Z} = \{\ldots,-2,-1,0,1,2,\ldots\}$$

that are needed in our later work. The use of $\mathbf{Z}$ for the integers reflects the strong German influence on the modern development of algebra; $\mathbf{Z}$ comes from the German word for numbers, "Zahlen." Some of the computational techniques we study here will reappear numerous times in later chapters. Furthermore, we will construct some concrete examples that will serve as important building blocks for later work on groups, rings, and fields.

To give a simple illustration of how we will use elementary number theory, consider the matrix $A = \begin{bmatrix} 0 & 1 \\ -1 & 0 \end{bmatrix}$. The powers of A are

$$A^2 = \begin{bmatrix} -1 & 0 \\ 0 & -1 \end{bmatrix}, \qquad A^3 = \begin{bmatrix} 0 & -1 \\ 1 & 0 \end{bmatrix}, \qquad A^4 = \begin{bmatrix} 1 & 0 \\ 0 & 1 \end{bmatrix}, \qquad A^5 = \begin{bmatrix} 0 & 1 \\ -1 & 0 \end{bmatrix},$$

etc. Since A^4 is the identity matrix I, the powers begin to repeat at A^5, as we can see by writing $A^5 = A^4 \cdot A = I \cdot A = A$, $A^6 = A^4 \cdot A^2 = I \cdot A^2 = A^2$, etc.

What about A^{231}, for example? If we divide 231 by 4, we get 57 with remainder 3, so $231 = 4 \cdot 57 + 3$. This provides our answer, since

$$A^{231} = A^{4\cdot57+3} = A^{4\cdot57} \cdot A^3 = (A^4)^{57} \cdot A^3 = I^{57} \cdot A^3 = I \cdot A^3 = A^3.$$

We can see that two powers A^i and A^j agree precisely when i and j differ by a multiple of 4. Thus there are only the four powers:

$$\begin{bmatrix} 1 & 0 \\ 0 & 1 \end{bmatrix}, \quad \begin{bmatrix} 0 & 1 \\ -1 & 0 \end{bmatrix}, \quad \begin{bmatrix} -1 & 0 \\ 0 & -1 \end{bmatrix}, \quad \begin{bmatrix} 0 & -1 \\ 1 & 0 \end{bmatrix}.$$

In this chapter we will develop the notion of congruence modulo n. This is used to describe situations in which we need to consider numbers to be similar when they differ by a multiple of n. One familiar situation of this sort occurs when telling time, since on a clock we do not distinguish between times that differ by a multiple of 12 (or 24 in some cases).

We do not mean to give the impression that number theory is not important in its own right. Historically, almost all civilizations have developed the integers (at least the positive ones) for use in agriculture, commerce, etc. After the elementary operations (addition, subtraction, multiplication, and division) have been understood, human curiosity has taken over and individuals have begun to look for deeper properties that the integers may possess.

Nonmathematicians are often surprised that research is currently being done in mathematics. They seem to believe that all possible questions have already been answered. At this point an analogy may be useful. Think of all that is known as being contained in a ball. Adding knowledge enlarges the ball, and this means that the surface of the ball—the interface between known and unknown where research occurs—also becomes larger. In short, the more we know, the more questions there are to ask. In number theory, perhaps more than in any other branch of mathematics, there are still many unanswered questions that can easily be posed. In fact, it seems that often the simplest sounding questions require the deepest tools to resolve.

One aspect of number theory that has particular applications in algebra is the one that concerns itself with questions of divisibility and primality. Fortunately for our study of algebra, this part of number theory is easily accessible, and it is with these properties of integers that we will deal in this chapter. Number theory got its start with Euclid and much of what we do in the first two sections appears in his book *Elements*.

Our approach to number theory will be to study it as a tool for later use. By taking this opportunity to mention several important problems with which number theorists are concerned, we hope to indicate why the subject is so interesting in its own right.

Remember that an integer $p > 1$ is called *prime* if its only positive divisors are itself and 1. In one sense, these are the basic building blocks in number theory, since every positive integer can be written (essentially uniquely) as a product of primes. Euclid considered primes and proved that there are infinitely many. When we look at the sequence of primes

$$2,3,5,7,11,13,17,19,23,29,31,\ldots$$

we observe that except for 2, all primes are odd. Any two odd primes on the list must differ by at least 2, but certain pairs of "twin primes" that differ by the minimal amount 2 do appear, for example,

$$(3,5), (5,7), (11,13), (17,19), (29,31), (41,43), \ldots.$$

Are there infinitely many "twin prime" pairs? The answer to this innocent question is unknown.

Although any positive integer is a product of primes, what about sums? Another open question is attributed to Goldbach (1690–1764). He asks whether every even integer greater than 2 can be written as the sum of two primes. (Since the sum of two odd primes is even, the only way to write an odd integer as a sum of two primes is to use an odd prime added to 2. That means that the only odd integers that can be represented as a sum of two primes are the ones that occur as the larger prime in a pair of "twin primes.") We invite you to experiment in writing some even integers as sums of two primes.

A beautiful theorem proved by Lagrange (1736–1813) in 1770 states that every positive integer can be written as the sum of four squares (where an integer of the form n^2 is called a square). Could we get by with fewer than four squares? The answer is no; try representing 7 as a sum of three squares. This naturally leads to the question of which positive integers can be written as the sum of three squares. The answer is that n can be written as a sum of three squares if and only if n is not of the form $4^m(8k + 7)$, where m, k are any nonnegative integers. This theorem was first correctly proved by Gauss (1777–1855) and appears in his famous book *Disquisitiones Arithmeticae* (1801).

This raises the question of which positive integers can be written as the sum of two squares. The answer in this case is slightly more complicated. It is that n can be written as the sum of two squares if and only if when we factor n as a product of primes, all those primes that give a remainder of 3 upon division by 4 have even exponents. The first published proof of this fact (dating from 1749) is due to Euler (1707–1783). Around 1640 Fermat (1601–1665) had stated, without proof, all three of these theorems on the representation of n as a sum of squares.

Our fourth and final topic deals with another statement of Fermat, namely the famous "Fermat's last theorem," which, since we still have no proof, is more accurately if less romantically known as "Fermat's last conjecture." The ancient Greeks (the Pythagoreans, in particular) knew that for certain triples (x, y, z) of nonzero integers we have $x^2 + y^2 = z^2$, for example, $(3,4,5), (5,12,13), \ldots.$ Fermat considered a generalization of this and asked whether for any integer $n > 2$ there exists a triple (x, y, z) such that $x^n + y^n = z^n$. In the margin of his copy of a number theory text he stated that he had a wonderful proof that there exists no such triple for $n \geq 3$, but he went on to say that the margin was not wide enough to write it out. Mathematicians have spent the last 300 years looking for a proof! Several books and a whole branch of number theory have been devoted to this search. In fact, at least two books are devoted to the question of whether Fermat could have had a proof.

Whether Fermat actually had a proof or not, he is clearly the first modern (post-Greek) number theorist, and he deserves much of the credit for the subject as we know it today. Another important milestone in modern number theory is Gauss's *Disquisitiones Arithmeticae,* which changed number theory from a "hodge-podge" of results into a coherent subject. The material on congruences in Section 1.3 first appeared there and contributed much to the systematic organization of number theory.

1.1 DIVISORS

Obviously, at the beginning of the book we must decide where to start mathematically. We would like to give a careful mathematical development, including proofs of virtually everything we cover. However, that would take us farther into the foundations of mathematics than we believe is profitable in a beginning course in abstract algebra. As a compromise, we have chosen to assume a knowledge of basic set theory and some familiarity with the set of integers.

For the student who is concerned about how the integers can be described formally and how the basic properties of the integers can be deduced, we have provided some very sketchy information in the appendices at the end of this chapter. Even there we have taken a naive approach, rather than formally treating the basic notions of set theory as undefined terms and giving the axioms that relate them. We have included a list of the Peano postulates, which use concepts and axioms of set theory to characterize the natural numbers. We then sketch how the integers, and larger sets of numbers, can be logically developed.

In the beginning sections of this chapter we will assume some familiarity with the set of integers, and we will simply take for granted some of the basic arithmetic and order properties of the integers. (These properties should be familiar from elementary school arithmetic. They are listed in detail in Appendix C at the end of the chapter.) The set of *integers* $\{0, \pm 1, \pm 2, \ldots\}$ will be denoted by $\mathbf{Z}$ throughout the text, while we will use $\mathbf{N}$ for the set of *natural numbers* $\{0, 1, 2, \ldots\}$.

Our first task is to study divisibility. We will then develop a theory of prime numbers based on our work with greatest common divisors. The fact that exact division is not always possible within the set of integers should not be regarded as a deficiency. Rather, it is one source of the richness of the subject of number theory and leads to many interesting and fundamental propositions about the integers.

1.1.1 Definition.   An integer a is called a *multiple* of an integer b if $a = bq$ for some integer q. In this case we also say that b is a *divisor* of a, and we use the notation $b | a$.

In the above case we can also say that b is a *factor* of a, or that a is *divisible* by b. If b is not a divisor of a, meaning that $a \neq bq$ for all $q \in \mathbf{Z}$, then we write $b \nmid a$. The set of all multiples of an integer a will be denoted by $a\mathbf{Z}$.

We note some elementary facts about divisors. If $a \neq 0$ and $b|a$, then $|b| \leq |a|$ since $|b| \leq |b||q| = |a|$ for some nonzero integer q. It follows from this observation that if $b|a$ and $a|b$, then $|b| = |a|$ and so $b = \pm a$. Therefore, if $b|1$, then since it is always true that $1|b$, we must have $b = \pm 1$.

Note that the only multiple of 0 is 0 itself. On the other hand, for any integer a we have $0 = a \cdot 0$, and thus 0 is a multiple of any integer. With the notation we have introduced, the set of all multiples of 3 is $3\mathbf{Z} = \{0, \pm 3, \pm 6, \pm 9, \ldots\}$. To describe $a\mathbf{Z}$ precisely, we can write

$$a\mathbf{Z} = \{m \in \mathbf{Z} | m = aq \text{ for some } q \in \mathbf{Z}\}.$$

Suppose that a is a multiple of b. Then every multiple of a is also a multiple of b, and in fact we can say that a is a multiple of b if and only if every multiple of a is also a multiple of b. In symbols we can write $b|a$ if and only if $a\mathbf{Z} \subseteq b\mathbf{Z}$. Exercise 10 asks for a more detailed proof of this statement.

Before we study divisors and multiples of a fixed integer, we need to state an important property of the set of natural numbers, which we will take as an axiom.

1.1.2 Axiom (Well-Ordering Principle).   Every nonempty set of natural numbers contains a smallest element.

Let S be a nonempty set of integers that has a lower bound. That is, there is an integer b such that $b \leq n$ for all $n \in S$. If $b \geq 0$, then S is actually a set of natural numbers, so it contains a smallest element by the well-ordering principle. If $b < 0$, then adding $|b|$ to each integer in S produces a set of natural numbers, since $|b| + n \geq 0$ for all $n \in S$. This new set must contain a smallest element, say s, and it is easy to see that $s - |b|$ is the smallest element of S. This allows us to use, if necessary, a somewhat stronger version of the well-ordering principle: Every set of integers that is bounded below contains a smallest element.

The first application of the well-ordering principle will be to prove the division algorithm. In familiar terms, the division algorithm states that dividing an integer a by a positive integer b gives a quotient q and nonnegative remainder r, such that r is less than b. You might write this as

$$\frac{a}{b} = q + \frac{r}{b}.$$

To simplify our work by reducing the number of operations, we prefer to use only addition and multiplication in the statement, writing instead

$$a = bq + r.$$

For example, if $a = 29$ and $b = 8$, then

$$29 = 8 \cdot 3 + 5,$$

so the quotient q is 3 and the remainder r is 5. You must be careful when a is a negative number, since the remainder must be nonnegative. Simply changing signs in the previous equation, we have

$$-29 = (8)(-3) + (-5),$$

which does not give an appropriate remainder. Rewriting this in the form

$$-29 = (8)(-4) + 3$$

gives the correct quotient $q = -4$ and remainder $r = 3$.

     Solving for r in the equation $a = bq + r$ shows that $r = a - bq$, and r must be the smallest nonnegative integer that can be written in this form, since $0 \le r < b$. This observation forms the basis of our proof that the division algorithm can be deduced from the well-ordering principle. Another way to see that the quotient depends on the remainder is to notice that you could find the remainder and quotient by repeatedly subtracting b from a. You know that you have the correct remainder when you obtain a nonnegative integer less than b.

     1.1.3 Theorem (Division Algorithm).   For any integers a and b, with $b > 0$, there exist unique integers q (the *quotient*) and r (the *remainder*) such that $a = bq + r$, with $0 \le r < b$.

     Proof. Consider the set $R = \{a - bq \mid q \in \mathbf{Z}\}$. The elements of R are the potential remainders, and among these we need to find the smallest nonnegative one. We want to apply the well-ordering principle to the set R^+ of nonnegative integers in R, so we must first show that R^+ is nonempty. Since $b \ge 1$, the number $a - b(-|a|) = a + b \cdot |a|$ is nonnegative and belongs to R^+.

     Now by the well-ordering principle, R^+ has a smallest element, and we will call this element r. We will show that $a = bq + r$, with $0 \le r$ and $r < b$. By definition, $r \ge 0$, and since $r \in R^+$, we must have $r = a - bq$ for some integer q. If $r \ge b$, then $s = r - b = a - b(q + 1) \in R^+$, and so $s \ge 0$ with $s < r$. This contradicts the way r was defined, and so we must have $r < b$. We have now proved the existence of r and q satisfying the conditions $a = bq + r$ and $0 \le r < b$.

     To show that q and r are unique, suppose that we can also write $a = bp + s$ for integers p and s with $0 \le s < b$. We have $0 \le r < b$ and $0 \le s < b$, and this implies that $|s - r| < b$. But $bp + s = bq + r$ and so $s - r = b(q - p)$, which shows that $b \mid (s - r)$. The only way that b can be a divisor of a number with smaller absolute value is if that number is 0, and so we must have $s - r = 0$, or $s = r$. Then $bp = bq$, which implies that $p = q$ since $b > 0$. Thus the quotient and remainder are unique, and we have completed the proof of the theorem. $\square$

     Given integers a and b, with $b > 0$, we can use the division algorithm to write $a = bq + r$, with $0 \le r < b$. Since $b \mid a$ if and only if there exists $q \in \mathbf{Z}$ such that

$a = bq$, we see that $b|a$ if and only if $r = 0$. This simple observation gives us a useful tool in doing number theoretic proofs. To show that $b|a$ we can use the division algorithm to write $a = bq + r$ and then show that $r = 0$. This technique makes its first appearance in the proof of Theorem 1.1.4.

A set of multiples $a\mathbf{Z}$ has the property that the sum or difference of two integers in the set is again in the set, since $aq_1 \pm aq_2 = a(q_1 \pm q_2)$. We say that the set $a\mathbf{Z}$ is *closed under addition and subtraction*. This will prove to be a very important property in our later work. The next theorem shows that this property characterizes sets of multiples, since a nonempty set of integers is closed under addition and subtraction if and only if it is a set of the form $a\mathbf{Z}$, for some nonnegative integer a.

1.1.4 Theorem.   Let I be a nonempty set of integers that is closed under addition and subtraction. Then I either consists of zero alone or else contains a smallest positive element, in which case I consists of all multiples of its smallest positive element.

Proof. Since I is nonempty, either it consists of 0 alone, or else it contains a nonzero integer a. In the first case we are done. In the second case, if I contains the nonzero integer a, then it must contain the difference $a - a = 0$, and hence the difference $0 - a = -a$, since I is assumed to be closed under subtraction. Now either a or $-a$ is positive, so I contains at least one positive integer. Having shown that the set of positive integers in I is nonempty, we can apply the well-ordering principle to guarantee that it contains a smallest member, say b.

Next we want to show that I is equal to the set $b\mathbf{Z}$ of all multiples of b. To show that $I = b\mathbf{Z}$, we will first show that $b\mathbf{Z} \subseteq I$, and then show that $I \subseteq b\mathbf{Z}$.

Any nonzero multiple of b is given by just adding b (or $-b$) to itself a finite number of times, so since I is closed under addition, it must contain all multiples of b. Thus $b\mathbf{Z} \subseteq I$.

On the other hand, to show that $I \subseteq b\mathbf{Z}$ we must take any element c in I and show that it is a multiple of b, or equivalently, that $b|c$. (Now comes the one crucial idea in the proof.) Using the division algorithm we can write $c = bq + r$, for some integers q and r with $0 \le r < b$. Since I contains bq and is closed under subtraction, it must also contain $r = c - bq$. But this is a contradiction unless $r = 0$, because b was chosen to be the smallest positive integer in I and yet $r < b$ by the division algorithm. We conclude that $r = 0$, and therefore $c = bq$, so $b|c$ and we have shown that $I \subseteq b\mathbf{Z}$.

This completes the proof that $I = b\mathbf{Z}$. $\square$

One of the main goals of Chapter 1 is to develop some properties of prime numbers (see Section 1.2). To do so we need information about relatively prime numbers, and this in turn depends on the notion of the greatest common divisor of two numbers. Our definition is given in terms of divisibility, not in terms of size. See Exercise 15 for an equivalent definition.

1.1.5 Definition.   A positive integer d is called the *greatest common divisor* of the nonzero integers a and b if

 (i) d is a divisor of both a and b, and

 (ii) any divisor of both a and b is also a divisor of d.

We will use the notation $\gcd(a,b)$ or (a,b) for the greatest common divisor of a and b.

The fact that we have written down a definition of the greatest common divisor does not guarantee that there is such a number. Furthermore, the use of the word "the" has to be justified, since it implies that there can be only one greatest common divisor. The next theorem will guarantee the existence of the greatest common divisor, and the question of uniqueness is easily answered: If d_1 and d_2 are greatest common divisors of a and b, then the definition requires that $d_1 | d_2$ and $d_2 | d_1$, so $d_1 = \pm d_2$. Since both d_1 and d_2 are positive, we have $d_1 = d_2$.

If a and b are integers, then we will refer to any integer of the form $ma + nb$ as a *linear combination* of a and b. The next theorem gives a very useful connection between greatest common divisors and linear combinations.

1.1.6 Theorem.   Any two nonzero integers a and b have a greatest common divisor, which can be expressed as the smallest positive linear combination of a and b.

Moreover, an integer is a linear combination of a and b if and only if it is a multiple of their greatest common divisor.

Proof. Let I be the set of all linear combinations of a and b, that is,

$$I = \{ma + nb \,|\, m, n \in \mathbf{Z}\}.$$

The set I is nonempty since it contains $a = 1 \cdot a + 0 \cdot b$ and $b = 0 \cdot a + 1 \cdot b$, and it is closed under addition and subtraction since

$$(m_1 a + n_1 b) \pm (m_2 a + n_2 b) = (m_1 \pm m_2)a + (n_1 \pm n_2)b.$$

By Theorem 1.1.4, the set I consists of all multiples of the smallest positive integer it contains, say d. Since $d \in I$, $d = ma + nb$ for some integers m and n.

Since we already know that d is positive, to show that $d = (a,b)$ we must show that (i) $d | a$ and $d | b$ and (ii) if $c | a$ and $c | b$, then $c | d$. First, d is a divisor of every element in I, so $d | a$ and $d | b$ since $a, b \in I$. Secondly, if $c | a$ and $c | b$, say $a = cq_1$ and $b = cq_2$, then

$$d = ma + nb = m(cq_1) + n(cq_2) = c(mq_1 + nq_2),$$

which shows that $c | d$.

The second assertion follows from the fact that I, the set of all linear combinations of a and b, is equal to $d\mathbf{Z}$, the set of all multiples of d.   □

The greatest common divisor of two numbers can be computed by using a procedure known as the *Euclidean algorithm*. (Our proof of the existence of the greatest common divisor did not include an explicit method for finding it.) Before discussing the Euclidean algorithm, we need to note some properties of the greatest common divisor. First, if $a \neq 0$ and $b|a$, then $(a,b) = |b|$.

The next observation provides the basis for the Euclidean algorithm. If $a = bq + r$, then $(a,b) = (b,r)$. This can be shown by noting first that a is a multiple of (b,r) since it is a linear combination of b and r. Then $(b,r)|(a,b)$ since b is also a multiple of (b,r). A similar argument using the equality $r = a - bq$ shows that $(a,b)|(b,r)$, and it follows that $(a,b) = (b,r)$.

Given integers $a > b > 0$, the Euclidean algorithm uses the division algorithm repeatedly to obtain

$$a = bq_1 + r_1 \qquad \text{with} \qquad 0 \le r_1 < b$$

$$b = r_1q_2 + r_2 \qquad \text{with} \qquad 0 \le r_2 < r_1$$

$$r_1 = r_2q_3 + r_3 \qquad \text{with} \qquad 0 \le r_3 < r_2$$

$$\text{etc.}$$

Since $r_1 > r_2 > \ldots$, the remainders get smaller and smaller, and after a finite number of steps we obtain a remainder $r_{n+1} = 0$. The algorithm ends with the equation

$$r_{n-1} = r_nq_{n+1} + 0.$$

This gives us the greatest common divisor:

$$(a,b) = (b,r_1) = (r_1,r_2) = \ldots = (r_{n-1},r_n) = r_n.$$

Example 1.1.1.

In showing that $(24,18) = 6$, we have $(24,18) = (18,6)$ since $24 = 18 \cdot 1 + 6$, and $(18,6) = 6$ since $6|18$. Thus $(24,18) = 6$.

To show that $(126,35) = 7$, we first have $(126,35) = (35,21)$ since $126 = 35 \cdot 3 + 21$. Then $(35,21) = (21,14)$ since $35 = 21 \cdot 1 + 14$, and $(21,14) = (14,7)$ since $21 = 14 \cdot 1 + 7$. Finally, $(14,7) = 7$ since $14 = 7 \cdot 2$.

In finding $(83,38)$, we can arrange the work in the following manner:

$$(83,38) = (38,7) \qquad \text{since} \qquad 83 = 38 \cdot 2 + 7$$

$$(38,7) = (7,3) \qquad \text{since} \qquad 38 = 7 \cdot 5 + 3$$

$$(7,3) = (3,1) \qquad \text{since} \qquad 7 = 3 \cdot 2 + 1$$

$$(3,1) = 1 \qquad \text{since} \qquad 3 = 3 \cdot 1.$$

If you only need to find the greatest common divisor, stop as soon as you can compute it in your head. In finding $(83,38)$, note that since 7 has no positive divisors except 1 and 7 and is not a divisor of 38, it is clear immediately that $(38,7) = 1$. □

Example 1.1.2.

Sometimes it is necessary to find the linear combination of a and b that gives (a,b). In finding $(126,35)$ we had the following equations:

$$a = bq_1 + r_1 \qquad 126 = 35 \cdot 3 + 21$$
$$b = r_1q_2 + r_2 \qquad 35 = 21 \cdot 1 + 14$$
$$r_1 = r_2q_3 + d \qquad 21 = 14 \cdot 1 + 7$$
$$r_2 = dq_4 + 0 \qquad 14 = 7 \cdot 2 + 0.$$

We first solve for the remainder in each equation except for the final one:

$$r_1 = a + (-q_1)b \qquad 21 = 1 \cdot 126 + (-3) \cdot 35$$
$$r_2 = b + (-q_2)r_1 \qquad 14 = 1 \cdot 35 + (-1) \cdot 21$$
$$d = r_1 + (-q_3)r_2 \qquad 7 = 1 \cdot 21 + (-1) \cdot 14$$

We then substitute r_1 into the second equation, and collect terms to write r_2 as a linear combination of a and b:

$$14 = 1 \cdot 35 - 1 \cdot (126 - 3 \cdot 35) = (-1) \cdot 126 + 4 \cdot 35.$$

The next equation, which will give us d, involves both r_1 and r_2. Since we have already expressed each of these as a linear combination of a and b, we can substitute and obtain d as a linear combination of a and b:

$$7 = 1 \cdot 21 + (-1) \cdot 14 =$$
$$(126 - 3 \cdot 35) - (-1 \cdot 126 + 4 \cdot 35) = 2 \cdot 126 - 7 \cdot 35. \qquad \square$$

The technique introduced in the previous example can easily be extended to the general situation in which it is desired to express (a,b) as a linear combination of a and b. After solving for the remainder in each of the relevant equations, we obtain

$$r_1 = a + (-q_1)b$$
$$r_2 = b + (-q_2)r_1$$
$$r_3 = r_1 + (-q_3)r_2$$
$$r_4 = r_2 + (-q_4)r_3$$
$$\vdots$$

At each step, the expression for the remainder depends upon the previous two remainders. By substituting into the successive equations and then rearranging terms, it is possible to express each remainder (in turn) as a linear combination of a and b. The final step is to express (a,b) as a linear combination of a and b.

The Euclidean algorithm can be put into a convenient matrix format that keeps track of the remainders and linear combinations at the same time. To find (a,b), the idea is to start with the following system of equations:

$$x \qquad = a$$
$$y = b$$

and, by using elementary row operations, to find an equivalent system of the following form:

$$m_1 x + n_1 y = (a,b)$$
$$m_2 x + n_2 y = 0$$

Beginning with the matrix

$$\begin{bmatrix} 1 & 0 & a \\ 0 & 1 & b \end{bmatrix},$$

we use the division algorithm to write $a = bq_1 + r_1$. We then subtract q_1 times the bottom row from the top row, to get

$$\begin{bmatrix} 1 & -q_1 & r_1 \\ 0 & 1 & b \end{bmatrix}.$$

We next write $b = r_1 q_2 + r_2$, and subtract q_2 times the top row from the bottom row. This gives the matrix

$$\begin{bmatrix} 1 & -q_1 & r_1 \\ -q_2 & 1 + q_1 q_2 & r_2 \end{bmatrix}$$

and it can be checked that this algorithm produces rows in the matrix that give each successive remainder, together with the coefficients of the appropriate linear combination of a and b. The procedure is continued until one of the entries in the right-hand column is zero. Then the other entry in this column is the greatest common divisor, and its row contains the coefficients of the desired linear combination.

Example 1.1.3.

In computing $(126,35)$ we have the following matrices:

$$\begin{bmatrix} 1 & 0 & 126 \\ 0 & 1 & 35 \end{bmatrix} \rightsquigarrow \begin{bmatrix} 1 & -3 & 21 \\ 0 & 1 & 35 \end{bmatrix} \rightsquigarrow \begin{bmatrix} 1 & -3 & 21 \\ -1 & 4 & 14 \end{bmatrix} \rightsquigarrow$$

$$\begin{bmatrix} 2 & -7 & 7 \\ -1 & 4 & 14 \end{bmatrix} \rightsquigarrow \begin{bmatrix} 2 & -7 & 7 \\ -5 & 18 & 0 \end{bmatrix}.$$

Thus $(126,35) = 7$ and $(2)(126) + (-7)(35) = 7$.

In matrix form, the solution for (83,38) is the following:

$$\begin{bmatrix} 1 & 0 & 83 \\ 0 & 1 & 38 \end{bmatrix} \rightsquigarrow \begin{bmatrix} 1 & -2 & 7 \\ 0 & 1 & 38 \end{bmatrix} \rightsquigarrow \begin{bmatrix} 1 & -2 & 7 \\ -5 & 11 & 3 \end{bmatrix} \rightsquigarrow$$

$$\begin{bmatrix} 11 & -24 & 1 \\ -5 & 11 & 3 \end{bmatrix} \rightsquigarrow \begin{bmatrix} 11 & -24 & 1 \\ -38 & 83 & 0 \end{bmatrix}$$

Thus $(83,38) = 1$ and $(11)(83) + (-24)(38) = 1$. □

The number (a,b) can be written in many different ways as a linear combination of a and b. The matrix method gives a linear combination with $0 = m_1a + n_1b$, so if $(a,b) = ma + nb$, then adding the previous equation gives $(a,b) = (m + m_1)a + (n + n_1)b$. In fact, any multiple of the equation $0 = m_1a + n_1b$ could have been added, so there are infinitely many linear combinations of a and b that give (a,b).

EXERCISES: SECTION 1.1

Before working on the exercises, you must make sure that you are familiar with all of the definitions and theorems of this section. You also need to be familiar with the techniques of proof that have been used in the theorems and examples in the text. As a reminder, we take this opportunity to list several useful approaches.

—When working questions involving divisibility you may find it useful to go back to the definition. If you rewrite $b|a$ as $a = bq$ for some $q \in \mathbf{Z}$, then you have an equation involving integers, something concrete and familiar to work with.

—To show that $b|a$, try to write down an expression for a that has b as a factor.

—Another approach to proving that $b|a$ is to use the division algorithm to write $a = bq + r$, where $0 \le r < b$, and show that $r = 0$.

—Theorem 1.1.6 is extremely useful in questions involving greatest common divisors.

1. A number n is called perfect if it is equal to the sum of its proper positive divisors (those divisors different from n). The first perfect number is 6 since $1 + 2 + 3 = 6$. For each number between 6 and the next perfect number, make a list containing the number, its proper divisors, and their sum. Note: If you reach 40, you have missed the next perfect number.

2. Find the quotient and remainder when a is divided by b.
   (a) $a = 99, b = 17$
   (b) $a = -99, b = 17$
   (c) $a = 17, b = 99$
   (d) $a = -1017, b = 99$

3. Use the Euclidean algorithm to find the following greatest common divisors:
   (a) $(14,35)$
   (b) $(11,15)$

    (c) (180,252)

    (d) (513,187)

    (e) (1001,7655)

    (f) (2873,6643)

    (g) (4148,7684)

    (h) (26460,12600)

    (i) (6540,1206)

4. For each part of the previous question, find integers m and n such that (a,b) is expressed in the form $ma + nb$.

5. Let a, b, c be integers. Give a proof for these facts about divisors:
   (a) If $b \mid a$, then $b \mid ac$.
   (b) If $b \mid a$ and $c \mid b$, then $c \mid a$.
   (c) If $c \mid a$ and $c \mid b$, then $c \mid (ma + nb)$ for any integers m, n.

6. Let a, b, c be integers such that $a = b + c$. Show that if n is a positive integer that is a divisor of two of the three integers, then it is also a divisor of the third.

7. Let a, b, c be integers.
   (a) Show that if $b \mid a$ and $b \mid (a + c)$, then $b \mid c$.
   (b) Show that if $b \mid a$ and $b \nmid c$, then $b \nmid (a + c)$.

8. Let a, b, c be integers, with $c \neq 0$. Show that $bc \mid ac$ if and only if $b \mid a$.

9. Prove that the sum of the cubes of any three consecutive positive integers is divisible by 3.

10. Give a detailed proof of the statement in the text that if a and b are integers, then $b \mid a$ if and only if $a\mathbf{Z} \subseteq b\mathbf{Z}$.

11. Let a, b, c be integers, with $b > 0$, $c > 0$, and let q be the quotient and r the remainder when a is divided by b.
   (a) Show that q is the quotient and rc is the remainder when ac is divided by bc.
   (b) Show that if q' is the quotient when q is divided by c, then q' is the quotient when a is divided by bc. (Do not assume that the remainders are zero.)

12. Let a, b, n be integers with $n > 1$. Suppose that $a = nq_1 + r_1$ with $0 \leq r_1 < n$ and $b = nq_2 + r_2$ with $0 \leq r_2 < n$. Prove that $n \mid (a - b)$ if and only if $r_1 = r_2$.

13. Show that any nonempty set of integers that is closed under subtraction must also be closed under addition. (Thus part of the hypothesis of Theorem 1.1.4 is redundant.)

14. Let a, b, q, r be integers such that $a = bq + r$. Prove that $(a,b) = (b,r)$ by showing that (b,r) satisfies the definition of the greatest common divisor of a and b.

15. Perhaps a more natural definition of the greatest common divisor is the following: An integer d is called the greatest common divisor of the nonzero integers a and b if (i) d is a divisor of both a and b, and (ii) c is a divisor of both a and b implies $d \geq c$. Show that this definition is equivalent to Definition 1.1.5.

16. Show that if $a > 0$, then $(ab,ac) = a(b,c)$.

17. Find all integers x such that $3x + 7$ is divisible by 11.

18. Develop a theory of integer solutions x, y of equations of the form $ax + by = c$, where a, b, c are integers. That is, when can an equation of this form be solved, and if it can be solved, how can all solutions be found? Test your theory on these equations:

$$60x + 36y = 12, \qquad 35x + 6y = 8, \qquad 12x + 18y = 11.$$

Finally, give conditions on a and b under which $ax + by = c$ has solutions for every integer c.

19. Formulate a definition of the greatest common divisor of three positive integers a, b, c. With the appropriate definition you should be able to prove that the greatest common divisor is a linear combination of a, b, and c.

1.2 PRIMES

The main focus of this section is on prime numbers. Our method will be to investigate the notion of two integers which are relatively prime, that is, those which have no common divisors except ± 1. Using some facts which we will prove about them, we will be able to prove the prime factorization theorem, which states that every nonzero integer can be expressed as a product of primes. Finally, we will be able to use prime factorizations to learn more about greatest common divisors and least common multiples.

1.2.1 Definition. The nonzero integers a and b are said to be *relatively prime* if $(a,b) = 1$.

1.2.2 Proposition. Let a, b be nonzero integers. Then $(a,b) = 1$ if and only if there exist integers m, n such that $ma + nb = 1$.

Proof. If a and b are relatively prime, then by Theorem 1.1.6 integers m and n can be found for which $ma + nb = 1$. To prove the converse, we only need to note that if there exist integers m and n with $ma + nb = 1$, then 1 must be the smallest positive linear combination of a and b, and thus $(a,b) = 1$, again by Theorem 1.1.6. $\square$

Proposition 1.2.2 will be used repeatedly in the proof of the next result. A word of caution—it is often tempting to jump from the equation $d = ma + nb$ to the conclusion that $d = (a,b)$. For example, $16 = 2 \cdot 5 + 3 \cdot 2$, but obviously $(5,2) \neq 16$. The most that it is possible to say (using Theorem 1.1.6) is that d is a multiple of (a,b). Of course, if $ma + nb = 1$, then Proposition 1.2.2 implies that $(a,b) = 1$.

1.2.3 Proposition. Let a, b, c be integers.
(a) If $b|ac$, then $b|(a,b) \cdot c$.
(b) If $b|ac$ and $(a,b) = 1$, then $b|c$.
(c) If $b|a$, $c|a$ and $(b,c) = 1$, then $bc|a$.
(d) $(a,bc) = 1$ if and only if $(a,b) = 1$ and $(a,c) = 1$.

Proof. (a) Assume that $b|ac$. To show that $b|(a,b) \cdot c$, we will try to find an expression for $(a,b) \cdot c$ that has b as an obvious factor. We can write $(a,b) = ma + nb$ for some $m, n \in \mathbf{Z}$, and then multiplying by c gives

$$(a,b) \cdot c = mac + nbc.$$

Now b is certainly a factor of nbc, and by assumption it is also a factor of ac, so it is a factor of mac and therefore of the sum $mac + nbc$. Thus $b|(a,b) \cdot c$.

(b) Simply letting $(a,b) = 1$ in part (a) gives the result immediately.

(c) If $b|a$, then $a = bq$ for some integer q. If $c|a$, then $c|bq$, so if $(b,c) = 1$, it follows from part (b) that $c|q$, say with $q = cq_1$. Substituting for q in the equation $a = bq$ gives $a = bcq_1$, and thus $bc|a$.

(d) Suppose that $(a,bc) = 1$. Then $ma + n(bc) = 1$ for some integers m and n, and by viewing this equation as $ma + (nc)b = 1$ and $ma + (nb)c = 1$ we can see that $(a,b) = 1$ and $(a,c) = 1$.

Conversely, suppose that $(a,b) = 1$ and $(a,c) = 1$. Then $m_1a + n_1b = 1$ for some integers m_1 and n_1, and $m_2a + n_2c = 1$ for some integers m_2 and n_2. Multiplying these two equations gives

$$(m_1m_2a + m_1n_2c + m_2n_1b)a + (n_1n_2)bc = 1,$$

which shows that $(a,bc) = 1$. $\square$

1.2.4 Definition.   An integer $p > 1$ is called a *prime number* if its only divisors are ± 1 and $\pm p$. An integer $a > 1$ is called *composite* if it is not prime.

Euclid's lemma, the next step in our development of the fundamental theorem of arithmetic, is the one that requires our work on relatively prime numbers. We will use Proposition 1.2.3 (b) in a crucial way.

1.2.5 Lemma (Euclid).   An integer $p > 1$ is prime if and only if it satisfies the following property: If $p|ab$ for integers a and b, then either $p|a$ or $p|b$.

Proof. Suppose that p is prime and $p|ab$. If $a = 0$, then the result is clear. For any nonzero integer a, we know that either $(p,a) = p$ or $(p,a) = 1$, since (p,a) is always a divisor of p and p is prime. In the first case $p|a$ and we are done. In the second case, since $(p,a) = 1$, we can apply Proposition 1.2.3 (b) to show that $p|ab$ implies $p|b$. Thus we have shown that if $p|ab$, then either $p|a$ or $p|b$.

Conversely, suppose that p satisfies that given condition. If p were composite, then we could write $p = ab$ for some positive integers smaller than p. The condition would imply that either $p|a$ or $p|b$, which would be an obvious contradiction. $\square$

Example 1.2.1 (Sieve of Eratosthenes)

The primes less than a fixed positive integer a can be found by using the "sieve of Eratosthenes." List all positive integers less than a (except 1), and cross off every even number except 2. Then go to the first number that has not been crossed off, which will be 3, and cross off all higher multiples of 3. Continue this process to find all primes less than a. You can stop after you have crossed off all proper multiples of primes p for which $p < \sqrt{a}$, since you will have crossed off every number less than a that has a proper factor. (If b is composite, say $b = b_1b_2$, then either $b_1 \leq \sqrt{b}$ or $b_2 \leq \sqrt{b}$.) For example, we can find all primes less than 20 by just crossing off all multiples of 2 and 3, since $5 > \sqrt{20}$:

$$2 \quad 3 \quad \cancel{4} \quad 5 \quad \cancel{6} \quad 7 \quad \cancel{8} \quad \cancel{9} \quad \cancel{10}$$

$$11 \quad \cancel{12} \quad 13 \quad \cancel{14} \quad \cancel{15} \quad \cancel{16} \quad 17 \quad \cancel{18} \quad 19 \quad .$$

Similarly, the integers less than a and relatively prime to a can be found by crossing off the prime factors of a and all of their multiples. For example, the prime divisors of 36 are 2 and 3, and so the positive integers less than 36 and relatively prime to it can be found as follows:

$$1 \quad \cancel{2} \quad \cancel{3} \quad \cancel{4} \quad 5 \quad \cancel{6} \quad 7 \quad \cancel{8} \quad \cancel{9} \quad \cancel{10} \quad 11 \quad \cancel{12} \quad 13 \quad \cancel{14} \quad \cancel{15} \quad \cancel{16} \quad 17 \quad \cancel{18}$$

$$19 \quad \cancel{20} \quad \cancel{21} \quad \cancel{22} \quad 23 \quad \cancel{24} \quad 25 \quad \cancel{26} \quad \cancel{27} \quad \cancel{28} \quad 29 \quad \cancel{30} \quad 31 \quad \cancel{32} \quad \cancel{33} \quad \cancel{34} \quad 35 \quad . \quad \square$$

The next theorem, on prime factorization, is sometimes called the fundamental theorem of arithmetic. The naive way to prove that an integer a can be written as a product of primes is to note that either a is prime and we are done, or else a is composite, say $a = bc$. Then the same argument can be applied to b and c, and continued until a has been broken up into a product of primes. (This process must stop after a finite number of steps because of the well-ordering principle.) We also need to prove that any two factorizations of a number are in reality the same. The idea of the proof is to use Euclid's lemma to pair the primes in one factorization with those in the other.

1.2.6 Theorem (Fundamental Theorem of Arithmetic). Any integer $a > 1$ can be factored uniquely as a product of prime numbers, in the form

$$a = p_1^{\alpha_1}p_2^{\alpha_2} \cdots p_n^{\alpha_n},$$

where $p_1 < p_2 < \ldots < p_n$ and the exponents $\alpha_1, \alpha_2, \ldots, \alpha_n$ are all positive.

Proof. Suppose that there is some integer that cannot be written as a product of primes. Then the set of all integers $a > 1$ that have no prime factorization must be nonempty, so as a consequence of the well-ordering principle it must have a smallest member, say b. Now b cannot itself be a prime number since then it would have a prime factorization. Thus b is composite, and we can write $b = cd$ for positive integers c, d that are smaller than b. By assumption, both c and d must have factorizations into products of primes, and this shows that b also has such a factorization, which is a contradiction. Since multiplication is commutative, the prime factors can be ordered in the desired manner.

If there exists an integer > 1 for which the factorization is not unique, then by the well-ordering principle there exists a smallest such integer, say a. Assume that a has two factorizations $a = p_1^{\alpha_1} p_2^{\alpha_2} \cdots p_n^{\alpha_n}$ and $a = q_1^{\beta_1} q_2^{\beta_2} \cdots q_m^{\beta_m}$, where $p_1 < p_2 < \ldots < p_n$, $q_1 < q_2 < \ldots < q_m$, $\alpha_i > 0$ for $i = 1, \ldots, n$, and $\beta_i > 0$ for $i = 1, \ldots, m$. By Euclid's lemma, $q_1 | p_k$ for some k with $1 \leq k \leq n$ and $p_1 | q_j$ for some j with $1 \leq j \leq m$. Since all of the numbers p_i and q_i are prime, we must have $q_1 = p_k$ and $p_1 = q_j$. Then $j = k = 1$ since $q_1 \leq q_j = p_1 \leq p_k = q_1$. Hence we can let

$$s = \frac{a}{p_1} = \frac{a}{q_1} = p_1^{\alpha_1 - 1} p_2^{\alpha_2} \cdots p_n^{\alpha_n} = q_1^{\beta_1 - 1} q_2^{\beta_2} \cdots q_m^{\beta_m}.$$

If $s = 1$ then $a = p_1$ has a unique factorization, contrary to the choice of a. If $s > 1$, then since $s < a$ and s has two factorizations, we again have a contradiction to the choice of a. $\square$

If the prime factorization of an integer is known, then it is easy to list all of its divisors. If $a = p_1^{\alpha_1} p_2^{\alpha_2} \cdots p_n^{\alpha_n}$, then b is a divisor of a if and only if $b = p_1^{\beta_1} p_2^{\beta_2} \cdots p_n^{\beta_n}$, where $\beta_i \leq \alpha_i$ for all i. Thus we can list all possible divisors of a by systematically decreasing the exponents of each of its prime divisors.

Example 1.2.2

The positive divisors of 12 are 1, 2, 3, 4, 6, 12; the divisors of 8 are 1, 2, 4, 8; and the divisors of 36 are 1, 2, 3, 4, 6, 9, 12, 18, 36. In Figure 1.2.2, we have arranged the divisors so as to show the divisibility relations among them. There is a path (moving upward only) from a to b if and only if $a|b$. In constructing the first diagram in Figure 1.2.2, it is easiest to use the prime factorization of 12. Since $12 = 2^2 3$, we first divide 12 by 2 to get 6 and then divide again by 2 to get 3. This gives us the first part of the diagram, as shown in Figure 1.2.1. To complete the diagram we divide each number by 3.

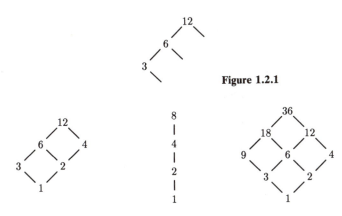

Figure 1.2.1

Figure 1.2.2

If the number has three different prime factors, then we would need a three-dimensional diagram. (Visualize the factors as if on the edges of a box.) With more than three distinct prime factors, the diagrams lose their clarity. □

The following proof, although easy to follow, is an excellent example of the austere beauty of mathematics.

1.2.7 Theorem (Euclid). There exist infinitely many prime numbers.

Proof. Suppose that there were only finitely many prime numbers, say p_1, $p_2, \ldots, p_n$. Then consider the number $a = p_1 p_2 \cdots p_n + 1$. By Theorem 1.2.6, the number a has a prime divisor, say p. Now p must be one of the primes we listed, so $p | (p_1 p_2 \cdots p_n)$, which shows that $p | (a - p_1 p_2 \cdots p_n)$. This is a contradiction since p cannot be a divisor of 1. □

Example 1.2.3

Consider the numbers $2^2 - 1 = 3, 2^3 - 1 = 7, 2^4 - 1 = 15, 2^5 - 1 = 31$, and $2^6 - 1 = 63$. The prime exponents each give rise to a prime, while the composite exponents each give a composite number. Is this true in general? Continuing to investigate prime exponents gives $2^7 - 1 = 127$, which is prime, but $2^{11} - 1 = 2047 = 23 \cdot 89$. Thus a prime exponent may or may not yield a prime number.

On the other hand, it is always true that a composite exponent yields a composite number. To prove this, let n be composite, say $n = ab$ (where a and b are integers greater than 1), and consider $2^n - 1 = 2^{ab} - 1$. We need to find a nontrivial factorization of $2^{ab} - 1 = (2^a)^b - 1$. We can look at this as $x^b - 1$, and then we have the factorization

$$x^b - 1 = (x - 1)(x^{b-1} + x^{b-2} + \ldots + x^2 + x + 1).$$

Substituting $x = 2^a$ shows that $2^a - 1$ is a factor of $2^n - 1$. Now $1 < 2^a - 1 < 2^n - 1$ since both a and b are greater than 1, and so we have found a nontrivial factorization of $2^n - 1$. □

The final concept we study in this section is the least common multiple of two integers. Its definition is parallel to that of the greatest common divisor. We can characterize it in terms of the prime factorizations of the two numbers, or by the fact that the product of two numbers is equal to the product of their least common multiple and greatest common divisor.

1.2.8 Definition.   A positive integer m is called the *least common multiple* of the nonzero integers a and b if

   (i) m is a multiple of both a and b, and

   (ii) any multiple of both a and b is also a multiple of m.

We will use the notation lcm$[a,b]$ or $[a,b]$ for the least common multiple of a and b.

1.2.9 Proposition.   Let a and b be positive integers with prime factorizations $a = p_1^{\alpha_1} p_2^{\alpha_2} \cdots p_n^{\alpha_n}$ and $b = p_1^{\beta_1} p_2^{\beta_2} \cdots p_n^{\beta_n}$, where $\alpha_i \geq 0$ and $\beta_i \geq 0$ for all i (allowing use of the same prime factors.)

For each i let $\delta_i = \min\{\alpha_i, \beta_i\}$ and let $\mu_i = \max\{\alpha_i, \beta_i\}$. Then we have the following factorizations:

   (a) $\gcd(a,b) = p_1^{\delta_1} p_2^{\delta_2} \cdots p_n^{\delta_n}$

   (b) lcm$[a,b] = p_1^{\mu_1} p_2^{\mu_2} \cdots p_n^{\mu_n}$.

Proof. The proof follows immediately from the fundamental theorem of arithmetic and the definitions of the least common multiple and greatest common divisor.   □

As an immediate corollary of Proposition 1.2.9, it is clear that the product of $\gcd(a,b)$ and lcm$[a,b]$ is ab. This can also be shown directly from the definitions, as we have noted in Exercise 12. For small numbers it is probably easiest to use their prime factorizations to find their greatest common divisor and least common multiple. It takes a great deal of work to find the prime factors of a large number, even on a computer making use of sophisticated algorithms. In contrast, the Euclidean algorithm is much faster, so its use is more efficient for finding the greatest common divisor of large numbers.

Example 1.2.4

In the previous section we computed $(126,35)$. To do this using Proposition 1.2.9 we need the factorizations $126 = 2^1 \cdot 3^2 \cdot 7^1$ and $35 = 5^1 \cdot 7^1$. We then add terms so that we have the same primes in each case, to get $126 = 2^1 \cdot 3^2 \cdot 5^0 \cdot 7^1$ and $35 = 2^0 \cdot 3^0 \cdot 5^1 \cdot 7^1$. Thus we obtain $(126,35) = 2^0 \cdot 3^0 \cdot 5^0 \cdot 7^1 = 7$ and $[126,35] = 2^1 \cdot 3^2 \cdot 5^1 \cdot 7^1 = 630$.   □

EXERCISES: SECTION 1.2

Unique factorization into primes is a powerful result, but 1.2.2, 1.2.3, or 1.2.5 should usually be tried first.

   1. Use the sieve of Eratosthenes to find all prime numbers less than 200.

2. For each composite number a, with $4 \le a \le 20$, find all positive numbers less than a that are relatively prime to a.

3. For each of the numbers 9, 15, 20, 24, and 100, give the lattice diagram of all divisors of the number. (See Example 1.2.1.)

4. Give the lattice diagram of all divisors of 60. Do the same for 1575.

5. Let a, b be positive integers. Show that if

$$h = \frac{a}{(a,b)} \quad \text{and} \quad k = \frac{b}{(a,b)},$$

then $(h,k) = 1$.

6. Let a, b, c be positive integers, and let $d = (a,b)$. Show that if $h = a/d$ and $a|bc$, then $h|c$.

7. Show that $a\mathbf{Z} \cap b\mathbf{Z} = [a,b]\mathbf{Z}$.

8. Let a, b be nonzero integers, and let p be a prime. Show that if $p|[a,b]$, then either $p|a$ or $p|b$.

9. Let a, b, c be nonzero integers. Show that $(a,b) = 1$ and $(a,c) = 1$ if and only if $(a, [b,c]) = 1$.

10. Let a, b be nonzero integers. Prove $(a,b) = 1$ if and only if $(a + b, ab) = 1$.

11. Let a, b be nonzero integers with $(a,b) = 1$. Compute $(a + b, a - b)$.

12. Let a and b be positive integers. Without using the prime factorization theorem, prove that $(a,b)[a,b] = ab$ by verifying that

$$\frac{ab}{(a,b)}$$

satisfies the necessary properties of $[a,b]$.

13. Show that $a > 1$ is a perfect square if and only if every component in its prime factorization is even.

14. Show that if a, b are positive integers such that $(a,b) = 1$ and ab is a perfect square, then a and b are also perfect squares.

15. Show that if the positive integer a is not a perfect square, then $a \ne b^2/c^2$ for integers b, c. Thus any positive integer that is not a perfect square must have an irrational square root.
Hint: Use the previous exercise to show that $ac^2 \ne b^2$.

16. Prove that if $a > 1$, then there is a prime p with $a < p \le a! + 1$.

17. Show that for any $n > 0$, there are n consecutive composite numbers.

18. Show that if n is a positive integer such that $2^n + 1$ is prime, then n is a power of 2.

19. Show that log 2/log 3 is not a rational number.

20. If a, b, c are positive integers such that $a^2 + b^2 = c^2$, then (a,b,c) is called a Pythagorean triple. Examples: (3,4,5), (5,12,13). Assume that (a,b,c) is a Pythagorean triple in which the only common divisors of a, b, c are ± 1.
  (a) Show that a and b cannot both be odd.
  (b) If a is even, show that b and c must be of the form $b = m^2 - n^2$, $c = m^2 + n^2$, where $a = 2mn$.
  Hint: Factor $c^2 - b^2$ and show that $(c + b, c - b) = 2$.

1.3 CONGRUENCES

For many problems involving integers, all of the relevant information is contained in the remainders obtained by dividing by some fixed integer n. Since only n different remainders are possible $(0, 1, \ldots, n - 1)$, having only a finite number of cases to deal with can lead to considerable simplifications. For small values of n it even becomes feasible to use trial-and-error methods.

Example 1.3.1

In the introduction to Chapter 1, we briefly mentioned the theorem of Lagrange that states that every positive integer can be written as sum of four squares. To illustrate the use of remainders in solving a number theoretic problem, we will show that any positive integer whose remainder is 7 when divided by 8 cannot be written as the sum of three squares. Therefore this theorem of Lagrange is as sharp as possible.

If $n = a^2 + b^2 + c^2$, then when both sides are divided by 8, the remainders must be the same. It follows from Proposition 1.3.3 that we can compute the remainder of $n = a^2 + b^2 + c^2$ by adding the remainders of a^2, b^2, and c^2 (and dividing by 8 if necessary.) By the same proposition, we can compute the remainders of a^2, b^2, and c^2 by squaring the remainders of a, b, and c (and dividing by 8 if necessary). The possible remainders for a, b, and c are $0, 1, \ldots, 7$, and squaring and taking remainders yields only the values 0, 1, and 4. To check the possible remainders for $a^2 + b^2 + c^2$ we only need to add together three such terms. (If we get a sum larger than 7 we replace it with its remainder.) A careful analysis of all of the cases shows that we cannot obtain 7 as a remainder for $a^2 + b^2 + c^2$. Thus we cannot express any integer n whose remainder is 7 when divided by 8 in the form $n = a^2 + b^2 + c^2$. □

Similar techniques can sometimes be used to show that a polynomial equation has no integer solution. For example, if $x = c$ is a solution of the equation $a_k x^k + \cdots + a_1 x + a_0 = 0$, then $a_k c^k + \cdots + a_1 c + a_0$ must be divisible by every integer n. If some n can be found for which $a_k x^k + \cdots + a_1 x + a_0$ is never divisible by n, then this can be used to prove that the equation has no integer solutions. For example, $x^3 + x + 1$ has no integer roots since $c^3 + c + 1$ is odd for all integers c, and thus is never divisible by 2.

A more familiar situation in which we carry out arithmetic after dividing by a fixed integer is the addition of hours on a clock (where the fixed integer is 12). The familiar rules "even plus even is even," "even times even is even," etc., are useful in other circumstances (where the fixed integer is 2). Gauss introduced the following congruence notation, which simplifies computations of this sort.

1.3.1 Definition.   Let n be a positive integer. Integers a and b are said to be *congruent modulo n* if they have the same remainder when divided by n. This is denoted by writing $a \equiv b \pmod{n}$.

If we use the division algorithm to write $a = nq + r$, where $0 \leq r < n$, then $r = n \cdot 0 + r$. It follows immediately from the previous definition that $a \equiv r$ (mod n). In particular, any integer is congruent modulo n to one of the integers 0, $1, 2, \ldots, n - 1$.

We feel that the definition we have given provides the best intuitive understanding of the notion of congruence, but in almost all proofs it will be easiest to use the characterization given by the next proposition. Using this characterization makes it possible to utilize the facts about divisibility that we have developed in the preceding sections of this chapter.

1.3.2 Proposition.   Let a, b, and $n > 0$ be integers. Then $a \equiv b$ (mod n) if and only if $n|(a - b)$.

Proof. If $a \equiv b$ (mod n), then a and b have the same remainder when divided by n, so the division algorithm gives $a = nq_1 + r$ and $b = nq_2 + r$. Solving for the common remainder gives $a - nq_1 = b - nq_2$. Thus $a - b = n(q_1 - q_2)$, and so $n|(a - b)$.

To prove the converse, assume that $n|(a - b)$. By the division algorithm, $a = q_1 n + r_1$ and $b = q_2 n + r_2$, where $0 \leq r_1 < n$ and $0 \leq r_2 < n$. Subtracting gives $a - b = (q_1 - q_2)n + (r_1 - r_2)$, and since $n|(a - b)$, we must have $n|(r_1 - r_2)$. This can only occur if $r_1 - r_2 = 0$, since $-n < r_1 - r_2 < n$ and the only multiple of n in this interval is 0. Thus we must have $r_1 = r_2$, and so $a \equiv b$ (mod n). $\square$

When working with congruence modulo n, the integer n is called the *modulus*. By the preceding proposition, $a \equiv b$ (mod n) if and only if $a - b = nq$ for some integer q. We can write this in the form $a = b + nq$, for some integer q. This observation gives a very useful method of replacing a congruence with an equation (over **Z**). On the other hand, Proposition 1.3.3 shows that any equation can be converted to a congruence modulo n by simply changing the $=$ sign to $\equiv$. In doing so, any term congruent to 0 can simply be omitted. Thus the equation $a = b + nq$ would be converted back to $a \equiv b$ (mod n).

Congruence behaves in many ways like equality. The following properties, which are obvious from the definition of congruence modulo n, are a case in point. Let a, b, c be integers. Then

   (i) $a \equiv a$ (mod n).
   (ii) If $a \equiv b$ (mod n), then $b \equiv a$ (mod n).
   (iii) If $a \equiv b$ (mod n) and $b \equiv c$ (mod n), then $a \equiv c$ (mod n).

The following theorem carries this analogy even further. Perhaps its most important consequence is that when doing any computations with congruences you may substitute any congruent integer. For example, to show that $99^2 \equiv 1$ (mod 100), it is easier to substitute -1 for 99 and just show that $(-1)^2 = 1$.

1.3.3 Proposition.   Let $n > 0$ be an integer. Then the following conditions hold for all integers a, b, c, d:

  (a) If $a \equiv c \pmod{n}$ and $b \equiv d \pmod{n}$, then $a \pm b \equiv c \pm d \pmod{n}$, and $ab \equiv cd \pmod{n}$.

  (b) If $a + c \equiv a + d \pmod{n}$, then $c \equiv d \pmod{n}$. If $ac \equiv ad \pmod{n}$ and $(a,n) = 1$, then $c \equiv d \pmod{n}$.

Proof. (a) If $a \equiv c \pmod{n}$ and $b \equiv d \pmod{n}$, then $n \mid (a - c)$ and $n \mid (b - d)$. Adding shows that $n \mid ((a + b) - (c + d))$, and subtracting shows that $n \mid ((a - b) - (c - d))$. Thus $a \pm b \equiv c \pm d \pmod{n}$.

Since $n \mid (a - c)$, we have $n \mid (ab - cb)$, and since $n \mid (b - d)$, we have $n \mid (cb - cd)$. Adding shows that $n \mid (ab - cd)$ and thus $ab \equiv cd \pmod{n}$.

(b) If $a + c \equiv a + d \pmod{n}$, then $n \mid ((a + c) - (a + d))$. Thus $n \mid (c - d)$ and so $c \equiv d \pmod{n}$.

If $ac \equiv ad \pmod{n}$, then $n \mid (ac - ad)$, and since $(n,a) = 1$, it follows from Proposition 1.2.3 (b) that $n \mid (c - d)$. Thus $c \equiv d \pmod{n}$. $\square$

The second part of this proposition must be applied with care when you want to cancel. You may divide both sides of a congruence by an integer a only if $(a,n) = 1$. For example, $30 \equiv 6 \pmod{8}$, but dividing both sides by 6 gives $5 \equiv 1 \pmod{8}$, which is certainly false. On the other hand, since 3 is relatively prime to 8, we may divide both sides by 3 to get $10 \equiv 2 \pmod{8}$.

Proposition 1.3.3 shows that the remainder upon division by n of $a + b$ or ab can be found by adding or multiplying the remainders of a and b when divided by n and then dividing by n again if necessary. For example, if $n = 8$, then 101 has remainder 5 and 142 has remainder 6 when divided by 8. Thus $101 \cdot 142 = 14{,}342$ has the same remainder as 30 (namely, 6) when divided by 8. Formally, $101 \equiv 5 \pmod{8}$ and $142 \equiv 6 \pmod{8}$, so it follows that $101 \cdot 142 \equiv 5 \cdot 6 \equiv 6 \pmod{8}$.

As a further example, we compute the powers of 2 modulo 7. Rather than computing each power and then dividing by 7, we reduce modulo 7 at each stage of the computations:

$$2^2 \equiv 4 \pmod{7},$$

$$2^3 \equiv 2^2 2 \equiv 4 \cdot 2 \equiv 1 \pmod{7},$$

$$2^4 \equiv 2^3 2 \equiv 1 \cdot 2 \equiv 2 \pmod{7},$$

$$2^5 \equiv 2^4 2 \equiv 2 \cdot 2 \equiv 4 \pmod{7}.$$

From the way in which we have done the computations, it is clear that the powers will repeat. In fact, since there are only finitely many remainders modulo n, the powers of any integer will eventually begin repeating modulo n.

1.3.4 Proposition.   Let a and $n > 1$ be integers. There exists an integer b such that $ab \equiv 1 \pmod{n}$ if and only if $(a,n) = 1$.

Proof. If there exists an integer b such that $ab \equiv 1 \pmod{n}$, then we have $ab = 1 + qn$ for some integer q. This can be rewritten to give a linear combination of a and n equal to 1, and so $(a,n) = 1$.

Conversely, if $(a,n) = 1$, then there exist integers s, t such that $sa + tn = 1$. Letting $b = s$ and reducing the equation to a congruence modulo n gives $ab \equiv 1 \pmod{n}$. $\square$

We are now ready to study congruences that involve unknowns. The previous proposition shows that the congruence

$$ax \equiv 1 \pmod{n}$$

has a solution if and only if $(a,n) = 1$. In fact, the proof of the theorem shows that the solution can be obtained by using the Euclidean algorithm to write $1 = ab + nq$ for some b, $q \in \mathbf{Z}$, since then $1 \equiv ab \pmod{n}$.

The next theorem shows how to systematically solve linear congruences of the form

$$ax \equiv b \pmod{n}.$$

Of course, if the numbers involved are small, it may be simplest just to use trial and error. For example, to solve $3x \equiv 2 \pmod 5$, we only need to substitute $x = 0$, 1, 2, 3, 4. Thus by trial and error we can find the solution $x \equiv 4 \pmod 5$.

In many ways, solving congruences is like solving equations. There are a few important differences, however. A linear equation over the integers has at most one solution. On the other hand, $2x \equiv 2 \pmod 4$ has the solution $x \equiv 1 \pmod 4$ or $x \equiv 3 \pmod 4$. Two solutions r and s to the congruence $x \equiv b \pmod n$ are considered to be distinct if and only if r and s are not congruent modulo n. Thus in the next theorem the statement "d distinct solutions modulo n" means that there are d solutions s_1, s_2, . . . , s_d such that if $i \neq j$, then s_i and s_j are not congruent modulo n.

1.3.5 Theorem.   The congruence $ax \equiv b \pmod n$ has a solution if and only if $d|b$, where $d = (a,n)$. If $d|b$, then there are d distinct solutions modulo n, and these solutions are congruent modulo n/d.

Proof. To prove the first statement, observe that $ax \equiv b \pmod n$ has a solution if and only if there exist integers s and q such that $as = b + nq$, or, equivalently, $as + (-q)n = b$. Thus there is a solution if and only if b can be expressed as a linear combination of a and n. By Theorem 1.1.6 the linear combinations of a and n are precisely the multiples of d, so there is a solution if and only if $d|b$.

To prove the second statement, assume that $d|b$, and let $m = n/d$. Suppose that x_1 and x_2 are solutions of the congruence $ax \equiv b \pmod n$, giving $ax_1 \equiv ax_2 \pmod n$. Then $n|a(x_1 - x_2)$, and so Proposition 1.2.3 (a) implies that $n|d(x_1 - x_2)$.

Thus $m|(x_1 - x_2)$, and so $x_1 \equiv x_2 \pmod{m}$. On the other hand, if $x_1 \equiv x_2 \pmod{m}$, then $m|(x_1 - x_2)$, and so $n|d(x_1 - x_2)$ since $n = dm$. Then since $d|a$ we can conclude that $n|a(x_1 - x_2)$, and so $ax_1 \equiv ax_2 \pmod{n}$.

We can choose the distinct solutions from among the n remainders $0, 1, \ldots,$ $n - 1$. Given one such solution, we can find all others in the set by adding multiples of n/d, giving a total of d distinct solutions.   □

We now describe an algorithm for solving linear congruences of the form $ax \equiv b \pmod{n}$. First compute $d = (a,n)$, and if $d|b$, then write the congruence $ax \equiv b \pmod{n}$ as an equation $ax = b + qn$. Now by writing $a_1 = a/d$, $b_1 = b/d$, and $m = n/d$ we get $a_1 x = b_1 + qm$, which yields the congruence $a_1 x \equiv b_1 \pmod{m}$. It follows immediately from Proposition 1.2.9 that since $d = (a,n)$, the numbers a_1 and m must be relatively prime. Thus by Proposition 1.3.4 we may apply the Euclidean algorithm to find an integer c such that $ca_1 \equiv 1 \pmod{m}$. Multiplying the congruence $a_1 x \equiv b_1 \pmod{m}$ by c gives the solution $x \equiv b_1 c \pmod{m}$.

Finally, since the original congruence was given modulo n, we should give our answer modulo n instead of modulo m. The solution modulo m determines d distinct congruence classes modulo n, each of which is a solution modulo n. The solutions have the form $s_0 + km$, where s_0 is any particular solution of $x \equiv b_1 c \pmod{m}$ and k is any integer.

Example 1.3.2

To solve the congruence $60x \equiv 90 \pmod{105}$, we first note that $(60,105) = 15$, and then that $15|90$. Dividing by 15, we obtain the congruence $4x \equiv 6 \pmod{7}$. In order to solve this congruence, we need an integer c with $c \cdot 4 \equiv 1 \pmod{7}$. We could use the Euclidean algorithm, but with such a small modulus, trial and error is quicker. Using trial and error, we find that $c = 2$ will work, and so multiplying by 2 we have $8x \equiv 12 \pmod{7}$, or $x \equiv 5 \pmod{7}$. Thus we obtain the solutions $\ldots, -2, 5, 12, 19, \ldots$. There are 15 solutions that represent distinct congruence classes modulo 105, namely, 5, 12, 19, 26, 33, 40, 47, 54, 61, 68, 75, 82, 89, 96, and 103. These are obtained by adding multiples of 7 to the particular solution $x_0 = 5$.   □

In the next theorem we show how to solve two simultaneous congruences over moduli that are relatively prime. The motivation for the proof of the next theorem is as follows: Assume that the congruences $x \equiv a \pmod{n}$ and $x \equiv b \pmod{m}$ are given. If we can find integers y and z with

$$y \equiv 1 \pmod{n} \qquad y \equiv 0 \pmod{m}$$

$$z \equiv 0 \pmod{n} \qquad z \equiv 1 \pmod{m}$$

then $x = ay + bz$ will be a solution, as can be seen by reducing modulo n and then modulo m.

1.3.6 Theorem (Chinese Remainder Theorem).   Let n and m be positive integers, with $(n,m) = 1$. Then the system of congruences

$$x \equiv a \ (\text{mod } n) \qquad x \equiv b \ (\text{mod } m)$$

has a solution. Moreover, any two solutions are congruent modulo mn.

Proof. Since $(n,m) = 1$, there exist integers r and s such that $rm + sn = 1$. Then $rm \equiv 1 \ (\text{mod } n)$ and $sn \equiv 1 \ (\text{mod } m)$. Following the suggestion in the preceding paragraph, we let $x = arm + bsn$. Then a direct computation verifies that $x \equiv arm \equiv a \ (\text{mod } n)$ and $x \equiv bsn \equiv b \ (\text{mod } m)$.

If x is a solution, then adding any multiple of mn is obviously still a solution. Conversely, if x_1 and x_2 are two solutions of the given system of congruences, then they must be congruent modulo n and modulo m. Thus $x_1 - x_2$ is divisible by both n and m, so it is divisible by mn since by assumption $(n,m) = 1$. Therefore $x_1 \equiv x_2$ $(\text{mod } mn)$.   □

Example 1.3.3

The proof of Theorem 1.3.6 actually shows how to solve the given system of congruences. For example, if we wish to solve the system

$$x \equiv 7 \ (\text{mod } 8) \qquad x \equiv 3 \ (\text{mod } 5)$$

we first use the Euclidean algorithm to write $2 \cdot 8 - 3 \cdot 5 = 1$. Then $x = 7(-3)(5) + 3(2)(8) = -57$ is a solution, and the general solution is $-57 + 40t$. The smallest nonnegative solution is therefore 23.   □

Another proof of the existence of a solution in Theorem 1.3.6 can be given as follows. In some respects this method of solution is more intuitive and provides a convenient algorithm for solving the congruences. Given the congruences

$$x \equiv a \ (\text{mod } n) \qquad x \equiv b \ (\text{mod } m)$$

we can rewrite the first congruence as an equation in the form $x = a + qn$ for some $q \in \mathbf{Z}$. To find a simultaneous solution, we only need to substitute this expression for x in the second congruence, giving $a + qn \equiv b \ (\text{mod } m)$, or

$$qn \equiv b - a \ (\text{mod } m).$$

This congruence can be solved for q, since $(n,m) = 1$. This gives the simultaneous solutions to the two congruences in the form $x = a + qn$, and so we can choose as a particular solution the smallest positive integer in this form. The general solution is obtained by adding multiples of mn.

Example 1.3.4

To illustrate the second method of solution, again consider the system

$$x \equiv 7 \ (\text{mod } 8) \qquad x \equiv 3 \ (\text{mod } 5).$$

The first congruence gives us the equation $x = 7 + 8q$, and then substituting we obtain $7 + 8q \equiv 3 \pmod 5$, or equivalently, $3q \equiv -4 \pmod 5$. Multiplying by 2, since $2 \cdot 3 \equiv 1 \pmod 5$, gives $q \equiv -8 \pmod 5$ or $q \equiv 2 \pmod 5$. This yields the particular solution $x = 7 + 2 \cdot 8 = 23$.   □

EXERCISES: SECTION 1.3

1. Write n as a sum of four squares for $1 \le n \le 20$.
2. Solve the following congruences:
   (a) $2x \equiv 1 \pmod 9$
   (b) $5x \equiv 1 \pmod{32}$
   (c) $19x \equiv 1 \pmod{36}$
3. Solve the following congruences:
   (a) $10x \equiv 5 \pmod{21}$
   (b) $10x \equiv 5 \pmod{15}$
   (c) $10x \equiv 4 \pmod{15}$
4. Solve the following congruence:
   $20x \equiv 12 \pmod{72}$
5. Find all integers x such that $3x + 7$ is divisible by 11.
   (New techniques are available for this problem, which was Exercise 17 in Section 1.1.)
6. Let a, b, n be positive integers. Show that if $a \equiv b \pmod n$, then $(a,n) = (b,n)$.
7. The smallest positive solution of the congruence $ax \equiv 0 \pmod n$ is called the *additive order* of a modulo n. Find the additive orders of each of the following elements, by solving the appropriate congruences:
   (a) 8 modulo 12
   (b) 7 modulo 12
   (c) 21 modulo 28
   (d) 12 modulo 18
8. Show that if p is a prime number and a is any integer such that $p \nmid a$, then the additive order of a modulo p is equal to p.
9. Prove that if $n > 1$ and $a > 0$ are integers and $d = (a,n)$, then the additive order of a modulo n is n/d.
10. Show that $7|(6! + 1)$ and $11|(10! + 1)$.
11. Show that $4 \cdot (n^2 + 1)$ is never divisible by 11.
12. Prove that the sum of the cubes of any three consecutive positive integers is divisible by 9. (Compare Exercise 9 of Section 1.1.)
13. Find the units digit of $3^{29} + 11^{12} + 15$.
    Hint: Work modulo n, after choosing an appropriate modulus.
14. Solve the following congruences by trial and error:
    (a) $x^2 \equiv 1 \pmod{16}$
    (b) $x^3 \equiv 1 \pmod{16}$
    (c) $x^4 \equiv 1 \pmod{16}$

15. Solve the following congruences by trial and error:

(a) $x^3 + 2x + 2 \equiv 0 \pmod 5$

(b) $x^4 + x^3 + x^2 + x + 1 \equiv 0 \pmod 2$

(c) $x^4 + x^3 + 2x^2 + 2x \equiv 0 \pmod 3$

16. List and solve all quadratic congruences modulo 3; that is, list and solve all congruences of the form $ax^2 + bx + c \equiv 0 \pmod 3$. The only coefficients you need to consider are 0, 1, 2.

17. Solve the following system of congruences:

$$x \equiv 15 \pmod{27} \qquad x \equiv 16 \pmod{20}$$

18. Solve the following system of congruences:

$$x \equiv 11 \pmod{16} \qquad x \equiv 18 \pmod{25}$$

19. Solve the following system of congruences:

$$2x \equiv 5 \pmod 7 \qquad 3x \equiv 4 \pmod 8$$

Hint: Solve an equivalent system of congruences in which the coefficients of x are both 1.

20. Extend the techniques of the Chinese remainder theorem to solve the following system of congruences:

$$2x \equiv 3 \pmod 7 \qquad x \equiv 4 \pmod 6 \qquad 5x \equiv 50 \pmod{55}$$

21. (Casting out nines) Show that the remainder of n when divided by 9 is the same as the remainder of the sum of its digits when divided by 9. For example, $7862 \equiv 7 + 8 + 6 + 2 \pmod 9$.

Hint: How do you use the digits of a number to express it in terms of powers of 10?

22. Find a result similar to casting out nines for the integer 11.

23. Let p be a prime number and let a, b be any integers. Prove that

$$(a + b)^p \equiv a^p + b^p \pmod p.$$

1.4 INTEGERS MODULO *n*

In working with congruences, we have established that in any computations we can consider congruent numbers to be interchangeable. In this section we will formalize this point of view. We will now consider entire congruence classes as individual entities, and we will work with these entities much as we do with ordinary numbers. The point of introducing the notation given below is to allow us to use our experience with ordinary numbers as a guide to working with congruence classes. Most of the laws of integer arithmetic hold for the arithmetic of congruence classes. The notable exception is that the product of two nonzero congruence classes may be zero. For example, $2 \not\equiv 0 \pmod 6$ and $3 \not\equiv 0 \pmod 6$, but $2 \cdot 3 \equiv 0 \pmod 6$.

1.4.1 Definition.   Let a and $n > 0$ be integers. The set of all integers which have the same remainder as a when divided by n is called the *congruence class of a modulo n*, denoted by $[a]_n$. The collection of all congruence classes modulo n is called the *set of integers modulo n*, denoted by $\mathbf{Z}_n$.

Thus $[a]_n = \{x \in \mathbf{Z} \mid x \equiv a \pmod{n}\}$. Note that $[a]_n = [b]_n$ if and only if $a \equiv b$ (mod n). When the modulus is clearly understood from the context, the subscript n can be omitted and $[a]_n$ can be written simply as $[a]$.

A given congruence class can be denoted in many ways. For example, $x \equiv 5$ (mod 3) if and only if $x \equiv 8$ (mod 3), since $5 \equiv 8$ (mod 3). This shows that $[5]_3 = [8]_3$. We sometimes say that an element of $[a]_n$ is a *representative* of the congruence class. Each congruence class $[a]_n$ has a unique nonnegative representative that is smaller than n, namely, the remainder when a is divided by n. This shows that there are exactly n distinct congruence classes modulo n. For example, the congruence classes modulo 3 can be represented by 0, 1, and 2.

$$[0]_3 = \{\ldots, -9, -6, -3, 0, 3, 6, 9, \ldots\}$$
$$[1]_3 = \{\ldots, -8, -5, -2, 1, 4, 7, 10, \ldots\}$$
$$[2]_3 = \{\ldots, -7, -4, -1, 2, 5, 8, 11, \ldots\}$$

Note that each integer belongs to exactly one congruence class modulo 3, since the remainder on division by 3 is unique. In general, each integer belongs to a unique congruence class modulo n. Hence we have

$$\mathbf{Z}_n = \{[0]_n, [1]_n, \ldots, [n-1]_n\}.$$

The set $\mathbf{Z}_2$ consists of $[0]_2$ and $[1]_2$, where $[0]_2$ is the set of even numbers and $[1]_2$ is the set of odd numbers. The familiar rules "even + even = even," "odd + even = odd," and "odd + odd = even" can be expressed as $[0]_2 + [0]_2 = [0]_2$, $[1]_2 + [0]_2 = [1]_2$, and $[1]_2 + [1]_2 = [0]_2$. Similarly, "even $\times$ even = even," "even $\times$ odd = even," and "odd $\times$ odd = odd" can be expressed as $[0]_2 \cdot [0]_2 = [0]_2$, $[0]_2 \cdot [1]_2 = [0]_2$, and $[1]_2 \cdot [1]_2 = [1]_2$.

These rules can be summarized by giving an addition table (Table 1.4.1) and a multiplication table (Table 1.4.2). To use the addition table, select an element a from the first column, and an element b from the top row. Read from left to right in the row to which a belongs, until reaching the column to which b belongs. The corresponding entry in the table is $a + b$.

Here, as we will sometimes do elsewhere, we will simplify our notation for congruence classes. We will omit even the brackets, as well as the subscript, in $[a]_n$.

TABLE 1.4.1. Addition in $\mathbf{Z}_2$

+	0	1
0	0	1
1	1	0

TABLE 1.4.2. Multiplication in $\mathbf{Z}_2$

$\cdot$	0	1
0	0	0
1	0	1

A similar addition and multiplication can be introduced in $\mathbf{Z}_n$, for any n. Given congruence classes in $\mathbf{Z}_n$, we add (or multiply) them by picking representatives of each congruence class. We then add (or multiply) the representatives, and find the congruence class to which the result belongs.

This can be written formally as follows:

Addition: $[a]_n + [b]_n = [a + b]_n$

Multiplication: $[a]_n \cdot [b]_n = [ab]_n$

In $\mathbf{Z}_{12}$, for example, we have $[8]_{12} = [20]_{12}$ and $[10]_{12} = [34]_{12}$. Adding congruence classes gives the same answer, no matter which representatives we use: $[8]_{12} + [10]_{12} = [18]_{12} = [6]_{12}$ and also $[20]_{12} + [34]_{12} = [54]_{12} = [6]_{12}$.

1.4.2 Proposition.   Let n be a positive integer, and let a, b be any integers. Then the addition and multiplication of congruence classes given below are well-defined:

$$[a]_n + [b]_n = [a + b]_n, \qquad [a]_n \cdot [b]_n = [ab]_n.$$

Proof. We must show that the given formulas do not depend on the integers a and b which have been chosen to represent the congruence classes with which we are concerned. Suppose that x and y are any other representatives of the congruence classes $[a]_n$ and $[b]_n$, respectively. Then $x \equiv a \pmod{n}$ and $y \equiv b \pmod{n}$, and so we can apply Proposition 1.3.3. It follows from that proposition that $x + y \equiv a + b \pmod{n}$ and $xy \equiv ab \pmod{n}$, and thus we have $[x]_n + [y]_n = [a + b]_n$ and $[x]_n \cdot [y]_n = [ab]_n$. Since the formulas we have given do not depend on the particular representatives chosen, we say that addition and multiplication are "well-defined." $\square$

The familiar rules for addition and multiplication carry over from the addition and multiplication of integers. A complete discussion of these rules will be given in Chapter 5, when we study ring theory. If $[a]_n, [b]_n \in \mathbf{Z}_n$ and $[a]_n + [b]_n = [0]_n$, then $[b]_n$ is called an *additive inverse* of $[a]_n$. By Proposition 1.3.3 (b), additive inverses are unique. We will denote the additive inverse of $[a]_n$ by $-[a]_n$. It is easy to see that $-[a]_n$ is in fact equal to $[-a]_n$, since $[a]_n + [-a]_n = [a - a]_n = [0]_n$.

For any elements $[a]_n$, $[b]_n$, $[c]_n$ in $\mathbf{Z}_n$, the following laws hold:

Associativity: $([a]_n + [b]_n) + [c]_n = [a]_n + ([b]_n + [c]_n)$

$([a]_n \cdot [b]_n) \cdot [c]_n = [a]_n \cdot ([b]_n \cdot [c]_n)$

Commutativity: $[a]_n + [b]_n = [b]_n + [a]_n$

$[a]_n \cdot [b]_n = [b]_n \cdot [a]_n$

Distributivity: $[a]_n \cdot ([b]_n + [c]_n) = [a]_n \cdot [b]_n + [a]_n \cdot [c]_n$

Identities: $\qquad\qquad\qquad [a]_n + [0]_n = [a]_n$

$$[a]_n \cdot [1]_n = [a]_n$$

Additive inverses: $\qquad [a]_n + [-a]_n = [0]_n$

We will give a proof of the distributive law and leave the proofs of the remaining properties as an exercise. If a, b, c, $\in \mathbf{Z}$, then

$$[a]_n \cdot ([b]_n + [c]_n) = [a]_n \cdot [b + c]_n = [a(b + c)]_n$$
$$= [ab + ac]_n = [ab]_n + [ac]_n$$
$$= [a]_n \cdot [b]_n + [a]_n \cdot [c]_n.$$

The steps in the proof depend on the definitions of addition and multiplication and the equality $a(b + c) = ab + ac$, which is the distributive law for $\mathbf{Z}$.

In doing computations in $\mathbf{Z}_n$, the one point at which particular care must be taken is the cancellation law, which no longer holds in general. Otherwise, in almost all cases your experience with integer arithmetic can be trusted when working with congruence classes. A quick computation shows that $[6]_8 \cdot [5]_8 = [6]_8 \cdot [1]_8$, but $[5]_8 \neq [1]_8$. It can also happen that the product of nonzero classes is equal to zero. For example, $[6]_8 \cdot [4]_8 = [0]_8$. If $[a]_n[b]_n = [0]_n$ for some nonzero congruence class $[b]_n$, then $[a]_n$ is called a *divisor of zero*. If $[a]_n$ is not a divisor of zero, then we may cancel $[a]_n$. To see this, if $[a]_n[b]_n = [a]_n[c]_n$, then $[a]_n([b]_n - [c]_n) = [0]_n$, and so $[b]_n - [c]_n$ must be zero since $[a]_n$ is not a divisor of zero, which means that $[b]_n = [c]_n$.

From this point on, if the meaning is clear from the context we will omit the subscript on congruence classes. For the next remarks we will assume that the given congruence classes all belong to $\mathbf{Z}_n$.

If $[a][b] = [1]$, then $[b]$ is called a *multiplicative inverse* of $[a]$ and is denoted by $[a]^{-1}$. The next proposition (which is just a restatement of Proposition 1.3.4) shows that a has a multiplicative inverse modulo n if and only if $(a,n) = 1$. In this case it follows from Proposition 1.3.3 (b) that there is a unique solution to the congruence $ax \equiv 1 \pmod{n}$, and so we are justified in referring to *the* multiplicative inverse of $[a]$, whenever it exists.

In $\mathbf{Z}_7$, each nonzero congruence class contains representatives which are relative prime to 7, and so each nonzero congruence class has a multiplicative inverse. We can list them as $[1]^{-1} = [1]$, $[2]^{-1} = [4]$, $[3]^{-1} = [5]$ and $[6]^{-1} = [6]$. We did not need to list $[4]^{-1}$ and $[5]^{-1}$ since, in general, if $[a]^{-1} = [b]$, then $[b]^{-1} = [a]$.

If $[a][b] = [1]$, then we say that $[a]$ and $[b]$ are *invertible* elements of $\mathbf{Z}_n$, or *units* of $\mathbf{Z}_n$.

We note that if $[a]$ has a multiplicative inverse, then it cannot be a divisor of zero, since $[a][b] = [0]$ implies $[b] = [a]^{-1}([a][b]) = [a]^{-1}[0] = [0]$.

1.4.3 Proposition. Let n be a positive integer.

(a) The congruence class $[a]_n$ has a multiplicative inverse in $\mathbf{Z}_n$ if and only if $(a,n) = 1$.

(b) A nonzero element of $\mathbf{Z}_n$ either has a multiplicative inverse or is a divisor of zero.

Proof. (a) If $[a]$ has a multiplicative inverse, say $[a]^{-1} = [b]$, then $[a][b] = [1]$. Therefore $ab \equiv 1 \pmod{n}$, which implies that $ab = 1 + qn$ for some integer q. Thus $ab + (-q)n = 1$, and so $(a,n) = 1$.

Conversely, if $(a,n) = 1$, then there exist integers b and q such that $ab + qn = 1$. Reducing modulo n shows that $ab \equiv 1 \pmod{n}$, and so $[b] = [a]^{-1}$.

(b) Assume that a represents a nonzero congruence class, so that $n \nmid a$. If $(a,n) = 1$, then $[a]$ has a multiplicative inverse. If not, then $(a,n) = d$, where $1 < d < n$. In this case, since $d \mid n$ and $d \mid a$, we can find integers k, b with $n = kd$ and $a = bd$. Then $[k]$ is a nonzero element of $\mathbf{Z}_n$, but

$$[a][k] = [ak] = [bdk] = [bn] = [0],$$

which shows that $[a]$ is a divisor of zero. $\square$

1.4.4 Corollary. The following conditions on the modulus $n > 0$ are equivalent:

(1) The number n is prime.

(2) $\mathbf{Z}_n$ has no divisors of zero, except $[0]_n$.

(3) Every nonzero element of $\mathbf{Z}_n$ has a multiplicative inverse.

Proof. Since n is prime if and only if every positive integer less than n is relatively prime to n, Corollary 1.4.4 follows from Proposition 1.4.3. $\square$

The proof of Proposition 1.4.3(a) shows that if $(a,n) = 1$, then the multiplicative inverse of $[a]$ can be computed by using the Euclidean algorithm.

Example 1.4.1

For example, to find $[11]^{-1}$ in $\mathbf{Z}_{16}$ we have the following matrix computation:

$$\begin{bmatrix} 1 & 0 & 16 \\ 0 & 1 & 11 \end{bmatrix} \rightsquigarrow \begin{bmatrix} 1 & -1 & 5 \\ 0 & 1 & 11 \end{bmatrix} \rightsquigarrow \begin{bmatrix} 1 & -1 & 5 \\ -2 & 3 & 1 \end{bmatrix} \rightsquigarrow \begin{bmatrix} 11 & -16 & 0 \\ -2 & 3 & 1 \end{bmatrix}.$$

Thus $16(-2) + 11 \cdot 3 = 1$, which shows that $[11]_{16}^{-1} = [3]_{16}$.

When the numbers are small, as in this case, it is often easier to use trial and error. The positive integers less than 16 and relatively prime to 16 are $1, 3, 5, 7, 9, 11, 13, 15$. It is easier to use the representatives $\pm 1, \pm 3, \pm 5, \pm 7$ since if $[a][b] = [1]$, then $[-a][-b] = [1]$, and so $[-a]^{-1} = -[a]^{-1}$. Now we observe that $3 \cdot 5 = 15 \equiv -1 \pmod{16}$, so $3(-5) \equiv 1 \pmod{16}$. Thus

$[3]_{16}^{-1} = [-5]_{16} = [11]_{16}$ and $[-3]_{16}^{-1} = [5]_{16}$. Finally, $7 \cdot 7 \equiv 1 \pmod{16}$, so $[7]_{16}^{-1} = [7]_{16}$ and $[-7]_{16}^{-1} = [-7]_{16} = [9]_{16}$.   □

Another way to find the inverse of an element $[a] \in \mathbf{Z}_n$ is to take successive powers of $[a]$. If $(a,n) = 1$, then $[a]$ is not a zero divisor, and so no power of $[a]$ can be zero. We let $[a]^0 = [1]$. The set of powers $[1], [a], [a]^2, [a]^3, \ldots$ must contain fewer than n distinct elements, so after some point there must be a repetition. Suppose that the first repetition occurs for the exponent m, say $[a]^m = [a]^k$, with $k < m$. Then $[a]^{m-k} = [a]^0 = [1]$ since we can cancel $[a]$ from both sides a total of k times. This shows that for the first repetition we must have had $k = 0$, so actually $[a]^m = [1]$. From this we can see that $[a]^{-1} = [a]^{m-1}$.

Example 1.4.2

To find $[11]_{16}^{-1}$, we can list the powers of $[11]_{16}$. We have $[11]^2 = [-5]^2 = [9]$, $[11]^3 = [11]^2[11] = [99] = [3]$, and $[11]^4 = [11]^3[11] = [33] = [1]$. Thus again we see that $[11]_{16}^{-1} = [3]_{16}$.   □

We are now ready to continue our study of equations in $\mathbf{Z}_n$. A linear congruence of the form $ax \equiv b \pmod{n}$ can be viewed as a linear equation $[a]_n[x]_n = [b]_n$ in $\mathbf{Z}_n$. If $[a]_n$ has a multiplicative inverse, then there is a unique congruence class $[x]_n = [a]_n^{-1}[b]_n$ that is the solution to the equation. Note that without the notation for congruence classes we would need to modify the statement regarding uniqueness to say that if x_0 is a solution of $ax \equiv b \pmod{n}$, then so is $x_0 + qn$, for any integer q.

Solving congruences of the form $a_k x^k + \ldots + a_1 x + a_0 \equiv 0 \pmod{n}$ is considerably harder. It can be shown that in solving congruences modulo n of degree greater than or equal to 1, the problem reduces to solving congruences modulo p^α for the prime factors of n. (This uses the Chinese remainder theorem.) In an elementary course in number theory, it is shown how to determine the solutions modulo a prime power p^α from the solutions modulo p. Then to determine the solutions modulo p we can proceed by trial and error, simply substituting each of $0, 1, \ldots, p-1$ into the congruence. Fermat's theorem (Corollary 1.4.10) can be used to reduce the problem to considering polynomials of degree at most $p - 1$.

We will prove this theorem of Fermat as a special case of a more general theorem due to Euler. Another proof will also be given in Section 3.2, which takes advantage of the concepts we will have developed by then. The statement of Euler's theorem involves a function of paramount importance in number theory and algebra, which we now introduce.

1.4.5 Definition.   Let n be a positive integer. The number of positive integers less than or equal to n which are relatively prime to n will be denoted by $\varphi(n)$. This function is called *Euler's φ-function*, or the *totient function*.

In Section 1.2 we gave a procedure for listing the positive integers less than n and relatively prime to n. However, in many cases we only need to determine the numerical value of $\varphi(n)$, without actually listing the numbers themselves. With the formula in Proposition 1.4.6, $\varphi(n)$ can be given in terms of the prime factorization of n. Note that $\varphi(1) = 1$.

1.4.6 Proposition.   If the prime factorization of n is $n = p_1^{\alpha_1} p_2^{\alpha_2} \cdots p_k^{\alpha_k}$, then

$$\varphi(n) = n \left(1 - \frac{1}{p_1}\right) \left(1 - \frac{1}{p_2}\right) \cdots \left(1 - \frac{1}{p_k}\right).$$

Proof. See Exercises 10 and 26. A proof of this result will also be presented in Section 3.5.   □

Example 1.4.3

Using the formula, we have

$$\varphi(10) = 10 \left(\frac{1}{2}\right) \left(\frac{4}{5}\right) = 4$$

and

$$\varphi(36) = 36 \left(\frac{1}{2}\right) \left(\frac{2}{3}\right) = 12.   □$$

1.4.7 Definition.   The set of units of $\mathbf{Z}_n$, the congruence classes $[a]$ such that $(a,n) = 1$, will be denoted by $\mathbf{Z}_n^\times$.

1.4.8 Proposition.   The set $\mathbf{Z}_n^\times$ of units of $\mathbf{Z}_n$ is closed under multiplication.

Proof. This can be shown either by using Proposition 1.2.3 (d) or by using the formula $([a][b])^{-1} = [b]^{-1}[a]^{-1}$.   □

The number of elements of $\mathbf{Z}_n^\times$ is given by $\varphi(n)$. The next theorem should be viewed as a result on powers of elements in $\mathbf{Z}_n^\times$, although it is phrased in the more familiar congruence notation.

1.4.9 Theorem (Euler).   If $(a,n) = 1$, then $a^{\varphi(n)} \equiv 1 \pmod{n}$.

Proof. In the set $\mathbf{Z}_n$, there are $\varphi(n)$ congruence classes which are represented by an integer relatively prime to n. Let these representatives be $\{a_1, \ldots, a_{\varphi(n)}\}$. Consider the congruence classes represented by the products $\{aa_1, \ldots, aa_{\varphi(n)}\}$, which are all distinct because $(a,n) = 1$. Since each of the products is still relatively prime to n, we must have a representative from each of the $\varphi(n)$ congruence classes we started with. Therefore

$$a_1 a_2 \cdots a_{\varphi(n)} \equiv (aa_1)(aa_2) \cdots (aa_{\varphi(n)}) \equiv a^{\varphi(n)} a_1 a_2 \cdots a_{\varphi(n)} \pmod{n}.$$

Since the product $a_1 \cdots a_{\varphi(n)}$ is relatively prime to n, we can cancel it in the congruence

$$a_1 a_2 \cdots a_{\varphi(n)} \equiv a^{\varphi(n)} a_1 a_2 \cdots a_{\varphi(n)} \ (\text{mod } n),$$

and so we have $a^{\varphi(n)} \equiv 1 \ (\text{mod } n)$.   □

1.4.10 Corollary (Fermat).   If p is prime, then for any integer a, $a^p \equiv a$ $(\text{mod } p)$.

Proof. If $p|a$, then trivially $a^p \equiv a \equiv 0 \ (\text{mod } p)$. If $p \nmid a$, then $(a,p) = 1$ and Euler's theorem gives $a^{\varphi(p)} \equiv 1 \ (\text{mod } p)$. Then since $\varphi(p) = p - 1$, we have $a^p \equiv a$ $(\text{mod } p)$.   □

It is instructive to include another proof of Fermat's theorem, one that does not depend on Euler's theorem. Expanding $(a + b)^p$ we obtain

$$(a + b)^p = a^p + pa^{p-1}b + \frac{p(p-1)}{1 \cdot 2} a^{p-2}b^2 + \ldots + pab^{p-1} + b^p.$$

For $k \neq 0$, $k \neq p$, each of the coefficients $p!/(k!(p - k)!)$ is an integer and has p as a factor, since p is a divisor of the numerator but not the denominator. Therefore $(a + b)^p \equiv a^p + b^p \ (\text{mod } p)$. Using induction, this can be extended to more terms, giving $(a + b + c)^p \equiv a^p + b^p + c^p \ (\text{mod } p)$, etc. Writing a as $(1 + 1 + \ldots + 1)$ shows that

$$a^p = (1 + 1 + \ldots + 1)^p \equiv 1^p + \ldots + 1^p \equiv a \ (\text{mod } p).$$

As a final remark we note that if $(a,n) = 1$, then the multiplicative inverse of $[a]_n$ can be given explicitly as $[a]_n^{\varphi(n)-1}$, since by Euler's theorem, $a \cdot a^{\varphi(n)-1} \equiv 1$ $(\text{mod } n)$. Note also that for a given n the exponent $\varphi(n)$ in Euler's theorem may not be the smallest exponent possible. For example, in $\mathbf{Z}_8$ the integers $\pm 1, \pm 3$, are relatively prime to 8, and Euler's theorem states that $a^4 \equiv 1 \ (\text{mod } 8)$ for each of these integers. In fact, $a^2 \equiv 1 \ (\text{mod } 8)$ for $a = \pm 1, \pm 3$.

EXERCISES: SECTION 1.4

1. Make addition and multiplication tables for $\mathbf{Z}_3$, $\mathbf{Z}_4$, $\mathbf{Z}_6$, and $\mathbf{Z}_{12}$.
2. Find the multiplicative inverses of the given elements (if possible):
   (a) $[14]$ in $\mathbf{Z}_{15}$
   (b) $[38]$ in $\mathbf{Z}_{83}$
   (c) $[351]$ in $\mathbf{Z}_{6669}$
   (d) $[91]$ in $\mathbf{Z}_{2565}$
3. Use Proposition 1.3.3 (b) to show directly that if $[b]$ and $[c]$ are both multiplicative inverses of $[a]$ in $\mathbf{Z}_n$, then $b \equiv c \ (\text{mod } n)$.
4. Let a and b be integers. Prove that in $\mathbf{Z}_n$ either $[a]_n \cap [b]_n = \varnothing$ or $[a]_n = [b]_n$.

5. Prove that each congruence class $[a]_n$ in $\mathbf{Z}_n$ has a unique representative r that satisfies $0 \le r < n$.

6. Let $(a,n) = 1$. The smallest positive integer k such that $a^k \equiv 1 \pmod{n}$ is called the *order* of $[a]$ in $\mathbf{Z}_n^\times$.
   (a) Find the orders of $[5]$ and $[7]$ in $\mathbf{Z}_{16}^\times$.
   (b) Find the orders of $[2]$ and $[5]$ in $\mathbf{Z}_{17}^\times$.

7. Let $(a,n) = 1$. If $[a]$ has order k in $\mathbf{Z}_n^\times$, show that $k|\varphi(n)$.

8. In $\mathbf{Z}_9^\times$ each element is equal to a power of $[2]$. (Verify this.) Can you find a congruence class in $\mathbf{Z}_8^\times$ such that each element of $\mathbf{Z}_8^\times$ is equal to some power of that class? Answer the same question for $\mathbf{Z}_7^\times$.

9. Generalizing the previous exercise, we say that the set of units $\mathbf{Z}_n^\times$ of $\mathbf{Z}_n$ is *cyclic* if it has an element of order $\varphi(n)$. Show that $\mathbf{Z}_{10}^\times$ and $\mathbf{Z}_{11}^\times$ are cyclic, but $\mathbf{Z}_{12}^\times$ is not.

10. Using the formula for $\varphi(n)$, compute $\varphi(24)$, $\varphi(81)$, and $\varphi(p^\alpha)$, where p is a prime number. Give a proof that the formula for $\varphi(n)$ is valid when $n = p^\alpha$, where p is a prime.

11. Show that if a and b are positive integers such that $a|b$, then $\varphi(a)|\varphi(b)$.

12. Find all integers $n > 1$ such that $\varphi(n) = 2$.

13. Show that $\varphi(1) + \varphi(p) + \ldots + \varphi(p^\alpha) = p^\alpha$ for any prime p and any positive integer α.

14. Show that if $n > 2$, then $\varphi(n)$ is even.

15. For $n = 12$ show that $\Sigma_{d|n}\, \varphi(d) = n$. Do the same for $n = 18$.

16. Show that if $n > 1$, then the sum of all positive integers less than n and relatively prime to n is $n\varphi(n)/2$; that is,

$$\sum_{0<a<n \text{ and } (a,n)=1} a = \frac{n}{2}\,\varphi(n).$$

17. An element $[a]$ of $\mathbf{Z}_n$ is said to be *nilpotent* if $[a]^k = [0]$ for some k. Show that $\mathbf{Z}_n$ has no nonzero nilpotent elements if and only if n has no factor that is a perfect square (except 1).

18. An element $[a]$ of $\mathbf{Z}_n$ is said to be *idempotent* if $[a]^2 = [a]$. If p is a prime, show that $[0]$ and $[1]$ are the only idempotents in $\mathbf{Z}_p$.

19. Find all idempotent elements of $\mathbf{Z}_6$, $\mathbf{Z}_{10}$, and $\mathbf{Z}_{12}$.

20. If n is not a prime power, show that $\mathbf{Z}_n$ has an idempotent element different from $[0]$ and $[1]$.
   Hint: Suppose that $n = bc$, with $(b,c) = 1$. Solve the simultaneous congruences $x \equiv 1 \pmod{b}$ and $x \equiv 0 \pmod{c}$.

21. Show that if p is a prime, then the congruence $x^2 \equiv 1 \pmod{p}$ has only the solutions $x \equiv 1$ and $x \equiv -1$.

22. Let a, b be integers, and let p be a prime number of the form $p = 2k + 1$. Show that if $p \nmid a$ and $a \equiv b^2 \pmod{p}$, then $a^k \equiv 1 \pmod{p}$.

23. Let $p = 2k + 1$ be a prime number. Show that if a is an integer such that $p \nmid a$, then either $a^k \equiv 1 \pmod{p}$ or $a^k \equiv -1 \pmod{p}$.

24. Prove Wilson's theorem, which states that if p is a prime, then $(p - 1)! \equiv -1 \pmod{p}$.
   Hint: (p − 1)! is the product of all elements of $\mathbf{Z}_p^\times$. Pair each element with its inverse, and use Exercise 21. For two special cases see Exercise 10 in Section 1.3.

25. Prove that if $(m,n) = 1$, then $n^{\varphi(m)} + m^{\varphi(n)} \equiv 1 \pmod{mn}$.

26. Prove that if m, n are positive integers with $(m,n) = 1$, then $\varphi(mn) = \varphi(m)\varphi(n)$.

Hint: Use the Chinese remainder theorem to show that each pair of elements $[a]_m$ and $[b]_n$ (in $\mathbf{Z}_m$ and $\mathbf{Z}_n$ respectively) corresponds to a unique element $[x]_{mn}$ in $\mathbf{Z}_{mn}$. Then show that under this correspondence, $[a]$ and $[b]$ are units if and only if $[x]$ is a unit.

27. Use Exercise 10 and Exercise 26 to prove Proposition 1.4.6.

28. Prove that the associative and commutative laws hold for addition and multiplication of congruence classes, as defined in Proposition 1.4.2.

APPENDIX A: SETS

We have assumed that the reader is familiar with the basic language of set theory, allowing us to begin our book with elementary number theory. This appendix is designed to serve two purposes: to establish our notation, and to provide a quick review of some basic facts.

If S is a set (or collection) of elements denoted by a, b, c, etc., then we indicate that a belongs to S by writing $a \in S$. If $a \in S$, we say that a is an *element* of S, that a is a *member* of S, or simply that a is *in* S. If a is not an element of S, we write $a \notin S$. A set A consisting entirely of elements of S is said to be a *subset* of S, denoted by $A \subseteq S$. More formally, $A \subseteq S$ if and only if $a \in S$ for all $a \in A$.

Two sets A and B are said to be equal if they contain precisely the same elements. Thus to show that $A = B$, it is necessary to show that each element of A is also an element of B and that each element of B is also an element of A. In practice, equality is often proved by showing that $A \subseteq B$ and $B \subseteq A$, since different hypotheses may apply in the two different situations. If $A \subseteq B$ but $A \neq B$, then we say that A is a *proper* subset of B, and we will use the notation $A \subset B$. The notations $B \supseteq A$ and $B \supset A$ have the obvious meaning.

Let A and B be subsets of a given set S. There are several useful ways of constructing new subsets from A and B. The *intersection* of A and B is the set of all elements which belong to both A and B, and is denoted by $A \cap B$. In symbols we would write

$$A \cap B = \{x \in S \mid x \in A \text{ and } x \in B\}.$$

This notation $\{ \mid \}$ requires some explanation. The braces $\{ \}$ are used to denote a set, and the vertical bar is read "such that." Thus

$$\{x \in S \mid x \in A \text{ and } x \in B\}.$$

is read "the set of all x in S such that x is an element of A and x is an element of B." The intersection of two sets may very well not contain any elements, and this points up the necessity of introducing the notation $\varnothing$ for the *empty* or *null* set that has no elements.

We also define the *union* of sets A and B in the following way:

$$A \cup B = \{x \in S \mid x \in A \text{ or } x \in B\}.$$

The word "or" is used in the way generally accepted by mathematicians; that is, we use it in an inclusive rather than an exclusive way, so that $A \cup B$ contains all elements either in A or in B or in both A and B. We also define the *difference* of the two sets as follows:

$$A - B = \{x \in A \,|\, x \notin B\}.$$

Thus $A - B$ is the set obtained by taking from A all elements which belong to B.

If S is the set we are working with, and A is a subset of S, then we call $S - A$ the *complement* of A in S, and denote it by $\overline{A}$. The two important identities given below are known as DeMorgan's laws:

$$\overline{A \cap B} = \overline{A} \cup \overline{B}, \qquad \overline{A \cup B} = \overline{A} \cap \overline{B}.$$

Example A.1

Let A and B be sets. We will show that $A \subseteq B$ if and only if $A \cup B = B$.

First assume that $A \subseteq B$. To show that $A \cup B = B$ we must show that $A \cup B \subseteq B$ and that $B \subseteq A \cup B$. For this purpose, let $x \in A \cup B$. Thus $x \in A$ or $x \in B$. Since $A \subseteq B$, in either case $x \in B$, and thus $A \cup B \subseteq B$. On the other hand, it is always true that B is contained in $A \cup B$.

Conversely, assume that $A \cup B = B$, and let $x \in A$. Since x is in A, it is in $A \cup B$ and hence in B. This shows that $A \subseteq B$, completing the proof. □

In several places in the book we need to work with Cartesian products of sets; that is, we need to consider ordered pairs of elements. If $a \in A$ and $b \in B$, the notion of an ordered pair distinguishes between the pairs (a,b) and (b,a). These are different from the set $\{a,b\}$, which is equal as a set to $\{b,a\}$. Since ordered pairs are familiar from calculus and linear algebra, we will not go into more detail. However, we should note that it is possible to put the concept of an ordered pair on a precise footing by representing the ordered pair (a,b) as the set $\{\{a\},\{a,b\}\}$. This ordered pair is then distinguished from the ordered pair (b,a), which is represented by the set $\{\{b\},\{a,b\}\}$.

Let A and B be any sets. The *Cartesian product* of A and B is formed from all ordered pairs whose first element is in A and whose second element is in B. Formally, we define the Cartesian product of A and B as

$$A \times B = \{(a,b) \,|\, a \in A \text{ and } b \in B\}.$$

We can extend this definition to n sets by considering n-tuples in which the ith entry belongs to the ith set. The n-dimensional vector space $\mathbf{R}^n$ is just the Cartesian product of $\mathbf{R}$ with itself n times, together with the algebraic structure that defines addition of vectors and scalar multiplication.

Example A.2

In $A \times B$, ordered pairs (a_1,b_1) and (a_2,b_2) are equal if and only if $a_1 = a_2$ and $b_1 = b_2$. For example, if $A = \{1,2,3\}$ and $B = \{u,v\}$, then the Cartesian product of A and B has a total of six distinct elements:

$$A \times B = \{(1,u),(1,v),(2,u),(2,v),(3,u),(3,v)\}.$$

The Cartesian product $B \times A$ is quite different:

$$B \times A = \{(u,1),(u,2),(u,3),(v,1),(v,2),(v,3)\}. \quad \square$$

We have listed below some of the important facts about sets. They will provide good exercises for the reader who needs some review.

Let A, B, C be subsets of a given set S:
If $A \subseteq B$ and $B \subseteq C$, then $A \subseteq C$.
$A \cap B \subseteq A$ and $A \cap B \subseteq B$.
$A \subseteq A \cup B$ and $B \subseteq A \cup B$.
If $A \subseteq B$, then $A \cup C \subseteq B \cup C$.
If $A \subseteq B$, then $A \cap C \subseteq B \cap C$.
$A \subseteq B$ if and only if $A \cap B = A$.
$A \cup B = (A \cap B) \cup (A - B) \cup (B - A)$.
$A \cup (B \cap C) = (A \cup B) \cap (A \cup C)$.
$A \cap (B \cup C) = (A \cap B) \cup (A \cap C)$.
$(A - B) \cup (B - A) = (A \cup B) - (A \cap B)$.
$(A \cup B) \times C = (A \times C) \cup (B \times C)$.

APPENDIX B: CONSTRUCTION OF THE NUMBER SYSTEMS

The purpose of this appendix is to provide an outline of the logical development of our number systems—the natural numbers, integers, rational numbers, real numbers, and complex numbers. At best, we hope to whet the reader's appetite. We will only rarely attempt to give proofs for our statements. However, elsewhere in the text we will study general constructions which include as special cases the construction of the rational numbers from the integers and the construction of the complex numbers from the real numbers.

In Chapter 1 we have taken a naive approach in working with the set of integers. We have assumed that the reader is willing to accept the familiar properties of the operations of addition and multiplication. However, it is possible to derive these properties from a very short list of postulates. They are called the "Peano postulates," formulated about 1900 by G. Peano (1858–1932), and they provide a description of the natural numbers (nonnegative integers 0, 1, 2, ...), denoted by **N.**

We have chosen to take the language and concepts of set theory as the starting point of the development of the number systems. This means that the Peano postulates must be stated in set theoretic terms alone. Intuitively, to describe the natural numbers we begin with 0 and then list successive numbers. The process that extends the set from one natural number to the next can be described as a function, which we denote by S in the postulates. We have in mind the formula $S(m) = m + 1$, although at this stage of development the formula does not yet make sense. The third postulate is a statement of the principle of mathematical induction (see Appendix D).

B.1 Axiom (Peano Postulates).   The system $\mathbf{N}$ of natural numbers is a set $\mathbf{N}$ with a distinguished element 0 and a function S from $\mathbf{N}$ into $\mathbf{N}$ which satisfies
  (i) $S(n) \neq 0$ for all members n of $\mathbf{N}$;
 (ii) $S(n_1) \neq S(n_2)$ for all members $n_1 \neq n_2$ of $\mathbf{N}$;
(iii) any subset $\mathbf{N}'$ of $\mathbf{N}$ which contains 0 and which contains $S(n)$ for all n in $\mathbf{N}'$ must be equal to $\mathbf{N}$.

The function S utilized in the Peano postulates is called the *successor function*. Since it has as its model the function $S(m) = m + 1$, we will use it below to define addition and multiplication of natural numbers. Note that the assumption that S is a function means that it is possible to define the composition of S with itself n times, which we denote by S^n. We define S^0 to be the identity function.

B.2 Definition.   With the notation of the Peano postulates, let $m, n \in \mathbf{N}$. We define operations of addition and multiplication on $\mathbf{N}$ as follows:

$$m + n = S^n(m) \qquad \text{and} \qquad m \cdot n = (S^m)^n(0).$$

We define $m \geq n$ if the equation $m = n + x$ has a solution $x \in \mathbf{N}$.

It is possible to derive the basic arithmetic and order properties of the natural numbers from the Peano postulates, but that is beyond the scope of what we have set out to do. After defining the integers $\mathbf{Z}$ in terms of $\mathbf{N}$, it is then possible to extend properties of $\mathbf{N}$ to $\mathbf{Z}$. This indication of how the properties which are listed in Appendix C can be proved is as much detail as we can provide without digressing.

In Section 1.1 we took the well-ordering principle to be an axiom. We now show that it is a direct consequence of the Peano postulates. In Appendix D we will show that the well-ordering principle implies the principles of mathematical induction, and so the well-ordering principle is logically equivalent to induction.

B.3 Proposition (Well-Ordering Principle).   Any nonempty set of natural numbers contains a smallest element.

Proof. Let T be a nonempty subset of $\mathbf{N}$ and let L be the set of natural numbers n such that $n \leq t$ for all $t \in T$. We cannot have $L = \mathbf{N}$ since there is some natural number t in T, and then $t + 1 = S(t)$ is not in L. (We are making use of the function S from the Peano postulates.) This means that L cannot satisfy the assumptions of postulate (iii), and since we certainly have $0 \in L$, there must be some n in L with $S(n) \notin L$. Thus we have $n \leq t$ for all $t \in T$, and to finish the proof we only need to show that $n \in T$. If this were not the case, then in fact $n < t$ for all $t \in T$, and this would imply that $S(n) \leq t$ for all $t \in T$, a contradiction. □

The next step is to use natural numbers to define the set of integers. We can do this by considering ordered pairs of natural numbers. We know that any negative integer can be expressed (in many ways) as a difference of natural numbers. To avoid the use of subtraction, which is as yet undefined, we consider the set $\mathbf{N} \times \mathbf{N}$, where an ordered pair (a,b) in $\mathbf{N} \times \mathbf{N}$ is thought of as representing $a - b$.

Just as with fractions, there are many ways in which a particular integer can be written as the difference of two natural numbers. For example, $(0,2)$, $(1,3)$, $(2,4)$, etc., all represent what we know should be -2. We need a notion of equivalence of ordered pairs, and since we know that we should have $a - b = c - d$ if and only if $a + d = c + b$, we can avoid the use of subtraction in the definition. The formulas given below for addition and multiplication are motivated by the fact that we know that we should get $(a - b) + (c - d) = (a + c) - (b + d)$ and $(a - b)(c - d) = (ac + bd) - (ad + bc)$. If we were going to prove all of our assertions, we would have to show that the definitions of addition and multiplication of integers do not depend on the particular ordered pairs of natural numbers which we choose to represent them. Furthermore, we would need to use the definition of an equivalence relation given in Section 2.2.

B.4 Definition.   The set of integers, denoted by **Z,** is defined via the set $\mathbf{N} \times \mathbf{N}$, where we specify that ordered pairs (a,b) and (c,d) are equivalent if and only if $a + d = b + c$.

We define addition and multiplication of ordered pairs as follows:

$$(a,b) + (c,d) = (a + c, b + d) \qquad \text{and} \qquad (a,b)(c,d) = (ac + bd, ad + bc).$$

It is possible to verify all of the properties of **Z** that are listed in Appendix C, using the above definitions and the properties of **N.** Furthermore, the set **N** can be identified with the set of ordered pairs (a,b) such that $a \geq b$, and so we can view **N** as the set of nonnegative integers. The well-ordering principle can easily be extended to the statement that any set of integers that is bounded below must contain a smallest element.

The next step is to construct the set of rational numbers **Q** from the set of integers. This is a special case of a general construction given in Section 5.4, where detailed proofs are provided.

B.5 Definition.   The set of rational numbers, denoted by **Q,** is defined via the set of ordered pairs (m,n) such that $m, n \in \mathbf{Z}$ and $n > 0$, where we agree that (a,b) is equivalent to (c,d) if and only if $ad = bc$. We define addition and multiplication as follows:

$$(a,b) + (c,d) = (ad + bc, bd) \quad \text{and} \quad (a,b)(c,d) = (ac, bd).$$

It is more difficult to describe the construction of the set of real numbers from the set of rational numbers. The Greeks used a completely geometric approach to real numbers, and initially considered numbers to be simply the ratios of lengths of line segments. However, the length of a diagonal of a square with sides of length 1 cannot be expressed as the ratio of two integer lengths, since $\sqrt{2}$ is not a rational number. This makes it necessary to introduce irrational numbers.

A sequence $\{a_n\}_{n=1}^{\infty}$ of rational numbers is said to be a *Cauchy sequence* if for each $\varepsilon > 0$ there exists N such that $|a_n - a_m| < 0$ for all $n, m > N$. It is then possible to define the set of real numbers **R** as the set of all Cauchy sequences of rational numbers, where such sequences are considered to be equivalent if the limit of the difference of the sequences is 0. To verify all of the properties of the real numbers is then quite an involved process.

We note only a few of the properties of real numbers: **R** is a field (see Section 4.1 for the definition and properties of a field) ordered by $\leq$. The set **Q** is dense in **R,** in the sense that between any two distinct real numbers there is a rational number. Any set of real numbers that has a lower bound has a greatest lower bound, and any set that has an upper bound has a least upper bound. The Archimedean property holds; i.e., for any two positive real numbers a, b there exists an integer n such that $na > b$.

Finally, the set **C** of complex numbers is described in Appendix E. One method of construction is to use Kronecker's theorem, Theorem 4.4.8. An alternative is to consider ordered pairs of real numbers. Then addition and multiplication are defined as follows:

$$(a,b) + (c,d) = (a + c, b + d) \quad \text{and} \quad (a,b) \cdot (c,d) = (ac - bd, ad + bc).$$

The ordered pair (a,b) is usually written $a + bi$, where $i^2 = -1$.

Detailed proofs of the assertions in this appendix can be found in various text books at a more advanced level. The construction of the real numbers from the rationals is usually viewed as a part of analysis rather than algebra.

APPENDIX C: BASIC PROPERTIES OF THE INTEGERS

We assume that the reader is familiar with the arithmetic and order properties of the integers, and indeed, we have been freely using these properties in Chapter 1. In the interest of completeness we will now explicitly list these properties, as well as their names.

C.1 Properties of Addition

1. Closure: Given any two integers a and b, there is a unique integer $a + b$.

2. Associativity: Given integers a, b, c, we have $(a + b) + c = a + (b + c)$.

3. Commutativity: Given integers a, b, we have $a + b = b + a$.
4. Zero element: There exists a unique integer 0 such that $a + 0 = a$ for any integer a.
5. Inverses: Given an integer a, there exists a unique integer, denoted by $-a$, such that $a + (-a) = 0$.

C.2 Properties of Multiplication

1. Closure: Given any two integers a and b, there is a unique integer $a \cdot b = ab$.
2. Associativity: Given integers a, b, c, we have $(a \cdot b) \cdot c = a \cdot (b \cdot c)$.
3. Commutativity: Given integers a, b, we have $a \cdot b = b \cdot a$.
4. Identity element: There exists a unique integer $1 (\neq 0)$ such that $a \cdot 1 = a$ for any integer a.

C.3 Joint Property of Addition and Multiplication

1. Distributivity: Given integers a, b, c, we have $a(b + c) = ab + ac$.

C.4 Properties of Order

There exists a subset $\mathbf{Z}^+ \subset \mathbf{Z}$, called the *set of positive integers,* which satisfies the following properties:

1. Closure under addition: If a, $b \in \mathbf{Z}^+$, then $a + b \in \mathbf{Z}^+$.
2. Closure under multiplication: If a, $b \in \mathbf{Z}^+$, then $ab \in \mathbf{Z}^+$.
3. Trichotomy: Given $a \in \mathbf{Z}$, exactly one of the following holds:
   (i) $a \in \mathbf{Z}^+$,　　　(ii) $a = 0$,　　　(iii) $-a \in \mathbf{Z}^+$.

　　A number of these properties are redundant. Our purpose is to provide a working knowledge of the system of integers, and so we have not given the most economical list of properties. Rather than investigating the foundations of the number systems, we will be content with simply making the following statement: Together with the well-ordering principle, the above list of thirteen properties completely characterizes the set of integers.
　　The following proposition lists some of the usual arithmetic properties of the set of integers. These properties hold in a more general setting, which will be studied in Chapter 5. We will use the notation $a - b$ for $a + (-b)$.

C.5 Proposition.　　Let a, b, $c \in \mathbf{Z}$.

(a) If $a + b = a + c$, then $b = c$.
(b) $-(-a) = a$.

(c) $a \cdot 0 = 0$

(d) $(-a)(-b) = ab$.

We introduce the usual order symbols as follows. We say that a is greater than b, denoted by $a > b$, if $a - b \in \mathbf{Z}^+$. For $a > b$ we also write $b < a$ (read "b is less than a"), and $a \geq b$ (read "a is greater than or equal to b") denotes that $a = b$ or $a > b$. Finally, the absolute value of a, denoted by $|a|$, is equal to a if $a \in \mathbf{Z}^+$ or $a = 0$ and is equal to $-a$ if $-a \in \mathbf{Z}^+$. The proof of the next proposition is left as an exercise.

C.6 Proposition. Let a, b, $c \in \mathbf{Z}$.

(a) If $a > b$ and $b > c$, then $a > c$.

(b) If $a > b$, then $a + c > b + c$.

(c) $a < 0$ if and only if $-a \in \mathbf{Z}^+$.

(d) If $a > b$ and $c > 0$, then $ac > bc$.

(e) If $a > b$ and $c < 0$, then $ac < bc$.

(f) $|a| \geq 0$, and $|a| = 0$ if and only if $a = 0$.

(g) If $a > 0$, then $|b| \leq a$ if and only if $-a \leq b \leq a$.

(h) $|ab| = |a||b|$.

(k) $|a + b| \leq |a| + |b|$.

(m) If $ab = ac$ and $a \neq 0$, then $b = c$.

APPENDIX D: INDUCTION

If one develops the natural numbers from the Peano postulates, then mathematical induction is taken to be one of the postulates. On the other hand, if one uses the list of properties given in Appendix C as a starting point, then the well-ordering principle is usually chosen as an axiom. We begin by showing that mathematical induction can be deduced from the well-ordering principle. We will let $\mathbf{Z}^+$ denote the positive integers $\{1, 2, 3, \ldots\}$.

D.1 Theorem. The well-ordering principle implies that if $S \subseteq \mathbf{Z}^+$ and (i) $1 \in S$ and (ii) $n + 1 \in S$ whenever $n \in S$, then $S = \mathbf{Z}^+$.

Proof. Suppose that $S \neq \mathbf{Z}^+$. Then the set $T = \{n \in \mathbf{Z}^+ \mid n \notin S\}$ is not empty and by the well-ordering principle has a least element k. Since $1 \in S$, $k \neq 1$, and so $k - 1 \in \mathbf{Z}^+$. Since $k - 1 < k$, we have $k - 1 \in S$. But by (ii), then we have $k = (k - 1) + 1 \in S$, a contradiction. Thus $T = \emptyset$ and $S = \mathbf{Z}^+$. $\square$

This theorem is applied in the principle of mathematical induction, which is of paramount importance. The principle of mathematical induction applies to statements which involve an arbitrary positive integer n. Examples of such statements are:

1. $1 + 2 + \ldots + n = n(n + 1)/2$.
2. $3 | (10^n - 1)$.
3. Let $a_1, \ldots, a_n$ be positive real numbers. Then

$$\sqrt[n]{a_1 a_2 \cdots a_n} \leq \frac{1}{n} \sum_{i=1}^{n} a_i.$$

4. $n^2 - n + 41$ is a prime number.

Observe that each statement depends on the positive integer n and becomes either true or false when some value is substituted for n. Example 4 is true when $n = 2$ but false when $n = 41$.

Suppose that P_n is a statement depending on the positive integer n. If P_n becomes true for each choice of n, then the principle of mathematical induction frequently allows us to establish this fact. Let us state the principle, prove that it holds, and then apply it to some examples. Note that we could begin numbering with any integer, say $P_0, P_1, \ldots$ or even $P_{-297}, P_{-296}, \ldots$.

D.2 Theorem (Principle of Mathematical Induction). Let $P_1, P_2, \ldots$ be a sequence of propositions. Suppose that
   (i) P_1 is true, and
   (ii) if P_k is true, then P_{k+1} is true for all positive integers k.
Then P_n is true for all positive integers n.

Proof. Let $S = \{n \in \mathbf{Z}^+ \mid P_n \text{ is true}\}$. By (i), $1 \in S$; and (ii), if $n \in S$, then $n + 1 \in S$. Apply Theorem D.1 to get $S = \mathbf{Z}^+$. Thus P_n is true for all positive integers. $\square$

Example D.1

To establish that $1 + 2 + \ldots + n = n(n + 1)/2$, let P_n be the statement $1 + 2 + \ldots + n = n(n + 1)/2$. Then $1 = 1(1 + 1)/2$, so P_1 is true. The next step is to show that P_k implies P_{k+1}. Assume that P_k is true, so that we have $1 + 2 + \ldots + k = k(k + 1)/2$. Add $k + 1$ to both sides of this equation to get

$$1 + 2 + \ldots + k + (k + 1) = \frac{k(k + 1)}{2} + (k + 1) = \frac{k(k + 1) + 2(k + 1)}{2}$$

$$= \frac{(k + 1)(k + 2)}{2} = \frac{(k + 1)[(k + 1) + 1]}{2}.$$

Thus P_{k+1} is true, and so by induction P_n holds for all $n \in \mathbf{Z}^+$. $\square$

This is a good point at which to emphasize that when we are using the principle of mathematical induction, we must establish the truth of P_1. However, when we establish the truth of P_{k+1}, we get to assume the truth of P_k without having to prove anything about the truth of P_k.

Example D.2

To prove that $3|(10^n - 1)$ for all positive integers n, let P_n be the statement $3|(10^n - 1)$. Now P_1 says $3|(10^1 - 1)$ or $3|9$, which is true. Assume that P_k is true, that is, that $3|(10^k - 1)$. Then since

$$10^{k+1} - 1 = 10 \cdot 10^k - 1 = 10 \cdot 10^k - 10 + 10 - 1 = 10 \cdot (10^k - 1) + 9$$

we have that $3|(10^{k+1} - 1)$ since $3|(10^k - 1)$ and $3|9$. Hence P_{k+1} is true. By the principle of mathematical induction, P_n holds for all positive integers n. □

A second form of mathematical induction is more useful for some purposes.

D.3 Theorem (Second Principle of Mathematical Induction). Let P_1, $P_2, \ldots$ be a sequence of propositions. Suppose that
   (i) P_1 is true, and
   (ii) if P_m is true for all $m \le k$, then P_{k+1} is true for all $k \in \mathbf{Z}^+$.
Then P_n is true for all positive integers n.

Proof. Let $S = \{n \in \mathbf{Z}^+ \mid P_n \text{ is true}\}$. If $S \ne \mathbf{Z}^+$, then the set $T = \{n \in \mathbf{Z}^+ \mid P_n$ is false} is nonempty. By the well-ordering principle, T has a least element k. Since $1 \notin T$, $k \ne 1$. Thus for all $m < k$ we have $m \in S$; that is, P_m is true for all $m \le k - 1$. By hypothesis P_k is true and so $k \notin T$, a contradiction. Thus P_n is true for all natural numbers n. □

Example D.3

Define a sequence of natural numbers as follows: Let $F_1 = F_2 = 1$, and $F_n = F_{n-1} + F_{n-2}$ for $n \ge 3$. Thus $F_3 = 2$, $F_4 = 3$, $F_5 = 5$, $F_6 = 8$, etc. The sequence $F_1, F_2, \ldots$ is called the *Fibonacci sequence*. We will show that $F_n < (7/4)^n$ for all positive integers n.
   Let P_n be the statement $F_n < (7/4)^n$. Then P_1 says $F_1 = 1 < (7/4)^1$, which is true. Assuming P_m for all $m \le k$, we have that

$$F_{k+1} = F_k + F_{k-1} < \left(\frac{7}{4}\right)^k + \left(\frac{7}{4}\right)^{k-1}$$

$$= \left(\frac{7}{4}\right)^{k-1} \left(\frac{7}{4} + 1\right) = \left(\frac{7}{4}\right)^{k-1} \left(\frac{11}{4}\right)$$

$$< \left(\frac{7}{4}\right)^{k-1} \left(\frac{49}{16}\right) = \left(\frac{7}{4}\right)^{k+1},$$

and so P_{k+1} holds. Thus by the second principle of mathematical induction, $F_n < (7/4)^n$ for all positive integers n.

EXERCISES: APPENDIX 1.D

Use the principle of mathematical induction to establish each of the following, where n is any positive integer:

1. $n < 2^n$
2. $1^2 + 2^2 + \ldots + n^2 = n(n + 1)(2n + 1)/6$
3. $1^3 + 2^3 + \ldots + n^3 = n^2(n + 1)^2/4$
4. $2 + 2^2 + \ldots + 2^n = 2^{n+1} - 2$
5. $x + 4x + 7x + \ldots + (3n - 2)x = n(3n - 1)x/2$
6. $10^{n+1} + 10^n + 1$ is divisible by 3.
7. $10^{n+1} + 3 \times 10^n + 5$ is divisible by 9.
8. $4 \times 10^{2n} + 9 \times 10^{2n-1} + 5$ is divisible by 99.
9. Let $a_1, \ldots, a_n$ be positive real numbers. Let

$$G_n = \sqrt[n]{a_1 a_2 \ldots a_n} \quad \text{and} \quad A_n = \frac{1}{n} \sum_{i=1}^{n} a_i.$$

G_n is called the *geometric mean* and A_n is called the *arithmetic mean*. We wish to show that $G_n \le A_n$.
   (i) Show that $G_2 \le A_2$.
   (ii) Show that $G_{2^n} \le A_{2^n}$ by using induction on n.
   (iii) Show that $G_n \le A_n$.
   Hint: Let m be such that $2^m \ge n$, and set $a_{n+1} = a_{n+2} = \ldots = a_{2^m} = A_n$ and apply part (ii).
10. Let a and b be real numbers. Prove the binomial theorem

$$(a + b)^n = \sum_{i=0}^{n} \binom{n}{i} a^i b^{n-i}$$

where

$$\binom{n}{i} = \frac{n!}{i!(n - i)!}$$

and $n! = n(n - 1) \cdots 2 \cdot 1$ for $n \ge 1$ and $0! = 1$.
   Hint: $\binom{m+1}{k} = \binom{m}{k} + \binom{m}{k-1}$.
11. Find a formula for the derivative of the product of n functions, and give a detailed proof by induction (assuming the product rule for the derivative of two functions).
12. Find a formula for the nth derivative of the product of two functions, and give a detailed proof by induction.

2

Functions

In studying mathematical objects, we need to develop ways of classifying them, and to do this, we must have various methods for comparing them. Since we will be studying algebraic objects which usually consist of a set together with additional structure, functions provide the most important means of comparison. In this chapter we will study functions in preparation for their later use, when we will utilize functions that preserve the relevant algebraic structure. Recall that in linear algebra the appropriate functions to work with are those that preserve scalar multiplication and vector addition, namely, linear transformations.

To give an example of such a comparison, we point out that the multiplicative structure of the set of all positive real numbers is algebraically similar to the additive structure of the set of all real numbers. The justification for this statement lies in the existence of the log and exponential functions, which provide one-to-one correspondences between the two sets and convert multiplication to addition (and back again).

When we need to show that two structures are essentially the same, we will need to use one-to-one correspondences. As a second example, consider the set

$$\mathbf{Z}_4 = \{[0]_4, [1]_4, [2]_4, [3]_4\}$$

using addition of congruence classes, and the set

$$\mathbf{Z}_5^\times = \{[1]_5, [2]_5, [3]_5, [4]_5\}$$

using multiplication of congruence classes. Although we are using addition in one case and multiplication in the other, there are clear similarities. In $\mathbf{Z}_4$ each congruence class can be written as a sum of $[1]_4$'s; in $\mathbf{Z}_5^\times$ each congruence class can be written as a product of $[2]_5$'s. We will see in Chapter 3 that this makes the two algebraic structures essentially the same.

To make the idea of similarity precise, we will work with one-to-one correspondences between the relevant structures. We will restrict ourselves to one-to-one correspondences that preserve the algebraic aspects of the structures. In the above example, if $[1]_4$ corresponds to $[2]_5$ and sums correspond to products, then we must have $[1]_4 \leftrightarrow [2]_5$, $[2]_4 \leftrightarrow [4]_5$, $[3]_4 \leftrightarrow [3]_5$ and $[0]_4 \leftrightarrow [1]_5$.

We can also obtain useful information from functions that are not necessarily one-to-one correspondences. In the introduction to Chapter 1 we discussed the problem of investigating the powers of the matrix $A = \begin{bmatrix} 0 & 1 \\ -1 & 0 \end{bmatrix}$. It is useful to consider the exponential function $f(n) = A^n$ that assigns to each integer n the corresponding power of the matrix A. The rules for exponents of matrices involve the relationship between addition of integers and taking powers of the matrix and show that the function respects the inherent algebraic structure. If we collect together the exponents that yield equal powers of A, we obtain the congruence classes of integers modulo 4. In fact, this function determines a natural one-to-one correspondence between $\mathbf{Z}_4$ and the powers of A.

In working with particular mathematical objects it is often useful to consider distinct objects to be essentially the same, just for the immediate purpose. In the previous example, we can consider two exponents to be the same if the corresponding powers of A are equal. In studying the Euclidean plane it is useful to consider all triangles that are congruent to each other to be "essentially the same." In Chapter 1 we studied the notion of congruence modulo n, where we did not differentiate between integers that had the same remainder on division by n. That led us to construct sets $\mathbf{Z}_n$ from the set of integers $\mathbf{Z}$.

This idea of collecting together similar objects leads to the important notion of a partition, or what amounts to the same thing—an equivalence relation. We partition a set into subsets consisting of those objects that we do not wish to distinguish from each other, and then we say that two objects belonging to the same subset of the partition are equivalent. Thus equivalence relations enable us to study objects by making distinctions between them that are no finer than those needed for the purpose at hand. Of course, for some other purpose we might need to preserve some other distinction that is lost in the equivalence relation we are using. In that case we would resort to a different equivalence relation, in which the subsets in the partition were smaller.

The notion of a one-to-one correspondence, when expressed in the concept of a permutation, can be used to describe symmetry of geometric objects and also symmetry in other situations. For example, the symmetry inherent in a square or equilateral triangle can be expressed in terms of various permutations of its vertices. Our study of permutations in this chapter will provide the motivation for a number of important concepts in later chapters.

Permutations are important in studying solvability by radicals. Certain "substitutions," or permutations, of the roots of a polynomial define its Galois group. Whether or not a particular permutation of the roots belongs to the Galois group depends on the coefficients of the equation, and is not generally easy to determine. We will see in Chapter 8 that the Galois group of an equation determines whether or not it is solvable by radicals.

We should emphasize that in this chapter we are developing important parts of our mathematical language which will be used throughout the remainder of the book. The reader needs to be thoroughly familiar with these ideas.

2.1 FUNCTIONS

The concept of a function should already be familiar from the calculus and linear algebra courses which we assume as a prerequisite for this book. At that level, it is standard to define a function f from a set S of real numbers into a set T of real numbers to be a "rule" that assigns to each real number x in S a unique real number y in T. An example would be the function $f(x)$ given by the rule $f(x) = x^2 + 3$, where f assigns to 1 the value 4, to 2 the value 7, and so on.

The graph of the function $f(x) = x^2 + 3$ is the set of points in the real plane described by $\{(x,y) \mid y = x^2 + 3\}$. Using the definition in the previous paragraph, a set of points in the plane is the graph of a function from the set of all real numbers into the set of all real numbers if and only if for each real number x there is a unique number y such that (x,y) belongs to the set. In a calculus course this is often expressed by saying that a set of points in the plane is the graph of a function if and only if every vertical line intersects the set in exactly one point.

In our development, we have chosen to take the concepts of set and element of a set as primitive (undefined) ideas. See Appendix A in Chapter 1 for a quick review of some basic set theory. The approach we will take is to define functions in terms of sets, and so we will do this by identifying a function with its graph.

It is convenient to introduce some notation for the sets we will be using. The symbol **R** will be used to denote the set of all real numbers. We will leave the precise development of the real numbers to a course in advanced calculus and simply view them as the set of all decimal numbers. They can be viewed as coordinates of points on a straight line, as in an introductory calculus course.

We will use the symbol **Q** to denote the set of ratios of integers, or rational numbers; that is

$$\mathbf{Q} = \left\{ \frac{m}{n} \mid m, n \in \mathbf{Z} \text{ and } n \neq 0 \right\}$$

where we must agree that m/n and p/q represent the same ratio if $mq = np$. Of course, we can view the set of integers $\mathbf{Z}$ as a subset of $\mathbf{Q}$ by identifying the integer m with the fraction $m/1$. The rational numbers can be viewed as a subset of $\mathbf{R}$, since fractions correspond to either terminating or repeating decimals.

The set $\mathbf{C} = \{a + bi \mid a, b \in \mathbf{R} \text{ and } i^2 = -1\}$ is called the set of *complex numbers*. Addition and multiplication of complex numbers are defined as follows:

$$(a + bi) + (c + di) = (a + c) + (b + d)i,$$

$$(a + bi)(c + di) = (ac - bd) + (ad + bc)i.$$

Note that $a + bi = c + di$ if and only if $a = c$ and $b = d$. See Appendix E at the end of this chapter for more details on the properties of complex numbers, which will be developed from a more rigorous point of view in Section 4.4.

We now return to the definition of a function. To describe the graph of a function, we need to consider ordered pairs. Let A and B be any sets. The *Cartesian product* of A and B is formed from ordered pairs of elements of A and B. Formally, we define the Cartesian product of A and B as

$$A \times B = \{(a,b) \mid a \in A \text{ and } b \in B\}.$$

In this set, ordered pairs (a_1,b_1) and (a_2,b_2) are equal if and only if $a_1 = a_2$ and $b_1 = b_2$. For example, if $A = \{1,2,3\}$ and $B = \{u,v,w\}$, then

$$A \times B = \{(1,u),(1,v),(1,w),(2,u),(2,v),(2,w),(3,u),(3,v),(3,w)\}.$$

Since the use of ordered pairs of elements should be familiar from calculus and linear algebra courses, we will not go into further detail on Cartesian products at this point.

2.1.1 Definition.   Let S and T be sets. A *function* from S into T is a subset F of $S \times T$ such that for each element $x \in S$ there is exactly one element $y \in T$ such that $(x,y) \in F$. The set S is called the *domain* of the function, and the set T is called the *codomain* of the function.

Many authors prefer to use the word *range* for what we have called the codomain of a function. The word range is also sometimes used for $\{y \in T \mid (x,y) \in F \text{ for some } x \in S\}$.

Example 2.1.1

Let $S = \{1,2,3\}$ and $T = \{u,v,w\}$. The subsets $F_1 = \{(1,u),(2,v),(3,w)\}$ and $F_2 = \{(1,u),(2,u),(3,u)\}$ of $S \times T$ both define functions since in both cases each element of S occurs exactly once among the ordered pairs. The subset $F_3 = \{(1,u),(3,w)\}$ does not define a function with domain S because the element $2 \in S$ does not appear as the first component of any ordered pair. Note that F_3 is a function if the domain is changed to the set $\{1,3\}$. Unlike the conventions used in calculus, the domain and codomain must be specified as well as the "rule of correspondence" (list of pairs) when you are presenting a function. The subset $F_4 = \{(1,u),(2,u),(2,v),(3,w)\}$ does not define a function since 2 appears as the first component of two ordered pairs. □

If $F \subseteq S \times T$ defines a function, then we will simply write $y = f(x)$ whenever $(x,y) \in F$. Thus f determines a "rule" that assigns to $x \in S$ the unique element $y \in T$, and we will call F the *graph* of f. We will use the familiar notation $f : S \to T$. We continue to emphasize the importance of the domain and codomain. In particular, we even distinguish between the functions $f : S \to T$ and $g : S \to T'$ where $T \subset T'$ and $\{(x,f(x)) \mid x \in S\} = \{(x,g(x)) \mid x \in S\}$.

Although the definition we have given provides a rigorous definition of a function, in the language of set theory, the more familiar definition and notation are usually easier to work with. But it is important to understand the precise definition. If you feel unsure of a concept involving functions, it is worth your time to try to rephrase it in terms of the formal definition using graphs. Be sure that you are comfortable with both the formal and informal definitions, so that you can use whichever one is appropriate.

For example, if we attempt to use the square root to define a function from **R** into **R,** we immediately run into a problem: The square root of a negative number cannot exist in the set of real numbers. This can be remedied in one of two ways. Either we can restrict the domain to the set $\mathbf{R}^+$ of all positive real numbers, or we can enlarge the codomain to the set **C** of all complex numbers. In either case the formula remains the same, but the functions are very different.

With the above notation, a function $f: S \to T$ must be determined by a rule or formula that assigns to each element $x \in S$ a unique element $f(x) \in T$. It is often the case that the uniqueness of $f(x)$ is in question. When checking that $f(x)$ is unique for all $x \in S$, we say that we are checking that the function f is "*well-defined.*" Problems arise when the element x can be described in more than one way, and the rule or formula for $f(x)$ depends on how x is written.

Example 2.1.2

Consider the formula $f(m/n) = m$, which might be thought to define a function $f : \mathbf{Q} \to \mathbf{Z}$. The difficulty is that a fraction has many equivalent representations, and the formula depends on one particular choice. For example, $f(1/2) = 1$, according to the formula, while $f(3/6) = 3$. Since we know that $1/2 = 3/6$ in the set of rational numbers, we are forced to conclude that the formula does not define a function from the set of rational numbers into the set of integers since it is not "well-defined."

On the other hand, the formula $f(m/n) = 2m/3n$ does define a function $f: \mathbf{Q} \to \mathbf{Q}$. To show that f is "well-defined," suppose that $m/n = p/q$. Then $mq = np$, and so multiplying both sides by 6 gives $2m \cdot 3q = 3n \cdot 2p$, which implies that $2m/3n = 2p/3q$, and thus $f(m/n) = f(p/q)$. □

Example 2.1.3

In defining functions on $\mathbf{Z}_n$ it is necessary to be very careful that the given formula is independent of the numbers chosen to represent each equivalence class.

On the one hand, the formula $f([x]_4) = [x]_6$ does not define a function from $\mathbf{Z}_4$ into $\mathbf{Z}_6$. To see this, we only need to note that although $[0]_4 = [4]_4$, the formula specifies that $f([0]_4) = [0]_6$, whereas $f([4]_4) = [4]_6$, giving two different values in $\mathbf{Z}_6$, since $[0]_6 \neq [4]_6$.

On the other hand, the formula $f([x]_4) = [3x]_6$ does define a function from $\mathbf{Z}_4$ into $\mathbf{Z}_6$. If $a \equiv b \pmod 4$, then $4|(a - b)$, and so multiplying by 3 shows that $12|3(a - b)$. But then of course $6|3(a - b)$, which implies that $3a \equiv 3b \pmod 6$. Thus $[a]_4 = [b]_4$ implies $f([a]_4) = f([b]_4)$, proving that f is well-defined.   □

Two subsets F and G of $S \times T$ define the same function from S into T if and only if the subsets are equal. In terms of the associated functions $f : S \to T$ and $g : S \to T$, this happens if and only if $f(x) = g(x)$ for all $x \in S$, giving the familiar condition for equality of functions, since f and g have the same domain and codomain. To see this, note that if $x \in S$, then $(x, f(x)) \in F$, and so we also have $(x, f(x)) \in G$, which means that $f(x) = g(x)$, since a given first component has exactly one possible second component in the graph of the function. On the other hand, if $f(x) = g(x)$ for all $x \in S$, then the ordered pairs $(x, f(x))$ and $(x, g(x))$ are equal for all $x \in S$, so we have $F = G$.

If $f : S \to T$ and $g : T \to U$ are functions, then the composition of f and g is defined by the formula $(g \circ f)(x) = g(f(x))$ for all $x \in S$. Its graph is the subset of $S \times T$ defined by

$$\{(x, z) \mid (x, y) \in F \text{ and } (y, z) \in G \text{ for some } y \in T\},$$

where F and G are the graphs of f and g, respectively. We note that $g \circ f$ is indeed a function, for given $x \in S$, there is a unique element $y \in T$ such that $y = f(x)$, and then there is a unique element $z \in U$ such that $z = g(y)$. Thus $z = g(f(x))$ is uniquely determined by x. This justifies the following formal definition.

2.1.2 Definition.   Let $f : S \to T$ and $g : T \to U$ be functions. The *composition* $g \circ f$ of f and g is the function from S to U defined by the formula $(g \circ f)(x) = g(f(x))$ for all $x \in S$.

The composition of two functions is defined only when the codomain of the first function is the same as the domain of the second function. Thus the composition of two functions may be defined in one order but not in the opposite order. (In calculus books, some authors allow the composition of two functions, $g \circ f$, to be defined when the codomain of f is merely a subset of the domain of g.)

In working with functions it can sometimes be helpful to use the analogy of a computer program. If $f : S \to T$ is a function, then we can think of the set S as consisting of the possible values which can be used as input for the program, and the set T as consisting of the potential output values. The conditions defining a function ensure that (i) any element of S can be used as an input and (ii) the output is uniquely determined by the input. If $g : T \to U$, then the composite function $g \circ f$ corresponds to a new program obtained by linking the two given programs, taking the output from the first program and using it as the input for the second program.

Example 2.1.4

Let $f : \mathbf{R} \to \mathbf{R}$ be given by $f(x) = x^2$ and $g : \mathbf{R} \to \mathbf{R}$ be given by $g(x) = x + 1$, for all $x \in \mathbf{R}.$ Then $g \circ f$ is given by

$$(g \circ f)(x) = g(f(x)) = g(x^2) = x^2 + 1$$

and $f \circ g$ is given by

$$(f \circ g)(x) = f(g(x)) = f(x + 1) = (x + 1)^2 = x^2 + 2x + 1.$$

This example shows that you should not expect $g \circ f$ and $f \circ g$ to be equal. □

Suppose that we are given three functions $f : S \to T$, $g : T \to U$, and $h : U \to V$. If we wish to compose these functions to obtain a function from S into V, then there are two ways to proceed. We could first form $g \circ f : S \to U$ and then compose with h to get $h \circ (g \circ f) : S \to V$, or we could first form $h \circ g : T \to V$ and then compose with f to get $(h \circ g) \circ f : S \to V$. These procedures define the same function, since in both cases the definition of the composition of functions leads to the expression $h(g(f(x)))$, for all $x \in S$. Thus we can say that composition of functions is *associative,* and write $h \circ (g \circ f) = (h \circ g) \circ f$. In practice, this allows us to ignore the use of parentheses, so that we can just write hgf for the composition.

2.1.3 Proposition. Composition of functions is associative.

Proof. Let $f : S \to T$, $g : T \to U$, and $h : U \to V$ be functions. Then for each $x \in S$, we have

$$(h \circ (g \circ f))(x) = (h(g \circ f))(x) = h(g(f(x)))$$

and also

$$((h \circ g) \circ f)(x) = (h \circ g)(f(x)) = h(g(f(x))).$$

This shows that $h \circ (g \circ f)$ and $(h \circ g) \circ f$ are equal as functions. □

In the definition of a function $f : S \to T$, every element of the domain S must appear as the first entry of some ordered pair in the graph of f, but nothing is said about the necessity of each element of the codomain T appearing as the second entry of some ordered pair. For example, in the function $f : \mathbf{R} \to \mathbf{R}$ defined by $f(x) = x^2$ for all $x \in \mathbf{R}$, no negative number appears as the second coordinate of any point on the graph of f. This particular example also points out the fact that two different elements of the domain may be assigned to the same element of the codomain of a function. This is important enough to warrant some definitions.

2.1.4 Definition.   Let $f : S \to T$ be a function. Then f is said to map S *onto* T if for each element $y \in T$ there exists an element $x \in S$ with $f(x) = y$.

   If $f(x_1) = f(x_2)$ implies $x_1 = x_2$ for all elements $x_1, x_2 \in S$, then f is said to be a *one-to-one* function.

   If f is both one-to-one and onto, then it is called a *one-to-one correspondence* from S to T.

Some other terminology is also in common use. An onto function is said to be *surjective* or is called a *surjection*. Similarly, a one-to-one function is said to be *injective* or is called an *injection*. In this terminology, which comes from the French, a one-to-one correspondence is a *bijection*.

   It may be helpful to think of onto functions in the following terms: If $f : S \to T$, then f is onto if and only if for each $y \in T$ the equation $y = f(x)$ has a solution $x \in S$. With this point of view, to show that $f : \mathbf{R} \to \mathbf{R}$ defined by $f(x) = x^3 + 1$ is onto, we need to show that for each $y \in \mathbf{R}$ we can solve the equation $y = x^3 + 1$. All we have to do is give the solution explicitly: $x = \sqrt[3]{y - 1}$.

   The definition we have given for a one-to-one function is perhaps not the most intuitive one. For $f : S \to T$ to be one-to-one we want to know that $f(x_1) \neq f(x_2)$ whenever $x_1 \neq x_2$ (for $x_1, x_2 \in S$). However, if we try to apply this definition to show that the function $f(x) = x^3 + 1$ of the previous paragraph is one-to-one, we would need to work with inequalities. Working with equalities is much more familiar, so it is useful to reformulate the definition. Rewriting it to state "if $x_1 \neq x_2$, then $f(x_1) \neq f(x_2)$" shows that an equivalent statement is "if $f(x_1) = f(x_2)$, then $x_1 = x_2$." Using the second statement, which we have taken as the definition, to show that f is one-to-one we only need to show that if $(x_1)^3 + 1 = (x_2)^3 + 1$, then $x_1 = x_2$. This is easy to do by just subtracting 1 from both sides and taking the cube root, which yields a unique value.

Example 2.1.5

   Let $S = \{1,2,3\}$ and $T = \{4,5,6\}$. The function $f : S \to T$ defined by $f(1) = 4$, $f(2) = 5$, and $f(3) = 6$ is a one-to-one correspondence because it is both one-to-one and onto. The function $g : S \to T$ defined by $g(1) = 4$, $g(2) = 4$, and $g(3) = 4$ is neither one-to-one nor onto. With these sets we cannot give an example that is one-to-one but not onto, and we cannot give an example that is onto but not one-to-one. (As we will see, this happens since S and T are finite and have the same number of elements.)

Let $S = \{1,2,3\}$ and $T = \{4,5\}$. The function $f : S \to T$ defined by $f(1) = 4$, $f(2) = 5$, and $f(3) = 5$ is onto but not one-to-one. The function $g : T \to S$ defined by $g(4) = 1$ and $g(5) = 2$ is one-to-one but not onto. $\square$

2.1.5 Proposition.   Let $f : S \to T$ be a function. Suppose that S and T are finite sets with the same number of elements. Then f is one-to-one if and only if it is onto.

Proof. First assume that f is one-to-one. Let $S = \{x_1, x_2, \ldots, x_n\}$, and consider the subset

$$B = \{f(x_1), f(x_1), \ldots, f(x_n)\} \subseteq T.$$

Since f is one-to-one, the elements $f(x_i)$ are distinct, for $i = 1, 2, \ldots, n$, and so B contains n elements. Since B is a subset of T and T also has n elements, we must have $B = T$, showing that f is onto.

Conversely, assume that f is onto. Suppose that $T = \{y_1, y_2, \ldots, y_n\}$ and $f(z) = y_i$ and $f(z') = y_i$ for some $z \neq z'$ in S and some i with $1 \leq i \leq n$. Since f is onto, for each $j \neq i$ (with $1 \leq j \leq n$) there exists an element z_j such that $f(z_j) = y_j$. Consider the subset

$$A = \{z, z', z_1, \ldots, z_{i-1}, z_{i+1}, \ldots, z_n\} \subseteq S.$$

The elements z_j are distinct since f is a function. Thus A is a subset that has $n + 1$ elements, which is impossible. Hence having $f(z) = f(z')$ for distinct elements z and z' is impossible. We conclude that f is one-to-one. $\square$

Example 2.1.6

Define $f : \mathbf{Z} \to \mathbf{Z}$ by $f(n) = 2n$, for all $n \in \mathbf{Z}$. Then f is one-to-one but not onto. On the other hand, if $g : \mathbf{Z} \to \mathbf{Z}$ is defined by letting $g(n) = n$ if n is odd and $g(n) = n/2$ if n is even, then g is onto but not one-to-one.

In fact, this is a property that characterizes infinite sets. A set S is infinite if and only if there exists a one-to-one correspondence between S and a proper subset of S. $\square$

2.1.6 Proposition.   Let $f : S \to T$ and $g : T \to U$ be functions.
 (a) If f and g are one-to-one, then $g \circ f$ is one-to-one.
 (b) If f and g are onto, then $g \circ f$ is onto.

Proof. (a) Assume that f and g are one-to-one functions. Let $x_1, x_2 \in S$. If $(g \circ f)(x_1) = (g \circ f)(x_2)$, then $g(f(x_1)) = g(f(x_2))$, and so $f(x_1) = f(x_2)$ since g is one-to-one. Then since f is one-to-one, we have $x_1 = x_2$. This shows that $g \circ f$ is one-to-one.

(b) Assume that f and g are onto functions, and let $z \in U$. Since g is onto, there exists $y \in T$ such that $g(y) = z$. Since f is onto, there exists $x \in S$ such that

$f(x) = y$. Hence $(g \circ f)(x) = g(f(x)) = g(y) = z$, and this shows that $g \circ f$ is onto.   □

We now want to study one-to-one correspondences in more detail. We first note that Proposition 2.1.6 implies that the composition of two one-to-one correspondences is a one-to-one correspondence. One of the most important properties that we want to show is that any one-to-one correspondence $f : S \rightarrow T$ is "reversible."

Let us return to our earlier analogy between functions and programs. If we are given a function $f : S \rightarrow T$, which we think of as a program, then we want to be able to construct another program (a new function $g : T \rightarrow S$) that is capable of taking any output from the program f and retrieving the corresponding input. Another way to express this is to say that if the two programs are linked via composition, they would end up doing nothing: that is, $g(f(x)) = x$ for all $x \in S$ and $f(g(y)) = y$ for all $y \in T$.

We need some formal notation at this point.

2.1.7 Definition.   Let S be a set. The *identity* function $1_S : S \rightarrow S$ is defined by the formula $1_S(x) = x$ for all $x \in S$.

If $f : S \rightarrow T$ is a function, then a function $g : T \rightarrow S$ is called an *inverse* for f if $g \circ f = 1_S$ and $f \circ g = 1_T$.

An identity function has the following important property: If $f : S \rightarrow T$, then for all $x \in S$ we have $f(1_S(x)) = f(x)$, showing that $f \circ 1_S = f$. Similarly, we have $1_T \circ f = f$. In particular, if $f : S \rightarrow S$, then $f \circ 1_S = f$ and $1_S \circ f = f$. Thus the identity function 1_S plays the same role for composition of functions as the number 1 does for multiplication.

If g is an inverse for f, then the definition shows immediately that f is an inverse for g. The next proposition shows that a function has an inverse precisely when it is one-to-one and onto. It also implies that the inverse function is again one-to-one and onto.

Suppose that $g, h : T \rightarrow S$ are both inverses for $f : S \rightarrow T$. Then on the one hand, $(g \circ f) \circ h = 1_S \circ h = h$, while on the other hand, $g \circ (f \circ h) = g \circ 1_T = g$. Since composition of functions is associative, the two expressions must be equal, so $h = g$. This shows that inverses are unique, and so the use of the notation f^{-1} for the inverse of f is justified.

2.1.8 Proposition.   Let $f : S \rightarrow T$ be a function. If f has an inverse, then it must be one-to-one and onto. Conversely, if f is one-to-one and onto, then it has a unique inverse.

Proof. First assume that f has an inverse $g : T \rightarrow S$ such that $g \circ f = 1_S$ and $f \circ g = 1_T$. Given any element $y \in T$, we have $y = 1_T(y) = f(g(y))$, and so f maps $g(y)$ onto y, showing that f is onto. If $x_1, x_2 \in S$ with $f(x_1) = f(x_2)$, then applying g

gives $g(f(x_1)) = g(f(x_2))$, and so we must have $x_1 = x_2$ since $g \circ f = 1_S$. Thus f is one-to-one.

Next, assume that f is one-to-one and onto. We will define a function $g : T \to S$ as follows. For each $y \in T$, there exists an element $x \in S$ with $f(x) = y$ since f is onto. Furthermore, there is only one such $x \in S$ since f is one-to-one. This allows us to define $g(y) = x$, and it follows immediately from this definition that $g(f(x)) = x$ for all $x \in S$. For any $y \in T$, we have $g(y) = x$ for the element $x \in S$ for which $f(x) = y$. Thus $f(g(y)) = f(x) = y$ for all $y \in T$, showing that g is an inverse for f.

As in the remarks preceding the proposition, suppose that $h : T \to S$ is also an inverse for f. Then

$$h = h \circ 1_T = h \circ (f \circ g) = (h \circ f) \circ g = 1_S \circ g = g$$

and the uniqueness is established. $\quad \square$

It is instructive to prove that a one-to-one and onto function has an inverse by using the graph of the function. Assume that $f : S \to T$ is one-to-one and onto, and let F denote the graph of f. We will define an inverse $g : S \to T$ by giving its graph G. Let $G = \{(y,x) \mid (x,y) \in F\}$. Then we have clearly defined a subset of $S \times T$. Since f is onto, for each $y \in T$ there exists $x \in S$ such that $(x,y) \in F$, and so then $(y,x) \in G$. Furthermore, the element x is uniquely determined by y, since f is one-to-one, showing that for each $y \in T$ there is only one ordered pair in G with first component y. This shows that G does in fact define a function. The graph of $g \circ f$ in $S \times S$ is $\{(x,x) \mid x \in S\}$, and the graph of $f \circ g$ in $T \times T$ is $\{(y,y) \mid y \in T\}$, and so g is the inverse of f.

EXERCISES: SECTION 2.1

We use $\mathbf{R}^+$ to denote the set of positive real numbers.

1. In each of the following parts, determine whether the given function is one-to-one or onto:
   (a) $f : \mathbf{R} \to \mathbf{R}; f(x) = x + 3$
   (b) $f : \mathbf{C} \to \mathbf{C}; f(x) = x^2 + 2x + 1$
   (c) $f : \mathbf{R} \to \mathbf{R}; f(x) = x^2$
   (d) $f : \mathbf{C} \to \mathbf{C}; f(x) = x^2$
   (e) $f : \mathbf{R}^+ \to \mathbf{R}^+; f(x) = x^2$
   (f) $f : \mathbf{Z}_n \to \mathbf{Z}_n; f([x]_n) = [mx + t]_n$, where $m, t \in \mathbf{Z}$
   (g) $f : \mathbf{R}^+ \to \mathbf{R}; f(x) = \ln x$
   (h) $f : \mathbf{R}^+ \to \mathbf{R}^+; f(x) = \begin{cases} x & \text{if } x \text{ is rational} \\ x^2 & \text{if } x \text{ is irrational} \end{cases}$

2. For each one-to-one and onto function in the previous question, find the inverse of the function.

Hint: It might not hurt to review the section on inverse functions in your calculus book.

3. Let $S = \{1,2,3\}$ and $T = \{a,b\}$.
   (a) How many functions are there from S into T? from T into S?
   (b) How many of the functions in part (a) are one-to-one? How many are onto?

4. **(a)** Does the formula $f(x) = 1/(x^2 + 1)$ define a function $f : \mathbf{R} \to \mathbf{R}$?
   (b) Does the formula given in part (a) define a function $f : \mathbf{C} \to \mathbf{C}$?

5. Which of the following formulas define functions from the set of rational numbers into itself? (Assume in each case that n, m are integers and that n is nonzero.)

   (a) $f\left(\dfrac{m}{n}\right) = \dfrac{m + 1}{n + 1}$

   (b) $g\left(\dfrac{m}{n}\right) = \dfrac{2m}{2n}$

   (c) $h\left(\dfrac{m}{n}\right) = \dfrac{m + n}{n^2}$

   (d) $p\left(\dfrac{m}{n}\right) = \dfrac{4m^2}{7n^2} - \dfrac{m}{n}$

   (e) $q\left(\dfrac{m}{n}\right) = \dfrac{m + 1}{m}$

6. Show that each of the following formulas yields a well-defined function:
   (a) $f : \mathbf{Z}_8 \to \mathbf{Z}_8$ defined by $f([x]_8) = [mx]_8$, for any $m \in \mathbf{Z}$
   (b) $g : \mathbf{Z}_8 \to \mathbf{Z}_{12}$ defined by $g([x]_8) = [6x]_{12}$
   (c) $h : \mathbf{Z}_{12} \to \mathbf{Z}_4$ defined by $h([x]_{12}) = [x]_4$
   (d) $p : \mathbf{Z}_{10} \to \mathbf{Z}_5$ defined by $p([x]_{10}) = [x^2 + 2x - 1]_5$
   (e) $q : \mathbf{Z}_4 \to \mathbf{Z}_{12}$ defined by $q([x]_4) = [9x]_{12}$

7. In each of the following cases, give an example to show that the formula does not define a function:
   (a) $f : \mathbf{Z}_8 \to \mathbf{Z}_{10}$ defined by $f([x]_8) = [6x]_{10}$
   (b) $g : \mathbf{Z}_2 \to \mathbf{Z}_5$ defined by $g([x]_2) = [x]_5$
   (c) $h : \mathbf{Z}_4 \to \mathbf{Z}_{12}$ defined by $h([x]_4) = [x]_{12}$
   (d) $p : \mathbf{Z}_{12} \to \mathbf{Z}_5$ defined by $p([x]_{12}) = [2x]_5$

8. Consider the formula $f : \mathbf{Z}_n \to \mathbf{Z}_m$, for positive integers m, n, defined by $f([x]_n) = [kx]_m$. Show that f defines a function if and only if $m | kn$.

9. Show that the function in the previous problem defines a one-to-one correspondence if and only if $m = n$ and $(k,n) = 1$.

10. Let $f : A \to B$ be a function, and let $f(A) = \{f(a) \mid a \in A\}$. Show that f is onto if and only if $f(A) = B$.

11. Let $f : A \to B$ and $g : B \to C$ be one-to-one and onto. Show that $(gf)^{-1}$ exists and that $(gf)^{-1} = f^{-1}g^{-1}$.

12. Let $f : A \to B$ and $g : B \to C$ be functions. Prove that if $g \circ f$ is one-to-one, then f is one-to-one, and that if $g \circ f$ is onto, then g is onto.

13. Let $f : A \to B$ be a function. Prove that f is onto if and only if there exists a function $g : B \to A$ such that $f \circ g = 1_B$.

14. Let $f : A \rightarrow B$ be a function. Prove that f is onto if and only if for every set C and all functions $h : B \rightarrow C$ and $k : B \rightarrow C$, if $h \circ f = k \circ f$, then $h = k$.

15. Let $f : A \rightarrow B$ be a function. Prove that f is one-to-one if and only if there exists a function $g : B \rightarrow A$ such that $g \circ f = 1_A$.

16. Let $f : A \rightarrow B$ be a function. Prove that f is one-to-one if and only if for every set C and all functions $h : C \rightarrow A$ and $k : C \rightarrow A$, if $f \circ h = f \circ k$ then $h = k$.

17. Define $f : \mathbf{Z}_{mn} \rightarrow \mathbf{Z}_m \times \mathbf{Z}_n$ by $f([x]_{mn}) = ([x]_m, [x]_n)$. Show that f is a function and that f is onto if and only if $\gcd(m, n) = 1$.

2.2 EQUIVALENCE RELATIONS

In working with functions that are not one-to-one, it is very useful to introduce the notion of an equivalence relation. The basic idea of an equivalence relation is to collect together elements that are related in some way, and then treat them as a single entity. This approach can be taken in many different situations and is fundamental in abstract algebra. We have already used this idea in Chapter 1, when we collected together integers congruent modulo n and constructed a new object $\mathbf{Z}_n$. In this section we will adopt the point of view that studying equivalence relations is simply a continuation of our study of functions.

Suppose that we are working with a function $f : S \rightarrow T$. If f is one-to-one, then for elements $x_1, x_2 \in S$ we know that $f(x_1) = f(x_2)$ if and only if $x_1 = x_2$. Of course, this is not true in general. For example, if $f : \mathbf{Z} \rightarrow \mathbf{Z}$ is the function that assigns to each integer its remainder when divided by 10, then infinitely many integers are mapped to each of the possible remainders $0, 1, \ldots, 9$. In this example we might want to consider two integers as equivalent whenever the function maps them to the same value. This is just saying that two integers are identified as the same by our function if they are congruent modulo 10. In general, for a function $f : S \rightarrow T$, we can say that two elements $x_1, x_2 \in S$ are equivalent with respect to f if $f(x_1) = f(x_2)$.

As with the definition of a function, for the sake of precision we will use set theory to give our formal definition of an equivalence relation. We then immediately give an equivalent definition which we hope is more useful and more intuitive.

2.2.1 Definition.   Let S be a set. A subset R of $S \times S$ is called an *equivalence relation* on S if

(i) for all $a \in S$, $(a,a) \in R$;

(ii) for all $a, b \in S$, if $(a,b) \in R$ then $(b,a) \in R$;

(iii) for all $a, b, c \in S$, if $(a,b) \in R$ and $(b,c) \in R$, then $(a,c) \in R$.

We will write $a \sim b$ to denote the fact that $(a,b) \in R$.

Using the above definition and notation, it is clear that R is an equivalence relation if and only if for all $a, b, c \in S$ we have

 (i) [Reflexive law] $a \sim a$;

 (ii) [Symmetric law] if $a \sim b$, then $b \sim a$;

 (iii) [Transitive law] if $a \sim b$ and $b \sim c$, then $a \sim c$.

We will usually use these conditions rather than the formal definition.

The most fundamental equivalence relation is given by simple equality of elements; that is, for $a, b \in S$, define $a \sim b$ if $a = b$. For this relation the reflexive, symmetric, and transitive laws are understood to be part of the definition. Under this relation, for any $a \in S$ the only element equivalent to a is a itself. In fact, equality is the only equivalence relation for which each element is related only to itself. For other equivalence relations we are interested in all elements related to a given element, leading to the following definition. In reading the definition, keep in mind the idea of congruence classes modulo n.

2.2.2 Definition.   Let $\sim$ be an equivalence relation on the set S. For a given element $a \in S$, we define the *equivalence class* of a to be the set of all elements of S that are equivalent to a. We will use the notation $[a]$. In symbols,

$$[a] = \{x \in S \mid x \sim a\}.$$

The notation $S/\sim$ will be used for the collection of all equivalence classes of S under $\sim$.

Example 2.2.1

Let n be a positive integer. For integers a, b we define $a \sim b$ if $n|(a - b)$. This, of course, is equivalent to the definition of congruence modulo n. It is not difficult to check that congruence modulo n defines an equivalence relation on the set $\mathbf{Z}$ of integers. First, the reflexive property holds since for any integer a we certainly have $a \sim a$ since $n|(a - a)$. Next, the symmetric property holds since if $a \sim b$, then $n|(a - b)$ and so $n|(b - a)$, showing that $b \sim a$. Finally, the transitive property holds since if $a \sim b$ and $b \sim c$, then $n|(a - b)$ and $n|(b - c)$, so adding shows that $n|(a - c)$, and thus $a \sim c$. The equivalence classes of this relation are the familiar congruence classes $[a]_n$.

As an alternate proof, we could use our original definition of congruence modulo n, in which we said that integers a and b are congruent modulo n if we obtain equal remainders when using the division algorithm to divide both a and b by n. Then the reflexive, symmetric, and transitive properties really follow from the identical properties for equality. $\square$

Because of our work with congruences in Chapter 1, the previous example is one of the most familiar nontrivial equivalence relations. The next example is probably the most basic one. In fact, later in this section we will show that every equivalence relation arises in this way from a function.

Example 2.2.2

Let $f : S \to T$ be any function. For $x_1, x_2 \in S$ we define $x_1 \sim x_2$ if $f(x_1) = f(x_2)$. Then for all $x_1, x_2, x_3 \in S$ we have (i) $f(x_1) = f(x_1)$; (ii) if $f(x_1) = f(x_2)$, then $f(x_2) = f(x_1)$; and (iii) if $f(x_1) = f(x_2)$ and $f(x_2) = f(x_3)$, then $f(x_1) = f(x_3)$. This shows that we have defined an equivalence relation on the set S. The proof of this is easy because the equivalence relation is defined in terms of equality of the images $f(x)$, and equality is the most elementary equivalence relation. The collection of all equivalence classes of S under $\sim$ will be denoted by S/f.

Later in this section we will show that any function $f : S \to T$ naturally induces a one-to-one function $\bar{f} : S/f \to T$. Thus by introducing the set S/f we can study f by studying the equivalence relation it defines on S and the corresponding one-to-one function. $\square$

Example 2.2.3

Let A be the set of all integers and let B be the set of all nonzero integers. On the set $S = A \times B$ of ordered pairs, define $(m, n) \sim (p, q)$ if $mq = np$. This defines an equivalence relation, which we can show as follows. Certainly $(m, n) \sim (m, n)$, since $mn = nm$. If $(m, n) \sim (p, q)$, then $mq = np$ implies $pn = qm$, and so $(p, q) \sim (m, n)$. Finally, suppose that $(m, n) \sim (p, q)$ and $(p, q) \sim (s, t)$. Then $mq = np$ and $pt = qs$, and multiplying the first equation by t and the second by n gives $mqt = npt$ and $npt = nqs$. After equating mqt and nqs and cancelling q (which is nonzero by assumption), we obtain $mt = ns$, and so $(m, n) \sim (s, t)$. Thus we have verified the reflexive, symmetric, and transitive laws, showing that we have in fact defined an equivalence relation. The equivalence classes of this equivalence relation form the set $\mathbf{Q}$ of rational numbers.

As a passing remark, we note that in this example using the ordered pair formulation of the notion of an equivalence relation would lead to considering ordered pairs of ordered pairs. $\square$

Example 2.2.4

Consider the set of all differentiable functions from $\mathbf{R}$ into $\mathbf{R}$. For two such functions $f(x)$ and $g(x)$ we define $f \sim g$ if the derivatives $f'(x)$ and $g'(x)$ are equal. It can easily be checked directly that the properties defining an equivalence relation hold in this case. Furthermore, the equivalence class of the function f is the set of all functions of the form $f(x) + C$, for a constant C. Using a somewhat more sophisticated point of view, we can consider differentiation as a function, whose domain is the set of all differentiable functions and whose codomain is the set of all functions. Then the equivalence relation we have defined arises as in the previous Example 2.2.2.   □

Example 2.2.5

Let S be the set $\mathbf{R}^2$ of points in the Euclidean plane; that is, $S = \{(x, y)|x, y \in \mathbf{R}\}$. Define $(x_1, y_1) \sim (x_2, y_2)$ if $x_1 = x_2$. There are several ways to check that we have defined an equivalence relation. First, the reflexive, symmetric, and transitive laws are easy to check since $\sim$ is defined in terms of equality of the first components. As a second method we could use Example 2.2.2. If we define $f : \mathbf{R}^2 \to \mathbf{R}$ by $f(x, y) = x$, then $f(x_1, y_1) = f(x_2, y_2)$ if and only if $x_1 = x_2$. Thus the relation defined by f is the same as $\sim$, and it follows that $\sim$ is an equivalence relation.

Now let us find the equivalence classes of the equivalence relation we have defined. The equivalence class of (a, b) consists of all points in the plane that have the same first coordinate; that is, $[(a, b)]$ is just the line $x = a$, which we will denote by L_a. Then since distinct vertical lines are parallel, we see that $L_a \cap L_b = \emptyset$ for $a \neq b$. We can summarize by saying that each point in the plane belongs to exactly one of the equivalence classes L_a.   □

The following fact, observed at the end of Example 2.2.5, holds for all equivalence relations.

2.2.3 Proposition.   Let S be a set, and let $\sim$ be an equivalence relation on S. Then each element of S belongs to exactly one of the equivalence classes of S determined by $\sim$.

Proof. If $a \in S$, then $a \sim a$, and so $a \in [a]$. If $a \in [b]$ for some $b \in S$, then we claim that $[a] = [b]$. To show this, we must check that each element of $[a]$ belongs to $[b]$, and also that each element of $[b]$ must belong to $[a]$. If $x \in [a]$, then $x \sim a$ by definition. We have assumed that $a \in [b]$, so $a \sim b$, and then it follows from the transitive law that $x \sim b$, and thus $x \in [b]$. On the other hand, if $x \in [b]$, then $x \sim b$. By the symmetric law, $a \sim b$ implies $b \sim a$, and so this gives us $x \sim a$, showing that $x \in [a]$. We have thus shown that each element $a \in S$ belongs to exactly one of the equivalence classes determined by $\sim$.   □

2.2.4 Definition.   Let S be any set. A family $\mathcal{P}$ of subsets of S is called a *partition* of S if each element of S belongs to exactly one of the members of $\mathcal{P}$.

In this terminology, Proposition 2.2.3 implies that any equivalence relation on a set S determines a partition of the set. We will show that the converse is also true. We first give some additional examples of partitions.

Example 2.2.6

The equivalence relation that we considered in Example 2.2.5 partitions the Euclidean plane $\mathbf{R}^2$ by using the collection of all vertical lines.

Another interesting partition of $\mathbf{R}^2$ uses lines radiating from the origin. First, let S_0 be the x-axis, $y = 0$. Since the origin $(0, 0)$ belongs to S_0, it cannot belong to any other set in the partition, so for any nonzero real number a, let S_a be the line $y = ax$ with the origin removed. Let S_∞ be the y-axis with the origin removed. Then each point in the plane (except the origin) determines a unique line through the origin, and so each point belongs to exactly one of the sets we have defined.   □

Example 2.2.7

It is obvious that the family of all circles with center at the origin forms a partition of the plane $\mathbf{R}^2$. (We must allow the origin itself to be considered a degenerate circle.) There is a corresponding equivalence relation, which can be described algebraically by defining $(x_1, y_1) \sim (x_2, y_2)$ if $x_1^2 + y_1^2 = x_2^2 + y_2^2$.   □

2.2.5 Proposition.   Any partition of a set determines an equivalence relation.

Proof. Assume that we are given a partition $\mathcal{P}$ of the set S. Then the given partition yields an equivalence relation on S by defining $a \sim b$ if there is some subset in the partition $\mathcal{P}$ to which both a and b belong. To prove that we have defined an equivalence relation, we proceed as follows. For all $a \in S$ we have $a \sim a$ since a belongs to exactly one subset in the partition, and thus the reflexive law holds. It is obvious from the definition that the relation is symmetric. Finally, for $a, b, c \in S$ suppose that $a \sim b$ and $b \sim c$. Then a and b both belong to some subset of the partition, say S_1. Similarly, b and c both belong to some subset, say S_2. Since we are given a partition, the element b can belong to only one subset, so we have $S_1 = S_2$, and then this implies that a and c belong to the same subset, so we have $a \sim c$ and the transitive law holds.   □

It can be shown that there is a one-to-one correspondence between equivalence relations on a set and partitions of the set. Beginning with an equivalence relation, its equivalence classes define a partition, and then the equivalence relation corresponding to this partition is precisely the original equivalence relation. On the other hand, if a partition on a set is given, then the equivalence classes of the corresponding equivalence relation are the subsets in the original partition.

Example 2.2.8

Let S be any set, and let $\sim$ be an equivalence relation on S. Define a function $\pi : S \rightarrow S/ \sim$ by $\pi(x) = [x]$ for all $x \in S$. Proposition 2.2.3 shows that each element $x \in S$ belongs to a unique equivalence class, and so this shows that π is a function, which we refer to as the *natural projection* from S onto $S/ \sim$.

For any x_1, $x_2 \in S$ we have $x_1 \sim x_2$ if and only if $\pi(x_1) = \pi(x_2)$. This shows that π defines the original equivalence relation $\sim$, so that the families $S/\sim$ and S/π are identical. Thus we have just verified the remark in Example 2.2.2 that every equivalence relation can be realized as the equivalence relation defined in a natural way by a function. $\square$

Before stating the main result of this section we need a definition. Let $f : S \to T$ be a function, and let $A \subseteq S$. Then

$$f(A) = \{y \in T \mid y = f(a) \text{ for some } a \in A\}$$

is called the *image* of A under f.

2.2.6 Theorem.   If $f : S \to T$ is any function, and $\sim$ is the equivalence relation defined on S by letting $x_1 \sim x_2$ if $f(x_1) = f(x_2)$, for all $x_1, x_2 \in S$, then there is a one-to-one correspondence between the elements of the image $f(S)$ of S under f and the equivalence classes S/f of the relation $\sim$.

Proof.   Define $\bar{f} : S/f \to f(S)$ by $\bar{f}([x]) = f(x)$, for all $x \in S$. The function $\bar{f}$ is well-defined since if x_1 and x_2 belong to the same equivalence class in S/f, then by definition of the equivalence relation we must have $f(x_1) = f(x_2)$. In addition, $\bar{f}$ is onto since if $y \in f(S)$, then $y = f(x)$ for some $x \in S$, and thus we have $y = \bar{f}([x])$. Finally, $\bar{f}$ is one-to-one since if $\bar{f}([x_1]) = \bar{f}([x_2])$, then $f(x_1) = f(x_2)$, and so by definition $x_1 \sim x_2$, which implies that $[x_1] = [x_2]$. Thus we have shown that $\bar{f}$ is a one-to-one correspondence. $\square$

If $f : S \to T$ is a function and y belongs to the image $f(S)$, then $\{x \in S \mid f(x) = y\}$ is usually called the *inverse image* of y, denoted by $f^{-1}(y)$. The inverse images of elements of $f(S)$ correspond to the equivalence classes S/f. (Note carefully that we are not implying that f has an inverse function.)

In Theorem 2.2.6 we have really factored f into a composition of better-behaved functions, since $f = \iota \circ \bar{f} \circ \pi$, where π is the natural function from S to S/f defined by $\pi(x) = [x]$, $\iota : f(S) \to T$ is the inclusion mapping defined by $\iota(y) = y$ for all $y \in f(S)$, and $\bar{f}$ is a one-to-one correspondence. This is illustrated by the following diagram.

$$S \xrightarrow{\pi} S/f \xrightarrow{\bar{f}} f(S) \xrightarrow{\iota} T$$

This factorization of f can also be shown as in the following diagram.

$$
\begin{array}{ccc}
 & f & \\
S & \to & T \\
\pi \downarrow & & \uparrow \iota \\
S/f & \to & f(S) \\
 & \bar{f} &
\end{array}
$$

EXERCISES: SECTION 2.2

1. It is shown in Theorem 2.2.6 that if $f : S \to T$ is a function, then there is a one-to-one correspondence between the elements of $f(S)$ and the equivalence classes of S/f. For each of the following functions, find $f(S)$ and S/f and exhibit the one-to-one correspondence between them:
   (a) $f : \mathbf{Z} \to \mathbf{C}$ given by $f(n) = i^n$ for all $n \in \mathbf{Z}$
   (b) $g : \mathbf{Z} \to \mathbf{Z}_{12}$ given by $g(n) = [8n]_{12}$ for all $n \in \mathbf{Z}$

2. Repeat the previous exercise for each of the following functions:
   (a) $f : \mathbf{Z}_{12} \to \mathbf{Z}_{12}$ defined by $f([x]_{12}) = [4x]_{12}$
   (b) $g : \mathbf{Z}_{12} \to \mathbf{Z}_{12}$ defined by $g([x]_{12}) = [5x]_{12}$
   (c) $h : \mathbf{Z}_{12} \to \mathbf{Z}_{12}$ defined by $h([x]_{12}) = [9x]_{12}$
   (d) $p : \mathbf{Z}_{12} \to \mathbf{Z}_{24}$ defined by $p([x]_{12}) = [4x]_{24}$
   (e) $q : \mathbf{Z}_{24} \to \mathbf{Z}_{12}$ defined by $q([x]_{24}) = [4x]_{12}$

3. Let S be the set of all ordered pairs (m, n) of positive integers. For $(a_1, a_2) \in S$ and $(b_1, b_2) \in S$, define $(a_1, a_2) \sim (b_1, b_2)$ if $a_1 + b_2 = a_2 + b_1$. Show that $\sim$ is an equivalence relation.

4. On $\mathbf{R}^2$, define $(a_1, a_2) \sim (b_1, b_2)$ if $a_1^2 + a_2^2 = b_1^2 + b_2^2$. Check that this defines an equivalence relation. What are the equivalence classes?

5. Define an equivalence relation on the set $\mathbf{R}$ that partitions the real line into subsets of length 1.

6. In $\mathbf{R}^3$, consider the standard (x, y, z)-coordinate system. We can define a partition of $\mathbf{R}^3$ by using planes parallel to the (x, y)-plane. Describe the corresponding equivalence relation by giving conditions on the coordinates x, y, z.

7. For integers m, n, define $m \sim n$ if and only if $n|m^k$ and $m|n^j$ for some positive integers k and j.
   (a) Show that $\sim$ is an equivalence relation on $\mathbf{Z}$.
   (b) Determine the equivalence classes [1], [2], [6] and [12].
   (c) Give a characterization of the equivalence class $[m]$.

8. Let S be a set. A subset $R \subseteq S \times S$ is called a *circular relation* if (i) for each $a \in S$, $(a, a) \in R$ and (ii) for each a, b, $c \in S$, if $(a, b) \in R$ and $(b, c) \in R$, then $(c, a) \in R$. Show that any circular relation must be an equivalence relation.

9. Let S be a set and let $2^S = \{A \mid A \subseteq S\}$ be the collection of all subsets of S. Define $\sim$ on 2^S by letting $A \sim B$ if and only if there exists a one-to-one correspondence from A to B.
   (a) Show that $\sim$ is an equivalence relation on 2^S.
   (b) If $S = \{1, 2, 3, 4\}$, list the subsets in 2^S and find each equivalence class determined by $\sim$.

10. Let W be a subspace of a vector space V over $\mathbf{R}$ (that is, the scalars are assumed to be real numbers). We say that two vectors u, $v \in V$ are congruent modulo W if $u - v \in W$, written $u \equiv v \pmod{W}$.
   (a) Show that $\equiv$ is an equivalence relation.
   (b) Show that if r, s are scalars and u_1, u_2, v_1, v_2 are vectors in V such that $u_1 \equiv v_1 \pmod{W}$ and $u_2 \equiv v_2 \pmod{W}$, then $ru_1 + su_2 \equiv rv_1 + sv_2 \pmod{W}$.
   (c) Let $[u]_W$ denote the equivalence class of the vector u. Set $U = \{[u]_W \mid u \in V\}$. Define $+$ and $\cdot$ on U by $[u]_W + [v]_W = [u + v]_W$ and $r \cdot [u]_W = [ru]_W$ for all u, $v \in V$ and $r \in R$. Show that U is a vector space with respect to these operations.

(d) Let $V = \mathbf{R}^2$, and let $W = \{(x, 0) \mid x \in \mathbf{R}\}$. Describe the equivalence class $[(x, y)]_W$ geometrically. Show that $T : \mathbf{R} \to U$ defined by $T(y) = [(0, y)]_W$ is a linear transformation that is one-to-one and onto.

11. Proposition 2.2.3 shows that any equivalence relation on a set S determines a partition of the set. As in Example 2.2.6, this partition defines an equivalence relation. Show that this "new" equivalence relation is, in fact, the same equivalence relation as the one with which we started.

12. Example 2.2.6 shows that any partition of a set S determines an equivalence relation. Show that the equivalence classes of this equivalence relation are precisely the sets in the original partition.

13. Let $T = \{(x, y, z) \in \mathbf{R}^3 \mid (x, y, z) \neq (0, 0, 0)\}$. Define $\sim$ on T by $(x_1, y_1, z_1) \sim (x_2, y_2, z_2)$ if there exists a nonzero real number λ such that $x_1 = \lambda x_2$, $y_1 = \lambda y_2$, and $z_1 = \lambda z_2$.
   (a) Show that $\sim$ is an equivalence relation on T.
   (b) Give a geometric description of the equivalence class of (x, y, z).
   The set $T/\sim$ is called the *real projective plane* and is denoted by $\mathbf{P}^2$. The class of (x, y, z) is denoted by $[x, y, z]$ and is called a *point*.
   (c) Let $(a, b, c) \in T$, and suppose that $(x_1, y_1, z_1) \sim (x_2, y_2, z_2)$. Show that if $ax_1 + by_1 + cz_1 = 0$, then $ax_2 + by_2 + cz_2 = 0$. Conclude that

$$L = \{[x, y, z] \in \mathbf{P}^2 \mid ax + by + cz = 0\}$$

   is a well-defined subset of $\mathbf{P}^2$. Such sets L are called *lines*.
   (d) Show that the triples $(a_1, b_1, c_1) \in T$ and $(a_2, b_2, c_2) \in T$ determine the same line if and only if $(a_1, b_1, c_1) \sim (a_2, b_2, c_2)$.
   (e) Given two distinct points of $\mathbf{P}^2$, show that there exists exactly one line that contains both points.
   (f) Given two distinct lines, show that there exists exactly one point that belongs to both lines.
   (g) Show that the function $f : \mathbf{R}^2 \to \mathbf{P}^2$ defined by $f(x, y) = [x, y, 1]$ is a one-to-one function. This is one possible embedding of the "affine plane" into the projective plane. We sometimes say that $\mathbf{P}^2$ is the "completion" of $\mathbf{R}^2$.
   (h) Show that the embedding of part (g) takes lines to "lines."
   (i) If two lines intersect in $\mathbf{R}^2$, show that the image of their intersection is the intersection of their images (under the embedding defined in part (g)).
   (j) If two lines are parallel in $\mathbf{R}^2$, what happens to their images under the embedding into $\mathbf{P}^2$?

2.3 PERMUTATIONS

We will now study one-to-one correspondences in more detail, particularly for finite sets. Our emphasis in this section will be on computations with such functions. We need to develop some notation that will make it easier to work with such functions, especially when finding the composition of two functions. We will change our notation slightly, using Greek letters for permutations, and instead of writing $\sigma \circ \tau$ for the composition of two permutations, we will simply write $\sigma\tau$.

2.3.1 Definition.   Let S be a set. A function $\sigma : S \to S$ is called a *permutation* of S if σ is one-to-one and onto. The set of all permutations of S will be denoted by $\text{Sym}(S)$.

The set of all permutations of the finite set $S = \{1, 2, \ldots, n\}$ will be denoted by S_n.

Proposition 2.1.6 shows that the composition of two permutations in $\text{Sym}(S)$ is again a permutation. It is obvious that the identity function on S is one-to-one and onto. Proposition 2.1.8 shows that any permutation in $\text{Sym}(S)$ has an inverse function that is also one-to-one and onto. We can summarize these important properties as follows:

(i) If $\sigma, \tau \in \text{Sym}(S)$, then $\tau\sigma \in \text{Sym}(S)$;

(ii) $1_S \in \text{Sym}(S)$;

(iii) if $\sigma \in \text{Sym}(S)$, then $\sigma^{-1} \in \text{Sym}(S)$.

We also mention that the composition of permutations is associative, by Proposition 2.1.3.

We need to develop some notation for working with permutations in S_n. Given $\sigma \in S_n$, note that σ is completely determined as soon as we know $\sigma(1)$, $\sigma(2), \ldots, \sigma(n)$, and so we introduce the notation

$$\sigma = \begin{pmatrix} 1 & 2 & \cdot & \cdot & \cdot & n \\ \sigma(1) & \sigma(2) & \cdot & \cdot & \cdot & \sigma(n) \end{pmatrix}$$

where under each integer i we write the image of i.

For example, if $S = \{1, 2, 3\}$ and $\sigma : S \to S$ is given by $\sigma(1) = 2$, $\sigma(2) = 3$, $\sigma(3) = 1$, then we would write $\sigma = \begin{pmatrix} 1 & 2 & 3 \\ 2 & 3 & 1 \end{pmatrix}$.

Since any element σ in S_n is one-to-one and onto, in the above notation for σ each element of S must appear once and only once in the second row. Thus an element $\sigma \in S_n$ is completely determined once we know the order in which the elements of S appear in the second row.

To count the number of elements of S_n we only need to count the number of possible second rows. Since there are n elements in $S = \{1, 2, \ldots, n\}$, there are n choices for the first element $\sigma(1)$ of the second row. Since the element that is assigned to $\sigma(1)$ cannot be used again, we have $n - 1$ choices when we wish to assign a value to $\sigma(2)$, and thus there are $n \cdot (n - 1)$ ways to assign values to both $\sigma(1)$ and $\sigma(2)$. Now there are $n - 2$ choices for $\sigma(3)$ and a total of $n(n - 1)(n - 2)$ ways to assign values to $\sigma(1)$, $\sigma(2)$, and $\sigma(3)$. Continuing in this fashion (there is an induction argument here), we have a total of $n!$ ways to assign values to $\sigma(1)$, $\sigma(2), \ldots, \sigma(n)$. Thus S_n has $n!$ elements.

The notation we have introduced is useful for computing in S_n. Suppose that

$$\sigma = \begin{pmatrix} 1 & 2 & . & . & . & n \\ \sigma(1) & \sigma(2) & . & . & . & \sigma(n) \end{pmatrix} \quad \text{and} \quad \tau = \begin{pmatrix} 1 & 2 & . & . & . & n \\ \tau(1) & \tau(2) & . & . & . & \tau(n) \end{pmatrix}.$$

Then to compute the composition

$$\sigma\tau = \begin{pmatrix} 1 & 2 & . & . & . & n \\ \sigma(\tau(1)) & \sigma(\tau(2)) & . & . & . & \sigma(\tau(n)) \end{pmatrix}$$

we proceed as follows: To find $\sigma(\tau(i))$ we first look under i in τ to get $j = \tau(i)$, and then we find $\sigma\tau(i) = \sigma(j)$ by looking under j in σ.

Example 2.3.1

Let $\sigma = \begin{pmatrix} 1 & 2 & 3 & 4 \\ 4 & 3 & 1 & 2 \end{pmatrix}$ and $\tau = \begin{pmatrix} 1 & 2 & 3 & 4 \\ 2 & 3 & 4 & 1 \end{pmatrix}$. To compute $\sigma\tau$ we have $\tau(1) = 2$ and then $\sigma(2) = 3$, giving $\sigma\tau(1) = 3$. Next we have $\tau(2) = 3$ and $\sigma(3) = 1$, giving $\sigma\tau(2) = 1$. Continuing this procedure we obtain $\sigma\tau = \begin{pmatrix} 1 & 2 & 3 & 4 \\ 3 & 1 & 2 & 4 \end{pmatrix}$. A similar computation gives $\tau\sigma = \begin{pmatrix} 1 & 2 & 3 & 4 \\ 1 & 4 & 2 & 3 \end{pmatrix}$. $\square$

Given $\sigma = \begin{pmatrix} 1 & 2 & . & . & . & n \\ \sigma(1) & \sigma(2) & . & . & . & \sigma(n) \end{pmatrix}$ in S_n, it is easy to compute σ^{-1}. To find $\sigma^{-1}(j)$ we find j in the second row of σ, say $j = \sigma(i)$. The inverse of σ must reverse this assignment, and so under j we write i, giving $\sigma^{-1}(j) = i$. This can be accomplished easily by simply turning the two rows of σ upside down and then rearranging terms.

Example 2.3.2

If $\sigma = \begin{pmatrix} 1 & 2 & 3 & 4 \\ 4 & 3 & 1 & 2 \end{pmatrix}$, then

$$\sigma^{-1} = \begin{pmatrix} 4 & 3 & 1 & 2 \\ 1 & 2 & 3 & 4 \end{pmatrix} = \begin{pmatrix} 1 & 2 & 3 & 4 \\ 3 & 4 & 2 & 1 \end{pmatrix}.   \square$$

The double-row notation that we have introduced for permutations is transparent but rather cumbersome. We now want to introduce another notation that is more compact and also helps to convey certain information about the permutation. Consider the permutation $\sigma = \begin{pmatrix} 1 & 2 & 3 & 4 & 5 \\ 3 & 1 & 4 & 2 & 5 \end{pmatrix}$. We do not necessarily have to write the first row in numerical order, and in this case it is informative to change the order as follows. The first column expresses the fact that $\sigma(1) = 3$, and using this we interchange the second and third columns. Now $\sigma(3) = 4$, and so as the third column we choose the one with 4 in the first row. Continuing gives us

$\sigma = \begin{pmatrix} 1 & 3 & 4 & 2 & 5 \\ 3 & 4 & 2 & 1 & 5 \end{pmatrix}$. Now writing $(1, 3, 4, 2)$ would give us all of the necessary information to describe σ, since $\sigma(1) = 3$, $\sigma(3) = 4$, $\sigma(4) = 2$, and $\sigma(2) = 1$. In the new notation we do not need to mention $\sigma(5)$ since $\sigma(5) = 5$.

2.3.2 Definition.   Let S be a set, and let $\sigma \in \mathrm{Sym}(S)$. Then σ is called a *cycle of length* k if there exist elements $a_1, a_2, \ldots, a_k \in S$ such that $\sigma(a_1) = a_2$, $\sigma(a_2) = a_3, \ldots, \sigma(a_{k-1}) = a_k$, $\sigma(a_k) = a_1$, and $\sigma(x) = x$ for all other elements $x \in S$ with $x \neq a_i$ for $i = 1, 2, \ldots, k$.

In this case we write $\sigma = (a_1, a_2, \ldots, a_k)$.

We can also write $\sigma = (a_2, a_3, \ldots, a_k, a_1)$ or $\sigma = (a_3, \ldots, a_k, a_1, a_2)$, etc. The notation for a cycle of length k can thus be written in k different ways, depending on the starting point. The notation (1) is used for the identity permutation.

Example 2.3.3

In S_5 the permutation $\begin{pmatrix} 1 & 2 & 3 & 4 & 5 \\ 3 & 2 & 4 & 1 & 5 \end{pmatrix}$ is a cycle of length 3, written $(1, 3, 4)$. The permutation $\begin{pmatrix} 1 & 2 & 3 & 4 & 5 \\ 3 & 5 & 4 & 1 & 2 \end{pmatrix}$ is not a cycle. $\square$

Example 2.3.4

For any permutation of a finite set there is an associated diagram, found by representing the elements of the set as points and joining two points with an arrow if the permutation maps one to the other. Since any permutation is a one-to-one and onto function, at each point of the set there is one and only one incoming arrow and one and only one outgoing arrow. To find the diagram of a permutation such as

$$\begin{pmatrix} 1 & 2 & 3 & 4 & 5 & 6 & 7 & 8 & 9 & 10 & 11 & 12 \\ 8 & 2 & 10 & 11 & 5 & 9 & 4 & 6 & 1 & 3 & 12 & 7 \end{pmatrix}$$

we first rearrange the columns to give the following form:

$$\begin{pmatrix} 1 & 8 & 6 & 9 & 2 & 3 & 10 & 4 & 11 & 12 & 7 & 5 \\ 8 & 6 & 9 & 1 & 2 & 10 & 3 & 11 & 12 & 7 & 4 & 5 \end{pmatrix}$$

The associated diagram is given in Figure 2.3.1.

The diagram of $\sigma = \begin{pmatrix} 1 & 2 & 3 & 4 & 5 \\ 3 & 2 & 4 & 1 & 5 \end{pmatrix}$ is in Figure 2.3.2.

Note that the diagram of a cycle of length k would consist of a connected component with k vertices, while all other components of it would

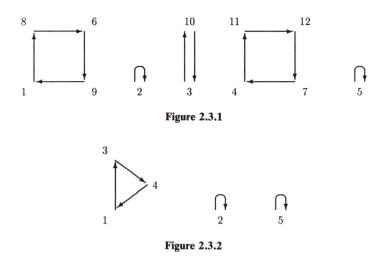

Figure 2.3.1

Figure 2.3.2

contain only one element. This diagram would clearly illustrate why such a
permutation is called a cycle.   □

Example 2.3.5

Let $(1, 4, 2, 5)$ and $(2, 6, 3)$ be cycles in S_6. Then

$$(1, 4, 2, 5) = \begin{pmatrix} 1 & 2 & 3 & 4 & 5 & 6 \\ 4 & 5 & 3 & 2 & 1 & 6 \end{pmatrix} \quad \text{and} \quad (2, 6, 3) = \begin{pmatrix} 1 & 2 & 3 & 4 & 5 & 6 \\ 1 & 6 & 2 & 4 & 5 & 3 \end{pmatrix}.$$

In computing the product of these two cycles we have

$$(1, 4, 2, 5)(2, 6, 3) = \begin{pmatrix} 1 & 2 & 3 & 4 & 5 & 6 \\ 4 & 5 & 3 & 2 & 1 & 6 \end{pmatrix} \begin{pmatrix} 1 & 2 & 3 & 4 & 5 & 6 \\ 1 & 6 & 2 & 4 & 5 & 3 \end{pmatrix}$$

$$= \begin{pmatrix} 1 & 2 & 3 & 4 & 5 & 6 \\ 4 & 6 & 5 & 2 & 1 & 3 \end{pmatrix}$$

$$= (1, 4, 2, 6, 3, 5),$$

which is again a cycle.

Note that it is not true in general that the product of two cycles is again
a cycle. In particular,

$$(1, 4, 2, 5)(1, 4, 2, 5) = \begin{pmatrix} 1 & 2 & 3 & 4 & 5 & 6 \\ 2 & 1 & 3 & 5 & 4 & 6 \end{pmatrix}$$

is not a cycle.   □

2.3.3 Definition.   Let $\sigma = (a_1, a_2, \ldots, a_k)$ and $\tau = (b_1, b_2, \ldots, b_m)$ be
cycles in Sym(S), for a set S. Then σ and τ are said to be *disjoint* if $a_i \neq b_j$ for
all i, j.

It often happens that $\sigma\tau \neq \tau\sigma$ for two permutations σ, τ. For example, in S_3 we have

$$\begin{pmatrix} 1 & 2 & 3 \\ 2 & 1 & 3 \end{pmatrix}\begin{pmatrix} 1 & 2 & 3 \\ 3 & 2 & 1 \end{pmatrix} = \begin{pmatrix} 1 & 2 & 3 \\ 3 & 1 & 2 \end{pmatrix}$$

but on the other hand

$$\begin{pmatrix} 1 & 2 & 3 \\ 3 & 2 & 1 \end{pmatrix}\begin{pmatrix} 1 & 2 & 3 \\ 2 & 1 & 3 \end{pmatrix} = \begin{pmatrix} 1 & 2 & 3 \\ 2 & 3 & 1 \end{pmatrix}.$$

It $\sigma\tau = \tau\sigma$, then we say that σ and τ *commute*. Using this terminology, the next proposition shows that disjoint cycles always commute.

2.3.4 Proposition.   Let S be any set. If σ and τ are disjoint cycles in Sym(S), then $\sigma\tau = \tau\sigma$.

Proof. Let $\sigma = (a_1, a_2, \ldots, a_k)$ and $\tau = (b_1, b_2, \ldots, b_m)$ be disjoint. If $i = a_j$ for some $j < k$, then

$$\sigma\tau(i) = \sigma(\tau(a_j)) = \sigma(a_j) = a_{j+1} = \tau(a_{j+1}) = \tau(\sigma(a_j)) = \tau\sigma(i)$$

because τ leaves $a_1, a_2, \ldots, a_k$ fixed. In case $j = k$, we use $\sigma(a_j) = a_1 = \tau(a_1)$. A similar computation can be given if $i = b_j$ for some j, since then σ leaves b_1, $b_2, \ldots, b_m$ fixed. If i appears in neither cycle, then both σ and τ leave it fixed, so $\sigma\tau(i) = \sigma(i) = i = \tau(i) = \tau\sigma(i)$. $\square$

Let σ be any permutation in Sym(S), for any set S. Taking the composition of σ with itself any number of times still gives us a permutation, and so for any positive integer i we define

$$\sigma^i = \sigma\sigma \cdots \sigma \qquad i \text{ times.}$$

Formally, we can define the powers of σ inductively by letting $\sigma^i = \sigma\sigma^{i-1}$ for $i \geq 2$. Then the following properties can be established by using induction. For positive integers m and n we have

$$\sigma^m\sigma^n = \sigma^{m+n} \qquad \text{and} \qquad (\sigma^m)^n = \sigma^{mn}.$$

To illustrate, we have

$$\sigma^2\sigma^3 = (\sigma\sigma)(\sigma\sigma\sigma) = \sigma^5 \quad \text{and} \quad (\sigma^2)^3 = (\sigma^2)(\sigma^2)(\sigma^2) = (\sigma\sigma)(\sigma\sigma)(\sigma\sigma) = \sigma^6.$$

To preserve these laws of exponents, we define $\sigma^0 = (1)$, where (1) is the identity element of Sym(S), and $\sigma^{-n} = (\sigma^n)^{-1}$. Using these definitions, it can be shown that

$$\sigma^m\sigma^n = \sigma^{m+n} \qquad \text{and} \qquad (\sigma^m)^n = \sigma^{mn}$$

for all integers m, n.

2.3.5 Theorem.   Every permutation in S_n can be written as a product of disjoint cycles. The cycles that appear in the product are unique.

Proof. Let $S = \{1, 2, \ldots, n\}$ and let $\sigma \in S_n = \text{Sym}(S)$. If we apply successive powers of σ to 1, we have the elements $1, \sigma(1), \sigma^2(1), \sigma^3(1), \ldots$, and after some point there must be a repetition since S has only n elements. Suppose that $\sigma^m(1) = \sigma^k(1)$ is the first repetition, with $m > k \geq 0$. If $k > 0$, then applying σ^{-1} to both sides of the equation a total of k times gives $\sigma^{m-k}(1) = 1$, which contradicts the choice of m. Thus the first time a repetition occurs it is 1 that is repeated.

If we let r be the exponent for which we first have $\sigma^r(1) = 1$, then the elements $1, \sigma(1), \sigma^2(1), \ldots \sigma^{r-1}(1)$ are all distinct, giving us a cycle of length r:

$$(1, \sigma(1), \sigma^2(1), \ldots, \sigma^{r-1}(1)).$$

If $r < n$, let a be the least integer not in $(1, \sigma(1), \ldots, \sigma^{r-1}(1))$ and form the cycle

$$(a, \sigma(a), \sigma^2(a), \ldots, \sigma^{s-1}(a))$$

in which s is the least positive integer such that $\sigma^s(a) = a$. (Such an exponent s exists by an argument similar to the one given above.) If $r + s < n$, then let b be the least positive integer not in the set

$$\{1, \sigma(1), \ldots, \sigma^{r-1}(1), a, \sigma(a), \ldots, \sigma^{s-1}(a)\}$$

and form the cycle beginning with b. We continue in this way until we have exhausted S. Then

$$\sigma = (1, \sigma(1), \ldots, \sigma^{r-1}(1))(a, \sigma(a), \ldots, \sigma^{s-1}(a)) \cdots$$

and we have written σ as a product of disjoint cycles.

In fact we have given an algorithm for finding the necessary cycles. Since the cycles are disjoint, by the previous proposition the product does not depend on their order. It is left as an exercise to show that a given permutation can be expressed as a product of disjoint cycles in only one way (if the order is disregarded).   □

Since any cycle of length one in S_n is the identity function, we will usually omit cycles of length one when we write a permutation as a product of disjoint cycles. Thus if a permutation $\sigma \in S_n$ is written as a product of disjoint cycles of length greater than or equal to two, then for any i $(1 \leq i \leq n)$ missing from these cycles we know that $\sigma(i) = i$.

Example 2.3.6

If $\sigma = \begin{pmatrix} 1 & 2 & 3 & 4 & 5 & 6 & 7 & 8 \\ 5 & 8 & 7 & 2 & 3 & 6 & 1 & 4 \end{pmatrix}$, then we have $\sigma = (1, 5, 3, 7)(2, 8, 4)$.   □

In general, if we wish to multiply (compose) two permutations in cycle notation, we do not have to return to the double-row format. We can use the algorithm in Theorem 2.3.5 to write the product (composition) as a product of disjoint cycles, remembering to work from right to left. The procedure is more difficult to describe in words than it is to carry out, but nonetheless we will now try to give a brief description of it.

Say we wish to find the composite function $(a_1, a_2, \ldots, a_k)(b_1, b_2, \ldots, b_m)$. To find the image of i we first see if i is equal to some b_s. If it is, then we know that $(b_1, b_2, \ldots, b_m)$ maps $i = b_s$ to b_{s+1}. (In case $s = m$, i is mapped to b_1.) We then look for b_{s+1} in $(a_1, a_2, \ldots, a_k)$. If it appears here, say $b_{s+1} = a_t$, then $(a_1, a_2, \ldots, a_k)$ maps b_{s+1} to a_{t+1}. (In case $t = k$, b_{s+1} is mapped to a_1.) Thus the composite function maps i to a_{t+1}. If b_{s+1} does not appear in $(a_1, a_2, \ldots, a_k)$, then $(a_1, a_2, \ldots, a_k)$ maps b_{s+1} to b_{s+1}, and so the composite function maps i to b_{s+1} in this case.

Likewise, if i does not appear in $(b_1, b_2, \ldots, b_m)$, then it leaves i fixed and we only need to look for i in $(a_1, a_2, \ldots, a_k)$. If $i = a_s$, then $(a_1, a_2, \ldots, a_k)$ maps i to a_{s+1} (in case $s = k$, i is mapped to a_1), and so the composite function maps i to a_{s+1}. If i does not appear in either $(a_1, a_2, \ldots, a_k)$ or $(b_1, b_2, \ldots, b_m)$, then i is left fixed by the composite function.

Thus if we want to write our product as a product of disjoint cycles, start with $i = 1$ and find the image of i, say j, as we have outlined above. Now, starting with j, repeat the procedure until we return to 1, completing the first cycle. Then, starting with the least integer that does not appear in that cycle, we apply the same procedure until the second cycle is complete. This is repeated until all entries of the two original cycles have been used.

Example 2.3.7

Let $(2,5,1,4,3)$ and $(4,6,2)$ be cycles in S_6. Then

$$(2,5,1,4,3)(4,6,2) = (1,4,6,5)(2,3)$$

and we note again that the product of two cycles need not be a cycle.   □

Since every positive power of σ must belong to S_n, while there are only finitely many elements in S_n, there must exist positive integers $i > j$ such that $\sigma^i = \sigma^j$. Taking the composition with σ^{-1} a total of j times shows that $\sigma^{i-j} = (1)$. Thus we have shown that there is a positive integer m such that $\sigma^m = (1)$.

If $\sigma = (a_1, a_2, \ldots, a_m)$ is a cycle of length m, then applying σ m times to any a_i, $i = 1, 2, \ldots, m$ gives a_i. Thus $\sigma^m = (1)$. Furthermore, m is the smallest positive power of σ that equals the identity, since $\sigma^k(a_1) = a_k$ for $1 \le k < m$. In terms of the following definition, we have just shown that a cycle of length m has order m.

2.3.6 Definition.   Let $\sigma \in S_n$. The least positive integer m such that $\sigma^m = (1)$ is called the *order* of σ.

2.3.7 Proposition.   Let $\sigma \in S_n$ have order m. Then for all integers i, j we have $\sigma^i = \sigma^j$ if and only if $i \equiv j \pmod{m}$.

Proof. By assumption m is the smallest positive exponent with $\sigma^m = (1)$. If $\sigma^i = \sigma^j$, for any integers i, j, then multiplying by $(\sigma^{-1})^j$ shows that $\sigma^{i-j} = (1)$. Using the division algorithm we can write $i - j = qm + r$ for integers q, r with $0 \leq r < m$. Then since

$$(1) = \sigma^{i-j} = (\sigma^m)^q \sigma^r = \sigma^r$$

we must have $r = 0$ because m is the least positive integer for which $\sigma^m = (1)$. Thus $m|(i - j)$ and so $i \equiv j \pmod{m}$.

Conversely, if $i \equiv j \pmod{m}$, then $i = j + mt$ for some integer t. Hence

$$\sigma^i = \sigma^{j+mt} = \sigma^j \sigma^{mt} = \sigma^j (\sigma^m)^t = \sigma^j. \quad \square$$

2.3.8 Proposition.   Let $\sigma \in S_n$ be written as a product of disjoint cycles. Then the order of σ is the least common multiple of the lengths of its cycles.

Proof. If $\sigma = (a_1, a_2, \ldots, a_m)$, then σ has order m. Furthermore, if $\sigma^k = (1)$, then $m|k$ by the previous proposition.

Next, if $\sigma = (a_1, a_2, \ldots, a_m)(b_1, b_2, \ldots, b_r)$ is a product of two disjoint cycles, then $\sigma^j = (a_1, a_2, \ldots, a_m)^j(b_1, b_2, \ldots, b_r)^j$ since $(a_1, a_2, \ldots, a_m)$ commutes with $(b_1, b_2, \ldots, b_r)$. If $\sigma^j = (1)$, then $(a_1, a_2, \ldots, a_m)^j = (1)$ and $(b_1, b_2, \ldots, b_r)^j = (1)$ since $(b_1, b_2, \ldots, b_r)^j$ leaves each a_i fixed and $(a_1, a_2, \ldots, a_m)^j$ leaves each b_i fixed. This happens if and only if $m|j$ and $r|j$, and then $\mathrm{lcm}[m, r]$ is a divisor of j. The smallest such j is thus $\mathrm{lcm}[m, r]$.

It should now be clear how to extend this argument to the general case. $\quad \square$

Example 2.3.8

The permutation $(1,5,3,7)(2,8,4)$ has order 12 in S_8. The permutation $(1,5,3)(2,8,4,6,9,7)$ has order 6 in S_9. $\quad \square$

If we wish to compute the inverse of a cycle, then we merely reverse the order of the cycle since

$$(a_1, a_2, \ldots, a_r)(a_r, a_{r-1}, \ldots, a_1) = (1).$$

The inverse of the product $\sigma\tau$ of two permutations is $\tau^{-1}\sigma^{-1}$ since

$$(\sigma\tau)(\tau^{-1}\sigma^{-1}) = \sigma(\tau\tau^{-1})\sigma^{-1} = \sigma(1)\sigma^{-1} = \sigma\sigma^{-1} = (1)$$

and similarly

$$(\tau^{-1}\sigma^{-1})(\sigma\tau) = (1).$$

Thus we have

$$[(a_1, \ldots, a_r)(b_1, \ldots, b_m)]^{-1} = (b_m, \ldots, b_1)(a_r, \ldots, a_1).$$

Note that if the cycles are disjoint, then they commute, and so the inverses do not need to be written in reverse order.

The simplest cycle, aside from (1), is one of the form (a_1, a_2). This represents an interchange of two elements. We now show that any permutation of a finite set can be obtained from a sequence of such interchanges. Whether the number of interchanges is even or odd is important in certain applications, for example in determining the signs of the elementary products used to compute the determinant of a matrix.

2.3.9 Definition.   A cycle (a_1, a_2) of length two is called a *transposition*.

2.3.10 Proposition.   Any permutation in S_n, where $n \geq 2$, can be written as a product of transpositions.

Proof. By Theorem 2.3.5 any permutation in S_n can be expressed as a product of cycles, and so we only need to show that any cycle can be expressed as a product of transpositions. The identity (1) can be expressed as (1,2)(1,2). For any other permutation, the proof is completed by just giving the explicit computation:

$$(a_1, a_2, \ldots, a_{r-1}, a_r) = (a_{r-1}, a_r)(a_{r-2}, a_r) \cdots (a_3, a_r)(a_2, a_r)(a_1, a_r). \quad \square$$

In the above proof, the expression we have given for $(a_1, a_2, \ldots, a_{r-1}, a_r)$ seems to be the most natural one, given that we read composition of permutations from right to left. It is also true that

$$(a_1, a_2, \ldots, a_{r-1}, a_r) = (a_1, a_2)(a_2, a_3) \cdots (a_{r-2}, a_{r-1})(a_{r-1}, a_r).$$

This expression may be easier for the student to remember. It also shows that representing a permutation as a product of transpositions is not unique.

Example 2.3.9

Applying Proposition 2.3.10 to S_3 gives us the following products. We have (1) = (1,2)(1,2), and (1,2), (1,3), and (2,3) are already expressed as transpositions. Finally, (1,2,3) = (2,3)(1,3) and (1,3,2) = (3,2)(1,2). We could also write (1,2,3) = (1,2)(1,3)(1,2)(1,3).

To give a longer example, we have (2,5,3,7,8) = (7,8)(3,8)(5,8)(2,8) as well as (2,5,3,7,8) = (2,5)(5,3)(3,7)(7,8), using the second method. $\square$

In the above example we have illustrated that writing (1,2,3) as a product of transpositions can be done in various ways, and in fact we have written it as a product of two transpositions in one case and four in another. In general, although the transpositions in the product are not uniquely determined, we do have a bit of uniqueness remaining, namely, the parity of the product. By this we mean that the number of transpositions in the product is either always even or always odd. This is the content of Theorem 2.3.11.

The proof of the theorem may appear to be rather complicated, so it seems to be worthwhile to comment on the general strategy of the proof. We wish to show that something cannot occur, and we do so by contradiction, assuming that a counterexample exists. If so, then a counterexample of minimal length exists. After modifying the counterexample without ever making it longer, we show that we can, in fact, produce an even shorter counterexample. This contradicts the minimality of the counterexample we started with. This general technique of proof goes back to Fermat and is often quite useful. We give another proof of the same fact in Section 3.6.

2.3.11 Theorem.   If a permutation is written as a product of transpositions in two ways, then the number of transpositions is either even in both cases or odd in both cases.

Proof. We will give a proof by contradiction. Suppose that the conclusion of the theorem is false. Then there exists a permutation σ that can be written as a product of an even number of transpositions and as a product of an odd number of transpositions, say

$$\sigma = \tau_1 \tau_2 \cdots \tau_{2m} = \delta_1 \delta_2 \cdots \delta_{2n+1}$$

for transpositions $\tau_1, \ldots, \tau_{2m}$ and $\delta_1, \ldots, \delta_{2n+1}$. Since $\delta_j = \delta_j^{-1}$ for $j = 1, \ldots, 2n + 1$, we have $\sigma^{-1} = \delta_{2n+1} \cdots \delta_1$, and so

$$(1) = \sigma \sigma^{-1} = \tau_1 \cdots \tau_{2m} \delta_{2n+1} \cdots \delta_1.$$

This shows that the identity permutation can be written as a product of an odd number of transpositions.

Next suppose that $(1) = \rho_1 \rho_2 \cdots \rho_k$ is the shortest product of an odd number of transpositions that is equal to the identity. Note that $k \geq 3$, and suppose that $\rho_1 = (a, b)$. We observe that a must appear in at least one other transposition, say ρ_i, with $i > 1$, since otherwise $\rho_1 \cdots \rho_k(a) = b$, a contradiction. Among all products of length k that are equal to the identity, and such that a appears in the transposition on the extreme left, we assume that $\rho_1 \rho_2 \cdots \rho_k$ has the fewest number of a's.

We now show that if ρ_i is the transposition of smallest index $i > 1$ in which a occurs, then ρ_i can be moved to the left without changing the number of transpositions or the number of times that a occurs in the product. Then combining ρ_i and ρ_1 will lead to a contradiction.

Let a, u, v, and r be distinct. Since $(u, v)(a, r) = (a, r)(u, v)$ and $(u, v)(a, v) = (a, u)(u, v)$, we can move a transposition with entry a to the second position without changing the number of a's that appear, and thus we may assume that ρ_2 is the next transposition in which a occurs, say $\rho_2 = (a, c)$ for some $c \neq a$. If $c = b$, then $\rho_1 \rho_2 = (1)$, and so $(1) = \rho_3 \cdots \rho_k$ is a shorter product of an odd number of transpositions; this gives us a contradiction. If $c \neq b$, then since $(a, b)(a, c) = (a, c)(b, c)$, we see that $(1) = (a, c)(b, c)\rho_3 \cdots \rho_k$ is a product of transpositions of length k with fewer a's; this again contradicts the choice of $\rho_1, \ldots, \rho_k$. Thus we have shown that (1) cannot be written as a product of an odd number of transpositions, completing the proof. $\square$

The point of Theorem 2.3.11 is that a given permutation is either even or odd, but not both. That result makes possible the following definition.

2.3.12 Definition.   A permutation σ is called *even* if it can be written as a product of an even number of transpositions, and *odd* if it can be written as a product of an odd number of transpositions.

Note that $(1,2)$ is odd and $(1,2,3) = (2,3)(1,3)$ is even. In remembering the parity of a cycle, it is important to note that in general a cycle of odd length is called even and a cycle of even length is called odd. Calling to mind the simplest case $(1,2)$ will remind you of this.

We should also note that the identity permutation is even. In addition, if σ is an even permutation, then so is the inverse of σ, since given σ as a product of transpositions, we only need to reverse the order of the transpositions to write σ^{-1} as a product of transpositions.

Finally, we note that the product of two even permutations is again an even permutation, and also that the product of two odd permutations is even, while the product of an odd permutation and an even permutation is odd. This remark follows from Theorem 2.3.11 and the fact that the sum of two even integers is even, the sum of two odd integers is even, and the sum of an odd and an even integer is odd.

EXERCISES: SECTION 2.3

1. Consider the following permutations in S_7:

$$\sigma = \begin{pmatrix} 1 & 2 & 3 & 4 & 5 & 6 & 7 \\ 3 & 2 & 5 & 4 & 6 & 1 & 7 \end{pmatrix} \quad \text{and} \quad \tau = \begin{pmatrix} 1 & 2 & 3 & 4 & 5 & 6 & 7 \\ 2 & 1 & 5 & 7 & 4 & 6 & 3 \end{pmatrix}$$

Compute the following products:
   (a) $\sigma\tau$ **(b)** $\tau\sigma$ **(c)** $\tau^2\sigma$ **(d)** σ^{-1} **(e)** $\tau^{-1}\sigma\tau$

2. Let $\sigma, \tau \in S_n$ be permutations such that $\sigma(k) = k$ and $\tau(k) = k$ for some k with $1 \le k \le n$. Show that $\sigma^{-1}(k) = k$ and $\rho(k) = k$, where $\rho = \sigma\tau$.

3. Write each of the permutations $\sigma\tau$, $\tau\sigma$, $\tau^2\sigma$, σ^{-1}, and $\tau^{-1}\sigma\tau$ in Exercise 1 as a product of disjoint cycles. Write σ and τ as products of transpositions.

4. Write $\begin{pmatrix} 1 & 2 & 3 & 4 & 5 & 6 & 7 & 8 & 9 & 10 \\ 3 & 4 & 10 & 5 & 7 & 8 & 2 & 6 & 9 & 1 \end{pmatrix}$ as a product of disjoint cycles and as a product of transpositions. Find its inverse, construct its associated diagram, and find its order.

5. Find the order of each of the following permutations:
   (a) $\begin{pmatrix} 1 & 2 & 3 & 4 & 5 & 6 \\ 6 & 4 & 5 & 3 & 2 & 1 \end{pmatrix}$

(b) $\begin{pmatrix} 1 & 2 & 3 & 4 & 5 & 6 & 7 & 8 \\ 4 & 6 & 7 & 5 & 1 & 8 & 2 & 3 \end{pmatrix}$

(c) $\begin{pmatrix} 1 & 2 & 3 & 4 & 5 & 6 & 7 & 8 & 9 \\ 5 & 9 & 8 & 7 & 3 & 4 & 6 & 1 & 2 \end{pmatrix}$

Hint: First write each permutation as a product of disjoint cycles.

6. List all of the cycles in S_4.

7. Find the number of cycles of each possible length in S_5. Then find all possible orders of elements in S_5. (Try to do this without having to write out all 120 possible permutations.)

8. Prove that in S_n, with $n \geq 3$, any even permutation is a product of cycles of length three.
   Hint: $(a,b)(b,c) = (a,b,c)$ and $(a,b)(c,d) = (a,b,c)(b,c,d)$.

9. Prove that (a,b) cannot be written as a product of two cycles of length three.

10. Let τ be the cycle $(1,2,\ldots,k)$ of length k.
    (a) Prove that if σ is any permutation, then $\sigma\tau\sigma^{-1} = (\sigma(1),\sigma(2),\ldots,\sigma(k))$. Thus $\sigma\tau\sigma^{-1}$ is a cycle of length k.
    (b) Let ρ be any cycle of length k. Prove that there exists a permutation σ such that $\sigma\tau\sigma^{-1} = \rho$.

11. Let S be any nonempty set, and let $\sigma \in \mathrm{Sym}(S)$. For $x, y \in S$ define $x \sim y$ if $\sigma^n(x) = y$ for some $n \in \mathbf{Z}$. Show that $\sim$ defines an equivalence relation on S.

12. For $\alpha, \beta \in S_n$, let $\alpha \sim \beta$ if there exists $\sigma \in S_n$ such that $\sigma\alpha\sigma^{-1} = \beta$. Show that $\sim$ is an equivalence relation on S_n.

APPENDIX E: COMPLEX NUMBERS

The equation $x^2 + 1 = 0$ has no real root since for any real number x we have $x^2 + 1 \geq 1$. The purpose of this appendix is to construct a set of numbers that extends the set of real numbers and includes a root of this equation. If we had a set of numbers that contained the set of real numbers, was closed under addition, subtraction, multiplication, and division, and contained a root i of the equation $x^2 + 1 = 0$, then it would have to include all numbers of the form $a + bi$ where a and b are real numbers. The addition and multiplication would be given by

$$(a + bi) + (c + di) = (a + c) + (b + d)i,$$

$$(a + bi)(c + di) = ac + (bc + ad)i + bdi^2 = (ac - bd) + (ad + bc)i.$$

Here we have used the fact that $i^2 = -1$ since $i^2 + 1 = 0$.

Our construction for the desired set of numbers is to invent a symbol i for which $i^2 = -1$, and then to consider all pairs of real numbers a and b, in the form $a + bi$. In Chapter 4 we will show how this construction can be done formally, by working with congruence classes of polynomials. At the end of this appendix, we also indicate how 2×2 matrices can be used to construct a set in which the equation $x^2 + 1 = 0$ has a solution. At this point, we simply ask the reader to accept the "invention" of the symbol i at an informal, intuitive level.

E.1 Definition.   The set $\mathbf{C} = \{a + bi \mid a, b \in \mathbf{R}$ and $i^2 = -1\}$ is called the *set of complex numbers*. Addition and multiplication of complex numbers are defined as follows:

$$(a + bi) + (c + di) = (a + c) + (b + d)i,$$

$$(a + bi)(c + di) = (ac - bd) + (ad + bc)i.$$

Note that $a + bi = c + di$ if and only if $a = c$ and $b = d$. If $c + di$ is nonzero, that is, if $c \neq 0$ or $d \neq 0$, then division by $c + di$ is possible:

$$\frac{a + bi}{c + di} = \frac{(a + bi)(c - di)}{(c + di)(c - di)} = \frac{ac + bd}{c^2 + d^2} + \frac{bd - ad}{c^2 + d^2}i.$$

A useful model for the set of complex numbers is a geometric model in which the number $a + bi$ is viewed as the ordered pair (a, b) in the plane. (See Figure 2.E.1.) Note that i corresponds to the pair $(0,1)$.

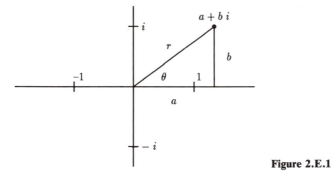

Figure 2.E.1

In polar coordinates, $a + bi$ is represented by (r, θ), where $r = \sqrt{a^2 + b^2}$ and $\cos \theta = a/r$, $\sin \theta = b/r$. The value r is called the *absolute value* of $a + bi$, and we write $|a + bi| = \sqrt{a^2 + b^2}$. This gives $a + bi = r(\cos \theta + i \sin \theta)$. In this form we can compute the product of two complex numbers as follows:

$r(\cos \theta + i \sin \theta) \cdot t(\cos \phi + i \sin \phi)$

$$= rt((\cos \theta \cos \phi - \sin \theta \sin \phi) + i(\sin \theta \cos \phi + \cos \theta \sin \phi))$$

$$= rt(\cos(\theta + \phi) + i \sin(\theta + \phi)).$$

This simplification of the product comes from the trigonometric formulas for the cosine and sine of the sum of two angles. Thus to multiply two complex numbers represented in polar form we multiply their absolute values and add their angles. A repeated application of this formula to $\cos \theta + i \sin \theta$ gives the following theorem.

E.2 Theorem (De Moivre).   For any positive integer n,

$$(\cos \theta + i \sin \theta)^n = \cos(n\theta) + i \sin(n\theta).$$

E.3 Corollary. For any positive integer n, the equation $z^n = 1$ has n distinct roots in the set of complex numbers.

Proof. For $k = 0, 1, \ldots, n - 1$, the values

$$\cos \frac{2k\pi}{n} + i \sin \frac{2k\pi}{n}$$

are distinct and

$$\left(\cos \frac{2k\pi}{n} + i \sin \frac{2k\pi}{n}\right)^n = \cos 2k\pi + i \sin 2k\pi = 1. \quad \square$$

The complex roots of $z^n = 1$ are called the *nth roots of unity.* When plotted in the complex plane, they form the vertices of a regular polygon with n sides inscribed in a circle of radius 1 with center at the origin.

Example E.1

The cube roots of unity are

$$1, \quad \cos \frac{2\pi}{3} + i \sin \frac{2\pi}{3}, \quad \text{and} \quad \cos \frac{4\pi}{3} + i \sin \frac{4\pi}{3},$$

or equivalently,

$$1, \quad \omega = \frac{-1}{2} + \frac{\sqrt{3}}{2} i, \quad \text{and} \quad \omega^2 = \frac{-1}{2} - \frac{\sqrt{3}}{2} i.$$

Note that $\omega^2 + \omega + 1 = 0$, since ω is a root of $z^3 - 1 = (z - 1)(z^2 + z + 1) = 0$. (See Figure 2.E.2.) $\quad \square$

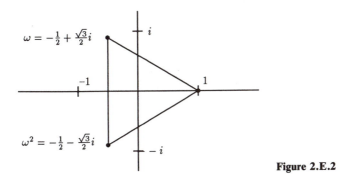

Figure 2.E.2

Example E.2

The fourth roots of unity are 1, i, $i^2 = -1$, and $i^3 = -i$. (See Figure 2.E.3.) $\quad \square$

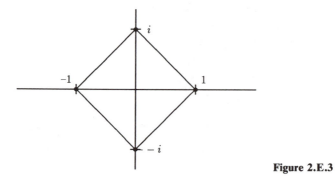

Figure 2.E.3

Example E.3

If $z^n = u$, then $(z\omega)^n = u$, where ω is any nth root of unity. Thus if all nth roots of unity are already known, it is easy to find the nth roots of any complex number. In general, the nth roots of $r(\cos\theta + i\sin\theta)$ are

$$r^{1/n}\left(\cos\frac{\theta + 2k\pi}{n} + i\sin\frac{\theta + 2k\pi}{n}\right), \quad \text{for} \quad 1 \le k \le n.$$

To find the square root of a complex number it may be helpful to use the formulas

$$\sin\frac{\theta}{2} = \pm\frac{\sqrt{1 - \cos\theta}}{2} \quad \text{and} \quad \cos\frac{\theta}{2} = \pm\frac{\sqrt{1 + \cos\theta}}{2}. \quad \square$$

We have noted that the powers of i are i, $i^2 = -1$, $i^3 = -i$, $i^4 = 1$. Since $i^4 = 1$, the powers repeat. For example, $i^5 = i^4 i = i$, $i^6 = i^4 i^2 = -1$, and so on. For any integer n, the power i^n depends on the remainder of n when divided by 4, since if $n = 4q + r$, then $i^n = i^{4q+r} = (i^4)^q i^r = i^r$. In particular, $i^{-1} = i^3 = -i$, and a similar computation can be given for any negative exponent.

We can use our knowledge of powers of i to give another way to express complex numbers. We need to recall the Taylor series expansion for e^x, $\sin(x)$, and $\cos(x)$:

$$e^x = 1 + x + \frac{x^2}{2} + \frac{x^3}{3!} + \frac{x^4}{4!} + \dots,$$

$$\sin(x) = 1 - \frac{x^2}{2} + \frac{x^4}{4!} - \frac{x^6}{6!} + \dots,$$

$$\cos(x) = x - \frac{x^3}{3!} + \frac{x^5}{5!} - \frac{x^7}{7!} + \dots.$$

If we formally substitute $i\theta$ into the series expansion for e^x, we can reduce the exponents of i modulo 4 and then group together those terms that contain i and those terms that do not contain i. This gives us the expression

$$e^{i\theta} = \cos(\theta) + i\sin(\theta).$$

Thus complex numbers can be written in polar coordinates in the form $re^{i\theta}$. The product $re^{i\theta} \cdot te^{i\phi}$ is then equal to $rte^{i(\theta+\phi)}$. Substituting $\theta = \pi$ yields the remarkable formula $e^{\pi i} = -1$.

The next theorem is usually referred to as the "fundamental theorem of algebra." It was discovered by D'Alembert in 1746, although he gave an incorrect proof. The first acceptable proof was given by Gauss in 1799. Using analytic techniques, a short proof is usually given in a beginning course in complex variables, and we will now provide an outline of this proof. In Theorem 8.3.10 we give a proof using Galois theory.

A function $f(z) : \mathbf{C} \to \mathbf{C}$ is called an *entire function* if it is differentiable at every point of $\mathbf{C}$. Liouville's theorem states that if f is an entire function and $|f(z)|$ is bounded for all $z \in \mathbf{C}$, then f is a constant function. The proof of Liouville's theorem requires the development of quite a bit of machinery, from which the fundamental theorem follows immediately. Suppose that $p(z)$ is a nonconstant complex polynomial that has no roots. If we let $f(z) = 1/p(z)$, then since the denominator is never zero, it follows that f is an entire function and hence must be constant. Simply computing $p(n)$ for all positive integers n shows that $p(z)$ cannot be constant, a contradiction.

E.4 Theorem (Fundamental Theorem of Algebra). Every polynomial of positive degree with complex coefficients has a complex root.

E.5 Corollary. Every polynomial $f(z)$ of degree $n > 0$ with complex coefficients can be expressed as a product of linear factors, in the form

$$f(z) = c(z - z_1)(z - z_2) \cdots (z - z_n).$$

Proof. We need to use the fact (proved in Chapter 4) that roots of $f(z)$ correspond to linear factors. A detailed proof would use Theorem E.4 and induction on the degree of $f(z)$. $\square$

Corollary E.5 can be used to give formulas relating the roots and coefficients of a polynomial. For example, if $f(z) = z^2 + a_1 z + a_0$ has roots z_1, z_2, then

$$z^2 + a_1 z + a_0 = (z - z_1)(z - z_2) = z^2 + (-z_1 - z_2)z + z_1 z_2.$$

Thus,

$$z_1 + z_2 = -a_1 \quad \text{and} \quad z_1 z_2 = a_0.$$

Similarly, if $f(z) = z^3 + a_2 z + a_1 z + a_0$ has roots z_1, z_2, z_3, then

$$z_1 + z_2 + z_3 = -a_2, \qquad z_1 z_2 + z_1 z_3 + z_2 z_3 = a_1, \qquad z_1 z_2 z_3 = -a_0.$$

This pattern can be extended easily to the general case.

If $z = a + bi$ is a complex number, then its *complex conjugate,* denoted by $\bar{z}$, is $\bar{z} = a - bi$. Note that $z\bar{z} = a^2 + b^2$ and $z + \bar{z} = 2a$ are real numbers, whereas $z - \bar{z} = (2b)i$ is a purely imaginary number. Furthermore, $z = \bar{z}$ if and only if z is a real number (i.e., $b = 0$). It can be checked that $\overline{(z + w)} = \bar{z} + \bar{w}$ and $\overline{(zw)} = \bar{z}\bar{w}$.

E.6 Proposition.   Let $f(x)$ be a polynomial with real coefficients. Then a complex number z is a root of $f(x)$ if and only if $\bar{z}$ is a root of $f(x)$.

Proof. If $f(x) = a_n x^n + \ldots + a_0$, then $a_n z^n + \ldots + a_0 = 0$ for any root z of $f(x)$. Taking the complex conjugate of both sides shows that

$$\bar{a}_n (\bar{z})^n + \ldots + \bar{a}_1 \bar{z} + \bar{a}_0 = a_n (\bar{z})^n + \ldots + a_1 \bar{z} + a_0 = 0$$

and thus $\bar{z}$ is a root of $f(x)$. Conversely, if $\bar{z}$ is a root of $f(x)$, then so is $z = \bar{\bar{z}}$. $\square$

E.7 Theorem.   Any polynomial with real coefficients can be factored into a product of linear and quadratic terms with real coefficients.

Proof. Let $f(x)$ be a polynomial with real coefficients, of degree n. By Corollary E.5 we can write $f(x) = c(x - z_1)(x - z_2) \cdots (x - z_n)$, where $c \in \mathbf{R}$. If z_i is not a real root, then by Proposition E.6, $\bar{z}_i$ is also a root, and so $x - \bar{z}_i$ occurs as one of the factors. But then

$$(x - z_i)(x - \bar{z}_i) = x^2 - (z_i + \bar{z}_i)x + z_i \bar{z}_i$$

has real coefficients. Thus if we pair each nonreal root with its conjugate, the remaining roots will be real, and so $f(x)$ can be written as a product of linear and quadratic polynomials each having real coefficients. $\square$

E.8 Corollary.   Any polynomial of odd degree that has real coefficients must have a real root.

Proof. By the previous theorem, such a polynomial must have a linear factor with real coefficients, and this factor yields a real root. $\square$

We will now construct a concrete model for the set of complex numbers, using 2×2 matrices over $\mathbf{R}$. We can identify the set $\mathbf{R}$ of real numbers with the set of all scalar 2×2 real matrices. That is, the real number a corresponds to the matrix $\begin{bmatrix} a & 0 \\ 0 & a \end{bmatrix}$. As you should recall from your study of linear algebra, scalar matrices are added and multiplied just like real numbers. We can now ask for a root of the equation $x^2 + 1 = 0$ in the set of 2×2 matrices, where we can identify 1 with the scalar matrix $\begin{bmatrix} 1 & 0 \\ 0 & 1 \end{bmatrix}$. In fact, the matrix $\begin{bmatrix} 0 & 1 \\ -1 & 0 \end{bmatrix}$ is one such root, since

$$\begin{bmatrix} 0 & 1 \\ -1 & 0 \end{bmatrix}^2 + \begin{bmatrix} 1 & 0 \\ 0 & 1 \end{bmatrix} = \begin{bmatrix} -1 & 0 \\ 0 & -1 \end{bmatrix} + \begin{bmatrix} 1 & 0 \\ 0 & 1 \end{bmatrix} = \begin{bmatrix} 0 & 0 \\ 0 & 0 \end{bmatrix}.$$

If we consider all matrices of the form

$$a\begin{bmatrix} 1 & 0 \\ 0 & 1 \end{bmatrix} + b\begin{bmatrix} 0 & 1 \\ -1 & 0 \end{bmatrix} = \begin{bmatrix} a & b \\ -b & a \end{bmatrix}$$

we have a set that is closed under addition, subtraction, and multiplication, since

$$\begin{bmatrix} a & b \\ -b & a \end{bmatrix} \pm \begin{bmatrix} c & d \\ -d & c \end{bmatrix} = \begin{bmatrix} a \pm c & b \pm d \\ -(b \pm d) & a \pm c \end{bmatrix}$$

and

$$\begin{bmatrix} a & b \\ -b & a \end{bmatrix} \cdot \begin{bmatrix} c & d \\ -d & c \end{bmatrix} = \begin{bmatrix} ac - bd & ad + bc \\ -(ad + bc) & ac - bd \end{bmatrix}.$$

Finally, if $\begin{bmatrix} a & b \\ -b & a \end{bmatrix}$ is not the zero matrix, then it has an inverse

$$\frac{1}{a^2 + b^2}\begin{bmatrix} a & -b \\ b & a \end{bmatrix}.$$

The set of 2×2 matrices of the form $\begin{bmatrix} a & b \\ -b & a \end{bmatrix}$ has the properties we are looking

for, and the correspondence $a + bi \leftrightarrow \begin{bmatrix} a & b \\ -b & a \end{bmatrix}$ preserves addition and multiplica-

tion. We have thus constructed a concrete model for the set of complex numbers, for those who were worried by our "invention" of the root i of $x^2 + 1 = 0$. The matrix representation for **C** is a convenient form in which to verify that the associative and distributive laws hold for the set of complex numbers, since these laws are known to hold for matrix multiplication.

EXERCISES: APPENDIX E

1. Compute each of the following:
   (a) $(1/\sqrt{2} + i/\sqrt{2})^6$
   (b) $(1 + i)^8$
   (c) $(\cos 20° + i \sin 20°)^9$
2. Find $(\cos \theta + i \sin \theta)^{-1}$.
3. (a) Find the 6th roots of unity.
   (b) Find the 8th roots of unity.

4. **(a)** Find the cube roots of $-8i$.
   (b) Find the cube roots of $-4\sqrt{2} + 4\sqrt{2}i$.
   (c) Find the cube roots of $2 + 2i$.
   (d) Find the fourth roots of $1 + i$.

5. Solve the equation $z^2 + z + (1 + i) = 0$.

6. Solve the equation $x^3 - 3x^2 - 6x - 20 = 0$, given that one root is $-1 + \sqrt{3}i$.

7. Use DeMoivre's theorem to find formulas for cos 3θ (in terms of cos θ) and sin 3θ (in terms of sin θ).

8. Verify each of the following, for complex numbers z and w:
   (a) $z\bar{z} = |z|^2$
   (b) $\overline{zw} = \bar{z}\bar{w}$
   (c) $|zw| = |z||w|$

9. Let a and b be integers, each of which can be written as the sum of two perfect squares. Show that ab has the same property.

   Hint: Use part (c) of the previous problem.

10. If η is an nth root of unity, it is called a *primitive* nth root of unity if it is not a root of $z^k - 1 = 0$ for any k such that $1 \le k < n$.
    (a) Show that $\cos(2\pi/n) + i\sin(2\pi/n)$ is a primitive nth root of unity.
    (b) If $\eta = \cos(2\pi/n) + i\sin(2\pi/n)$, show that η^m is a primitive nth root of unity if and only if $(n,m) = 1$.

3

Groups

Symmetry occurs frequently and in many forms in nature. Starfish possess rotational symmetry; the human body exhibits bilateral symmetry. A third sort of symmetry appears in some wallpaper or tile patterns that can be shifted in various directions without changing their appearance.

Each coefficient of a polynomial is a symmetric function of the polynomial's roots. To see what we mean by this, consider a monic cubic polynomial $f(x)$ with roots r_1, r_2, r_3. Then

$$f(x) = (x - r_1)(x - r_2)(x - r_3) = x^3 - bx^2 + cx - d$$

where $b = r_1 + r_2 + r_3$, $c = r_1r_2 + r_2r_3 + r_3r_1$, and $d = r_1r_2r_3$. Notice that if we permute r_1, r_2, r_3 by (for example) replacing r_1 by r_2, r_2 by r_3, and r_3 by r_1, then the coefficients b, c, d remain unchanged. In fact, the coefficients remain unchanged under any permutation of the roots, and so we say that they are symmetric functions of the roots.

The important feature of symmetry is the way that the shapes (or roots) can be changed while the whole figure (or the coefficients) remains unchanged. Geometrically, individual points move (or, algebraically, the roots interchange) while the figure as a whole (or the polynomial) remains the same. With respect to symmetry, geometrically the important thing is not the position of the points but the operation of moving them, and similarly, with respect to considering the roots of polynomials, it is the operation of shifting the roots among themselves that is most important and not the roots themselves. This was the key insight that enabled Galois to give a complete answer to the problem of solving polynomial equations by radicals.

The mathematical idea needed for the study of symmetry is that of a *group,* and in this chapter we introduce this important concept. The publication in 1870 of *Traité des substitutions et des équations algébriques* by Camille Jordan (1838–1922) represented a fundamental change in the character of group theory. Prior to its publication, various individuals had studied groups of permutations and groups of geometric transformations, but after 1870 the abstract notion of a group was developed in several steps. The modern definition of a group, using an axiomatic approach, was given in the commutative case in 1870 by Kronecker (1823–1891), and in the general noncommutative case in 1893 by Weber (1842–1913). Arthur Cayley (1821–1895) should be given credit for two papers he published in 1854 on the theory of groups. He introduced the concept of a group table, which is sometimes still referred to as a "Cayley table."

In this chapter we first discuss the abstract definition of a group, and attempt to clarify it for the reader by considering a wide variety of examples in considerable detail. We include sections on permutation groups, in which we consider groups of symmetries of some geometric objects, and cyclic groups, which we have already met in the guise of $\mathbf{Z}$ and $\mathbf{Z}_n$. We also study one-to-one correspondences that preserve the group structure (you may wish to reread the discussion in the introduction to Chapter 2). The last section deals with the notion of a group formed from equivalence classes of elements of a group. The procedure by which we formed $\mathbf{Z}_n$ from $\mathbf{Z}$ can be extended to any group. Such groups, which we call *factor groups,* play a crucial role in Chapter 8 in the study of Galois theory.

3.1 DEFINITION OF A GROUP

In describing the difference between arithmetic and algebra, it might be said that arithmetic deals exclusively with numbers, while algebra deals with letters that represent numbers. The next step in abstraction involves dealing with objects that may not even represent numbers. For example, in learning calculus it is necessary to develop an "arithmetic" for functions. To give another example, in working with matrices, it is again necessary to develop some rules for matrix operations, and these rules constitute an "arithmetic" for matrices. The common thread in these developments, from an algebraic point of view, is the idea of an operation. Thus, as operations, we have addition, multiplication, and composition of functions, together with addition and multiplication of matrices. When we write AB for a product of matrices, for example, we have created a notation that allows us to think in terms of ordinary multiplication, even though it represents a more complicated computation.

The operations we will study will be *binary* operations; that is, we will consider only operations which combine two elements at a time. A useful model to use is that of a computer program that allows two inputs and combines them in some way to give a single output. If we have an operation on a particular set, then we require that combining two inputs from the set will result in an output belonging to the same set. Furthermore, the output must depend only on the inputs, so that the answer is unique (when two inputs are specified).

A binary operation $*$ on a set S is a rule that assigns to each ordered pair (a, b) of elements of S a unique element $a * b$ of S. For example, the ordinary operations of addition, subtraction, and multiplication are binary operations on the set of real numbers. The operation of division is not a binary operation on the real numbers because it is not defined for all ordered pairs of real numbers (division by zero presents a problem). If we would restrict the set to the nonzero real numbers, then for any ordered pair (a, b) of real numbers, applying the operation $a \div b$ we get the quotient a/b, which is uniquely defined and is again a nonzero real number, showing that we have a binary operation. Although subtraction is a binary operation on the set of all real numbers, it is not a binary operation on the set of natural numbers, since, for example, $1 - 2$ is not in the set of natural numbers.

A binary operation permits us to combine only two elements, and so *a priori* $a * b * c$ does not make sense. But $(a * b) * c$ does make sense because we first combine a and b to get $a * b$ and then combine this element with c to get $(a * b) * c$. On the other hand, $a * (b * c)$ also makes sense because we combine b and c to get $b * c$ and then combine a with this element to get $a * (b * c)$. The point of requiring the *associative* law is that both of these options should yield the same result. As an example, if our binary operation is ordinary addition for real numbers and we are adding a column of three numbers, the associative law says that we can either add the column from top to bottom or from bottom to top.

3.1.1 Definition.   A *binary operation* $*$ on a set S is a function $*: S \times S \rightarrow S$ from the set $S \times S$ of all ordered pairs of elements in S into S.

The operation $*$ is said to be *associative* if $a * (b * c) = (a * b) * c$ for all $a, b, c \in S$.

An element $e \in S$ is called an *identity* element for $*$ if $a * e = a$ and $e * a = a$ for all $a \in S$.

If $*$ has an identity element e, and $a \in S$, then $b \in S$ is said to be an *inverse* for a if $a * b = e$ and $b * a = e$.

To illustrate these ideas, let S be the set of all functions from a set A into itself. If $\phi, \theta \in S$, then define $\phi * \theta$ by letting $\phi * \theta(a) = \phi(\theta(a))$ for all $a \in A$. This defines a binary operation on S, and the identity function is an identity element for the operation. Furthermore, composition of functions is associative, and the functions that have inverses are precisely the ones that are both one-to-one and onto.

The set $M_n(\mathbf{R})$ of all $n \times n$ matrices with entries from the real numbers $\mathbf{R}$ provides another good example. Matrix multiplication defines a binary operation on $M_n(\mathbf{R})$, and the identity matrix serves as an identity element. The proof that associativity holds is a laborious one if done directly from the definition. The appropriate way to remember why it holds is to use the correspondence between matrices and linear transformations, under which matrix multiplication corresponds to composition of functions. Then we only need to note that composition of functions is associative. Finally, recall that a matrix has a multiplicative inverse if and only if its determinant is nonzero.

Addition of matrices also defines an associative binary operation on $M_n(\mathbf{R})$, and in this case the identity element is the zero matrix. Each matrix has an inverse with respect to this operation, namely, its negative.

Since the definition of a binary operation involves a function, we sometimes face problems similar to those we have already encountered in checking that a function is well-defined. For example, consider the problem inherent in defining multiplication on the set of rational numbers

$$\mathbf{Q} = \left\{ \frac{m}{n} \middle| m, n \in \mathbf{Z} \text{ and } n \neq 0 \right\}$$

where m/n and p/q represent the same element if $mq = np$. If $a, b \in \mathbf{Q}$ with $a = m/n$ and $b = s/t$, then we use multiplication of integers to define $ab = ms/nt$. We must check that the product does not depend on how we choose to represent a and b. If we also have $a = p/q$ and $b = u/v$, then we must check that pu/qv is equivalent to ms/nt. Since m/n and p/q are assumed to be equivalent and s/t and u/v are assumed to be equivalent, we have $mq = np$ and $sv = tu$. Multiplying the two equations gives $(ms)(qv) = (nt)(pu)$, which shows that ms/nt is equivalent to pu/qv. This allows us to conclude that the given multiplication of rational numbers is well-defined.

3.1.2 Proposition.   Let $*$ be an associative, binary operation on a set S.
  (a) The operation $*$ has at most one identity element.
  (b) If $*$ has an identity element, then any element of S has at most one inverse.
  (c) If $*$ has an identity element and $a, b \in S$ have inverses a^{-1} and b^{-1}, respectively, then the inverse of a^{-1} exists and is equal to a, and the inverse of $a * b$ exists and is equal to $b^{-1} * a^{-1}$.

Proof. (a) Suppose that e and e' are identity elements for $*$. Since e is an identity element, we have $e * e' = e'$, and since e' is an identity element, we have $e * e' = e$. Therefore $e = e'$.

(b) Let e be the identity element for S relative to the operation $*$. Let b and b' be inverses for the element a. Then $b * a = e$ and $a * b' = e$, and so using the fact that $*$ is associative we have

$$b' = e * b' = (b * a) * b' = b * (a * b') = b * e = b.$$

(c) Let e be the identity element for S relative to the operation $*$. The equations $a * a^{-1} = e$ and $a^{-1} * a = e$ that state that a^{-1} is the inverse of a also show that a is the inverse of a^{-1}. Using the associative property for $*$, the computation

$$(a * b) * (b^{-1} * a^{-1}) = ((a * b) * b^{-1}) * a^{-1}$$
$$= (a * (b * b^{-1})) * a^{-1}$$
$$= (a * e) * a^{-1} = a * a^{-1} = e$$

and a similar computation with $(b^{-1} * a^{-1}) * (a * b)$ shows that the inverse of $a * b$ is $b^{-1} * a^{-1}$.   □

The general binary operations we work with will normally be denoted multiplicatively; that is, instead of writing $a * b$ we will just write $a \cdot b$, or simply ab. Since this may seem to imply that we only consider multiplication, some results will also be stated in additive notation, to help the student avoid confusion. In our symbolism, the previous proposition shows that $(a^{-1})^{-1} = a$ and $(ab)^{-1} = b^{-1}a^{-1}$, provided that a and b have inverses. As an elementary consequence of the proposition, we also note that $a = b$ if and only if $a^{-1} = b^{-1}$.

We now come to the main goal of this section—the definition of a group. Since definitions are the basic building blocks of abstract mathematics, the student will not progress without learning all definitions very thoroughly and carefully. Learning a definition should include associating with it several examples that will immediately come to mind to illustrate the important points of the definition. Following the definition we verify some elementary properties of groups and provide some broad classes of examples: groups of numbers, with familiar operations; groups of permutations, in which the operation is composition of functions; and groups of matrices, using matrix multiplication.

Our definition of a group contains some redundancy. Any binary operation must satisfy the condition called the closure property. By listing it separately as one of the conditions in the definition of a group, we hope to make it impossible for the student to forget to verify that the closure property holds when checking that a set with a given operation is a group.

3.1.3 Definition.   A *group* $(G, \cdot)$ is a nonempty set G together with a binary operation $\cdot$ on G such that the following conditions hold:
  (i) *Closure:* For all $a, b \in G$ the element $a \cdot b$ is a uniquely defined element of G.
 (ii) *Associativity:* For all $a, b, c \in G$, we have $a \cdot (b \cdot c) = (a \cdot b) \cdot c$.
(iii) *Identity:* There exists an *identity element* $e \in G$ such that $e \cdot a = a$ and $a \cdot e = a$ for all $a \in G$.
 (iv) *Inverses:* For each $a \in G$ there exists an *inverse* element $a^{-1} \in G$ such that $a \cdot a^{-1} = e$ and $a^{-1} \cdot a = e$.

In words, the above definition states that a group is a set G with an associative binary operation such that G has an identity element and each element of G has an inverse. Proposition 3.1.2 implies that the identity element is unique. Furthermore, each element has a unique inverse.

In the definition of a group G we do not require *commutativity*. That is, we do not assume that $a \cdot b = b \cdot a$ for all $a, b \in G$, since we want to allow the definition to include groups in which the operation is given by composition of functions or multiplication of matrices.

In the definition, note carefully one distinction between an identity element and an inverse element: An identity element satisfies a condition for all other elements of G, whereas an inverse element is defined relative to a single element of G. The order in which the axioms (iii) and (iv) are stated is important at least insofar as it is impossible to talk about an inverse of an element until an identity element is known to exist.

If G is a group and $a \in G$, then for any positive integer n we define a^n to be the product of a with itself n times. This can also be done inductively by letting $a^n = a \cdot a^{n-1}$. It is not difficult to show that the exponential laws

$$a^m a^n = a^{m+n} \quad \text{and} \quad (a^m)^n = a^{mn}$$

must hold for all positive exponents m, n. To illustrate, we have

$$a^2 \cdot a^3 = (a \cdot a) \cdot (a \cdot a \cdot a) = a^5$$

and

$$(a^2)^3 = (a^2) \cdot (a^2) \cdot (a^2) = (a \cdot a) \cdot (a \cdot a) \cdot (a \cdot a) = a^6.$$

To preserve these laws of exponents, we define $a^0 = e$, where e is the identity element of G, and $a^{-n} = (a^n)^{-1}$. Using these definitions, the above rules for exponents extend to all integers. The general proof is left as an exercise.

To begin our examples of groups, we now consider the set $\mathbf{R}$ of all real numbers, using as an operation the standard multiplication of real numbers. It is easy to check that the first three axioms for a group are satisfied, but the fourth axiom fails because 0 can never have a multiplicative inverse ($0 \cdot x = 1$ has no solution). Thus in order to define a group using the standard multiplication, we must reduce the set we work with.

Example 3.1.1 (Multiplicative Groups of Numbers)

Let $\mathbf{R}^\times$ denote the set of nonzero real numbers, with the operation given by standard multiplication. The first group axiom holds since the product of any two nonzero real numbers is still nonzero. The remaining axioms are easily seen to hold, with 1 playing the role of an identity element, and $1/a$ giving the inverse of an element $a \in \mathbf{R}^\times$. Thus $\mathbf{R}^\times$ is a group.

Similarly, we have the groups $\mathbf{Q}^\times$ of all nonzero rational numbers and $\mathbf{C}^\times$ of all nonzero complex numbers, under the operation of ordinary multiplication. If we attempt to form a multiplicative group from the integers $\mathbf{Z}$, we have to restrict ourselves to just ± 1, since these are the only integers that have multiplicative inverses in $\mathbf{Z}$.

The development of these number systems from first principles is outlined in Appendix B, as it is beyond the scope of this course to give a full development of them. To actually verify the group axioms in the course of such a development is a long and arduous task. We have chosen to take a naive approach, by assuming that the reader is familiar with the properties and is willing to accept that a careful development is possible.   □

The previous examples exhibit some of the most familiar groups. The next proposition deals with the most basic type of group. Recall that a permutation of a set S is a one-to-one function from S onto S. We will show in Section 3.6 that groups of permutations provide the most general models of groups. We recall our earlier definition from Section 2.3.

3.1.4 Definition.   The set of all permutations of a set S is denoted by Sym(S). The set of all permutations of the set $\{1, 2, \ldots, n\}$ is denoted by S_n.

3.1.5 Proposition.   If S is any nonempty set, then Sym(S) is a group under the operation of composition of functions.

Proof. The closure axiom is satisfied since by Proposition 2.1.6 the composition of two one-to-one and onto functions is again one-to-one and onto. Composition of functions is associative by Proposition 2.1.3. The identity function on S serves as an identity element for Sym(S). Finally, by Proposition 2.1.8 a function from S into S is one-to-one and onto if and only if it has an inverse function, and the inverse is again one-to-one and onto, so it belongs to Sym(S).   □

The group S_n is called the *symmetric group of degree n*. To see that S_n has $n!$ elements, let $S = \{1, 2, \ldots, n\}$. To define a permutation $\sigma : S \rightarrow S$, there are n choices for $\sigma(1)$. In order to make σ one-to-one, we must have $\sigma(2) \neq \sigma(1)$, and so there are only $n - 1$ choices for $\sigma(2)$. Continuing this analysis we can see that there will be a total of $n \cdot (n - 1) \cdots 2 \cdot 1 = n!$ possible distinct permutations of S.

Our next example will give the multiplication table for S_3. We will use cycle notation for the permutations. The function that leaves all three elements 1, 2, 3 fixed is the identity, which we will denote by (1). It is impossible for a permutation on three elements to leave exactly two elements fixed, so next we consider the functions that fix one element while interchanging the other two. The function that interchanges 1 and 2 will be denoted by (1, 2). Similarly, we have (1, 3) and (2, 3). Finally, there are two permutations that leave no element fixed. They are denoted by (1, 2, 3) for the permutation σ with $\sigma(1) = 2$, $\sigma(2) = 3$, and $\sigma(3) = 1$; and (1, 3, 2) for the permutation τ with $\tau(1) = 3$, $\tau(3) = 2$, and $\tau(2) = 1$. Remember that the products represent composition of functions and must be evaluated in the usual way functions are composed, from right to left. In Table 3.1.1, to find the product $\sigma\tau$, look in the row to the right of σ for the entry in the column below τ.

Example 3.1.2 (Symmetric Group on Three Elements)

In each row of Table 3.1.1, each element of the group occurs exactly once. The same is true in each column. To explain this, suppose we look at

TABLE 3.1.1. Multiplication in S_3

	(1)	(1,2,3)	(1,3,2)	(1,2)	(1,3)	(2,3)
(1)	(1)	(1,2,3)	(1,3,2)	(1,2)	(1,3)	(2,3)
(1,2,3)	(1,2,3)	(1,3,2)	(1)	(1,3)	(2,3)	(1,2)
(1,3,2)	(1,3,2)	(1)	(1,2,3)	(2,3)	(1,2)	(1,3)
(1,2)	(1,2)	(2,3)	(1,3)	(1)	(1,3,2)	(1,2,3)
(1,3)	(1,3)	(1,2)	(2,3)	(1,2,3)	(1)	(1,3,2)
(2,3)	(2,3)	(1,3)	(1,2)	(1,3,2)	(1,2,3)	(1)

the row corresponding to an element a. The entries in this row consist of all elements of the form ag, for $g \in G$. The next proposition, the cancellation law, implies that if $g_1 \neq g_2$, then $ag_1 \neq ag_2$. This guarantees that no group elements are repeated in the row. A similar argument applies to the column determined by a, which consists of elements of the form ga for $g \in G$. □

3.1.6 Proposition (Cancellation Property for Groups).   Let G be a group, and let $a, b, c \in G$.
  (a) If $ab = ac$, then $b = c$.
  (b) If $ac = bc$, then $a = b$.

Proof. Given $ab = ac$, multiplying on the left by a^{-1} (which exists since G is a group) gives $a^{-1}(ab) = a^{-1}(ac)$. Using the associative law, $(a^{-1}a)b = (a^{-1}a)c$. Then $eb = ec$, which shows that $b = c$. The proof of the second part of the proposition is similar. □

The next proposition provides some motivation for the study of groups. It shows that the group axioms are precisely the assumptions necessary to solve linear equations of the form $ax = b$.

3.1.7 Proposition.   If G is a group and $a, b \in G$, then the equations $ax = b$ and $xa = b$ have unique solutions.
Conversely, if G is a nonempty set with an associative binary operation in which the equations $ax = b$ and $xa = b$ have solutions for all $a, b \in G$, then G is a group.

Proof. Let G be a group, and let $a, b \in G$. Then a has an inverse, and substituting $x = a^{-1}b$ in the equation $ax = b$ gives

$$a(a^{-1}b) = (aa^{-1})b = eb = b,$$

showing that we have found a solution. If x' is any solution of $ax = b$, then multiplying both sides of the equation $ax' = b$ by a^{-1} shows (after some work) that $x' = a^{-1}b$, so we have in fact found a unique solution of $ax = b$. Similarly, $x = ba^{-1}$ can be shown to be the unique solution of the equation $xa = b$.

Conversely, suppose that G has an associative binary operation under which the equations $ax = b$ and $xa = b$ have solutions for all elements $a, b \in G$. There is at least one element $a \in G$, and so we first let e be a solution of the equation $ax = a$. Next, we will show that $be = b$ for all $b \in G$. Let $b \in G$ be given, and let c be a solution to the equation $xa = b$, so that $ca = b$. Then

$$be = (ca)e = c(ae) = ca = b.$$

Similarly, there exists an element $e' \in G$ such that $e'b = b$ for all $b \in G$. But then $e' = e'e$ and $e'e = e$, and so we conclude that $e' = e$. Thus e satisfies the conditions that show it to be an identity element for G.

Finally, given any element $b \in G$, we must find an inverse for b. Let c be a solution of the equation $bx = e$ and let d be a solution of the equation $xb = e$. Then

$$d = de = d(bc) = (db)c = ec = c$$

and so $d = c$. Thus $bc = e$ and $cb = e$, and so c is an inverse for b. This completes the proof that G is a group. $\square$

We have seen examples of groups of several different types. We already need some definitions to describe them. Groups that satisfy the commutative law are named in honor of Niels Abel, who was active in the early nineteenth century. He showed that such groups were important in the theory of equations, and gave a proof that the general fifth degree equation cannot be solved by radicals.

3.1.8 Definition.   A group G is said to be a *finite group* if the set G has a finite number of elements. In this case, the number of elements is called the *order* of G, denoted by $|G|$.

A group G is said to be *abelian* if $a \cdot b = b \cdot a$ for all $a, b \in G$.

In an abelian group G, the operation is very often denoted additively. The associative law then becomes $a + (b + c) = (a + b) + c$ for all $a, b, c \in G$. The identity element is then usually denoted by 0 and is called a *zero element,* and the additive inverse of an element a is denoted by $-a$.

Let G be an abelian group, and let $a \in G$. For a positive integer n, the sum of a with itself n times will be denoted by na. In additive notation, this replaces the exponential notation a^n. It is important to remember that this is not a multiplication that takes place in G, since n is not an element of G. For instance, G might be the group $M_n(\mathbf{R})$ of $n \times n$ matrices over $\mathbf{R}$, under addition of matrices. If $n \in \mathbf{Z}$ and A is a matrix in G, it is easy to see why nA makes sense, even though the operation is not a binary operation defined on G.

As with exponents in a group whose operation is denoted multiplicatively, we define $0 \cdot a = 0$ and $(-n)a = -(na)$. Notice that in the equation $0 \cdot a = 0$ the first 0 is the integer 0, while the second 0 is the identity of the group. The standard laws of exponents are then expressed as the following equations, which hold for all $a \in G$ and all $m, n \in \mathbf{Z}$:

$$ma + na = (m + n)a \qquad \text{and} \qquad m(na) = (mn)a.$$

When considering groups with additive notation, the most familiar examples are found in ordinary number systems. The set of integers $\mathbf{Z}$ is probably the most basic example. Results stated in Appendix C show that $\mathbf{Z}$ is closed under addition, that it satisfies the associative law, that 0 serves as the additive identity element, and that any integer n has an additive inverse $-n$.

Additional additive groups can be found by considering larger sets of numbers. In particular, the set of rational numbers $\mathbf{Q}$, the set of real numbers $\mathbf{R}$, and the set of complex numbers $\mathbf{C}$ all form groups using ordinary addition. Our next example will consist of an entire class of finite abelian groups. Before giving the example we will quickly review the notion of congruence for integers from Sections 1.3 and 1.4.

Let n be a positive integer, which we call the *modulus*. Then two integers a, b are *congruent modulo n*, written $a \equiv b$ (mod n), if a and b have the same remainder when divided by n. It can be shown that $a \equiv b$ (mod n) if and only if $a - b$ is divisible by n, and this condition is usually the easier one to work with. If we let $[a]_n$ denote the set of all integers that are congruent to a modulo n, then it is possible to define an addition for these *congruence classes*. Given $[a]_n$ and $[b]_n$ we define

$$[a]_n + [b]_n = [a + b]_n.$$

There is a question as to whether we have defined a binary operation, because the sum appears to depend on our choice of a and b. Proposition 1.4.2 shows that if $a_1 \equiv a_2$ (mod n) and $b_1 \equiv b_2$ (mod n), then it is true that $a_1 + b_1 \equiv a_2 + b_2$ (mod n), and so the sum is uniquely defined. The set of all congruence classes modulo n is denoted by $\mathbf{Z}_n$.

Example 3.1.3 (Group of Integers Modulo *n*)

Let n be a positive integer. The set $\mathbf{Z}_n$ of integers modulo n is an abelian group under addition of congruence classes. The group $\mathbf{Z}_n$ is finite and $|\mathbf{Z}_n| = n$.

Proposition 1.4.2 shows that addition of congruence classes defines a binary operation. The associative law holds since for all $a, b, c \in \mathbf{Z}$ we have

$$[a]_n + ([b]_n + [c]_n) = [a]_n + [b + c]_n = [a + (b + c)]_n = [(a + b) + c]_n$$

$$= [a + b]_n + [c]_n = ([a]_n + [b]_n) + [c]_n.$$

The commutative law holds since for all $a, b \in \mathbf{Z}$ we have

$$[a]_n + [b]_n = [a + b]_n = [b + a]_n = [b]_n + [a]_n.$$

Because $[a]_n + [0]_n = [a + 0]_n = [a]_n$ and $[a]_n + [-a]_n = [a - a]_n = [0]_n$, all of the necessary axioms are satisfied and $\mathbf{Z}_n$ is an abelian group. $\square$

Multiplication of congruence classes in $\mathbf{Z}_n$ is defined by

$$[a]_n \cdot [b]_n = [a \cdot b]_n.$$

As with addition of congruence classes, Proposition 1.4.2 implies that this multiplication is well-defined. Furthermore, a proof utilizing the corresponding properties of multiplication of integers can be given to show that multiplication of congruence classes is associative and commutative.

If a is an integer that is relatively prime to n, then there exist integers b, m such that $ab + mn = 1$. This gives $[a]_n \cdot [b]_n + [m]_n \cdot [n]_n = [1]_n$, or simply $[a]_n \cdot [b]_n = [1]_n$, since $[n]_n = [0]_n$. Thus $[a]_n$ has a multiplicative inverse. Conversely, if $[a]_n$ has a multiplicative inverse $[b]_n$, then $ab - 1$ must be divisible by n, and this implies that a and n are relatively prime.

The set of distinct congruence classes $[a]_n$ such that $(a,n) = 1$ is denoted by $\mathbf{Z}_n^{\times}$. Recall that the number of elements in this set is given by the Euler φ-function.

Example 3.1.4 (Group of Units Modulo n)

Let n be a positive integer. The set $\mathbf{Z}_n^{\times}$ of units modulo n is an abelian group under multiplication of congruence classes. The group $\mathbf{Z}_n^{\times}$ is finite and $|\mathbf{Z}_n^{\times}| = \varphi(n)$.

Proposition 1.4.8 shows that the multiplication of congruence classes is closed on $\mathbf{Z}_n^{\times}$. Another way to prove this is to use Proposition 1.2.3 (d) to show that ab is relatively prime to n if and only if both a and b are relatively prime to n. We have remarked that the associative and commutative laws can easily be checked. The element $[1]_n$ serves as an identity element. The set was defined so as to include all congruence classes that have multiplicative inverses. These multiplicative inverses are again in the set, so it follows that $\mathbf{Z}_n^{\times}$ is a group.

As a special case, we include the multiplication table of $\mathbf{Z}_8^{\times}$ (see Table 3.1.2). □

TABLE 3.1.2.   Multiplication in $\mathbf{Z}_8^{\times}$

	[1]	[3]	[5]	[7]
[1]	[1]	[3]	[5]	[7]
[3]	[3]	[1]	[7]	[5]
[5]	[5]	[7]	[1]	[3]
[7]	[7]	[5]	[3]	[1]

We next want to consider groups of matrices. In this section we will consider only matrices with entries from the set $\mathbf{R}$ of real numbers. We recall how to multiply 2×2 matrices:

$$\begin{bmatrix} a_{11} & a_{12} \\ a_{21} & a_{22} \end{bmatrix} \begin{bmatrix} b_{11} & b_{12} \\ b_{21} & b_{22} \end{bmatrix} = \begin{bmatrix} a_{11}b_{11} + a_{12}b_{21} & a_{11}b_{12} + a_{12}b_{22} \\ a_{21}b_{11} + a_{22}b_{21} & a_{21}b_{12} + a_{22}b_{22} \end{bmatrix}$$

A matrix $\begin{bmatrix} a & b \\ c & d \end{bmatrix}$ has an inverse if and only its determinant $ad - bc$ is nonzero, and the inverse can be found as follows:

$$\begin{bmatrix} a & b \\ c & d \end{bmatrix}^{-1} = \frac{1}{ad - bc} \begin{bmatrix} d & -b \\ -c & a \end{bmatrix}.$$

In general, if (a_{ij}) and (b_{ij}) are $n \times n$ matrices, then the product (c_{ij}) of the two matrices is defined as the matrix whose i,j-entry is

$$c_{ij} = \sum_{k=1}^{n} a_{ik} b_{kj}$$

and this product makes sense since in $\mathbf{R}$ the two operations of addition and multiplication are well-defined.

3.1.9 Definition.   The set of all invertible $n \times n$ matrices with entries in $\mathbf{R}$ is called the *general linear group of degree n over the real numbers,* and is denoted by $GL_n(\mathbf{R})$.

3.1.10 Proposition.   The set $GL_n(\mathbf{R})$ forms a group under matrix multiplication.

Proof. If A and B are invertible matrices, then the formulas $(A^{-1})^{-1} = A$ and $(AB)^{-1} = B^{-1}A^{-1}$ show that $GL_n(\mathbf{R})$ has inverses for each element and is closed under matrix multiplication. The identity matrix (with 1 in each diagonal entry and 0 in every other entry) is the identity element of the group. We assume that you have seen the proof that matrix multiplication is associative in a previous course in linear algebra. The proof is straightforward but rather unenlightening. $\square$

EXERCISES: SECTION 3.1

1. For each binary operation $*$ defined on a set below, determine whether or not $*$ gives a group structure on the set. If it is not a group, say which axioms fail to hold.
   (a) Define $*$ on $\mathbf{Z}$ by $a * b = ab$.
   (b) Define $*$ on $\mathbf{Z}$ by $a * b = \max\{a,b\}$.
   (c) Define $*$ on $\mathbf{Z}$ by $a * b = a - b$.
   (d) Define $*$ on $\mathbf{Z}$ by $a * b = |ab|$.
   (e) Define $*$ on $\mathbf{R}^+$ by $a * b = ab$.
   (f) Define $*$ on $\mathbf{Q}$ by $a * b = ab$.

2. Prove that multiplication of 2×2 matrices satisfies the associative law.

3. Is $GL_n(\mathbf{R})$ an abelian group? Support your answer by either giving a proof or a counterexample.

4. Write out the multiplication table for S_2.

5. Write out the addition table for $\mathbf{Z}_8$.

6. Write out the multiplication table for $\mathbf{Z}_7^\times$.

7. Write out the multiplication table for the following set of matrices over $\mathbf{Q}$:

$$\begin{bmatrix} 1 & 0 \\ 0 & 1 \end{bmatrix}, \begin{bmatrix} -1 & 0 \\ 0 & 1 \end{bmatrix}, \begin{bmatrix} 1 & 0 \\ 0 & -1 \end{bmatrix}, \begin{bmatrix} -1 & 0 \\ 0 & -1 \end{bmatrix}.$$

8. Show that the set $A = \{f : \mathbf{R} \to \mathbf{R} \mid f(x) = mx + b, m \neq 0\}$ of *affine functions* from $\mathbf{R}$ to $\mathbf{R}$ forms a group under composition of functions.

9. Show that the set of all 2×2 matrices over $\mathbf{R}$ of the form $\begin{bmatrix} m & b \\ 0 & 1 \end{bmatrix}$ with $m \neq 0$ forms a group under matrix multiplication.

10. In the group defined in the previous exercise, find all elements that commute with the element $\begin{bmatrix} 2 & 0 \\ 0 & 1 \end{bmatrix}$.

11. Let G be a group. We have shown that $(ab)^{-1} = b^{-1}a^{-1}$. Find a similar expression for $(abc)^{-1}$.

12. Let G be a group. If $g \in G$ and $g^2 = g$, then prove that $g = e$.

13. Let $S = \mathbf{R} - \{-1\}$. Define $*$ on S by $a * b = a + b + ab$. Show that $(S, *)$ is a group.

14. Show that a nonabelian group must have at least five distinct elements.

15. Let G be a group. Prove that $(ab)^n = a^n b^n$ for all $a, b \in G$ and all $n \in \mathbf{Z}$ if and only if G is abelian.

16. Let G be a group. Prove that $a^m a^n = a^{m+n}$ and $(a^m)^n = a^{mn}$ for all $m, n \in \mathbf{Z}$.

17. Let S be a nonempty finite set with a binary operation $*$ that satisfies the associative law. Show that S is a group if $a * b = a * c$ implies $b = c$ and $a * c = b * c$ implies $a = b$ for all $a, b, c \in S$. What can you say if S is infinite?

18. Let G be a finite set with an associative binary operation given by a table in which each element of the set G appears exactly once in each row and column. Prove that G is a group. How do you recognize the identity element? How do you recognize the inverse of an element?

19. Let G be a group. Prove that G is abelian if and only if $(ab)^{-1} = a^{-1}b^{-1}$ for all $a, b \in G$.

20. Let G be a group. Prove that if $x^2 = e$ for all $x \in G$, then G is abelian.
    Hint: Observe that $x = x^{-1}$ for all $x \in G$.

21. Show that if G is a finite group with an even number of elements, then there must exist an element $a \in G$ with $a \neq e$ such that $a^2 = e$.

3.2 SUBGROUPS

Verifying all of the group axioms in a particular example is often not an easy task. Many very useful examples, however, arise inside known groups. That is, we may want to consider a subset of a known group, and restrict the operation of the group to this subset. For example, the real numbers $+1$ and -1 form a group using ordinary multiplication. Of course, they form only a small subset of the full multiplicative group $\mathbf{R}^\times$.

In this section we introduce the notion of a subgroup, and present several conditions for checking that a subset of a given group is again a group. For example, we will show that if the closure axiom alone holds for a finite subset of a group, then that is enough to guarantee that the subset is a group. We will investigate one of the most important types of subgroup, that generated by a single element. Finally, we prove Lagrange's theorem, which relates the number of elements in any subgroup of a finite group to the total number of elements in the group.

If H is a subset of a group G, then for any $a, b \in H$ we can use the operation of G to define ab. This may not define a binary operation on H, unless we know that $ab \in H$ for all $a, b \in H$. In that case we say that we are using the operation on H *induced* by G.

3.2.1 Definition. Let G be a group, and let H be a subset of G. Then H is called a *subgroup* of G if H is itself a group, under the operation induced by G.

Example 3.2.1

As our first examples of subgroups we consider some groups of numbers. We already know that $\mathbf{Z}, \mathbf{Q}, \mathbf{R},$ and $\mathbf{C}$ are groups under ordinary addition. Furthermore, as sets we have

$$\mathbf{Z} \subseteq \mathbf{Q} \subseteq \mathbf{R} \subseteq \mathbf{C}$$

and each group is a subgroup of the next since the given operations are consistent. If we consider multiplicative groups of nonzero elements, we also have the subgroups

$$\mathbf{Q}^\times \subseteq \mathbf{R}^\times \subseteq \mathbf{C}^\times.$$

In the group $\mathbf{Z}$ of integers under addition consider the set of all multiples of a fixed positive integer n, denoted by

$$n\mathbf{Z} = \{x \in \mathbf{Z} \mid x = nk \text{ for some } k \in \mathbf{Z}\}.$$

Using the operation of addition induced from $\mathbf{Z}$, to show that $n\mathbf{Z}$ is a subgroup of $\mathbf{Z}$ we must check each of the axioms in the definition of a group. The closure axiom holds since if $a, b \in n\mathbf{Z}$, then we have $a = nq$ and $b = nk$ for some $q, k \in \mathbf{Z}$, and adding gives us $a + b = nq + nk = n(q + k)$. This shows that the sum of two elements in $n\mathbf{Z}$ has the correct form to belong to $n\mathbf{Z}$. The associative law holds for all elements in $\mathbf{Z}$, so in particular it holds for all elements in $n\mathbf{Z}$. The element 0 can be expressed in the form $0 = n \cdot 0$, and so it will work as an identity element for $n\mathbf{Z}$. Finally, the additive inverse of $x = nk$ has the correct form $-x = n(-k)$ to belong to $n\mathbf{Z}$, and so it also serves as an inverse in $n\mathbf{Z}$, since we have already observed that the identity elements of $\mathbf{Z}$ and $n\mathbf{Z}$ coincide.

For the final example of subgroups of sets of numbers, let $\mathbf{R}^+$ be the set of positive real numbers, and consider $\mathbf{R}^+$ as a subset of the multiplicative group $\mathbf{R}^\times$ of all nonzero real numbers. The product of two positive real numbers is again positive, so the closure axiom holds. The associative law holds for all real numbers, so in particular it holds for all real numbers in $\mathbf{R}^+$. The number 1 serves as an identity element, and the reciprocal of a positive number is still positive, so each element of $\mathbf{R}^+$ has a multiplicative inverse in $\mathbf{R}^+$. Thus we have shown that $\mathbf{R}^+$ is a subgroup of $\mathbf{R}^\times$. $\square$

Example 3.2.2

Recall our notation $GL_2(\mathbf{R})$ for the set of all 2×2 invertible matrices over the real numbers $\mathbf{R}$. The set of 2×2 matrices with determinant equal to 1 is a subgroup of $GL_2(\mathbf{R})$, which can be seen as follows: If $A, B \in GL_2(\mathbf{R})$ with $\det(A) = 1$ and $\det(B) = 1$, then we have $\det(AB) = \det(A)\det(B) = 1$. The associative law holds for all 2×2 matrices. The identity matrix certainly has determinant equal to 1, and if $\det(A) = 1$, then $\det(A^{-1}) = 1$.

The set of all $n \times n$ matrices over a field F with determinant equal to 1 is called the *special linear group over F,* denoted by $SL_n(F)$. Thus we have shown that $SL_2(\mathbf{R})$ is a subgroup of $GL_2(\mathbf{R})$. $\square$

In the examples of subgroups that we have considered, the associative law is always inherited by a subset of a known group. This means that at least one of the group axioms need not be checked. The next proposition and its corollaries are designed to give simplified conditions to use when checking that a subset is in fact a subgroup.

If a set S has a binary operation $\cdot$ defined on it, then the operation is said to be *closed* on a subset $X \subseteq S$ if $a \cdot b \in X$ for all $a, b \in X$. This is the reason that the first property in the definition of a group is called the *closure* property. (It is actually redundant in the definition of a group since it is a part of the definition of a binary operation. However, it is included for emphasis since it is very easy to forget the full implications of having a binary operation.)

The next proposition gives the conditions that are usually checked to determine whether or not a subset of a group is actually a subgroup. It can be summarized by saying that the subset must be closed under multiplication, must contain the identity, and must have inverses for each element.

3.2.2 Proposition.   Let G be a group with identity element e, and let H be a subset of G. Then H is a subgroup of G if and only if the following conditions hold:

(i) $ab \in H$ for all $a, b \in H$;

(ii) $e \in H$;

(iii) $a^{-1} \in H$ for all $a \in H$.

Proof. First, assume that H is a subgroup of G. Since H is a group under the operation of G, the closure axiom guarantees that ab must belong to H whenever a, b belong to H. There must be an identity element, say e', for H. Then considering the product in H, we have $e'e' = e'$. Now consider the same product as an element of G. Then we can write $e'e' = e'e$, and the cancellation law yields $e' = e$. Finally, if $a \in H$, then a must have an inverse b in H, with $ab = e$. But then in G we have $ab = e = aa^{-1}$, and the cancellation law implies that $a^{-1} = b$ is an element of H.

Conversely, suppose that H is a subset of G that satisfies the given conditions. Condition (i) shows that the operation of G defines a binary operation on H, and so the closure axiom holds. If $a, b, c \in H$, then in G we have the equation $a(bc) = (ab)c$, and so by considering this as an equation in H we see that H inherits the associative law. Conditions (ii) and (iii) assure that H has an identity element, and that every element of H has an inverse in H, since these elements have the same properties in H as they do when viewed as elements of G. $\square$

3.2.3 Corollary.   Let G be a group and let H be a nonempty subset of G. Then H is a subgroup if and only if $ab^{-1} \in H$ for all $a, b \in H$.

Proof. First assume that H is a subgroup of G. Using condition (ii) of the previous proposition, we see that H is nonempty since $e \in H$. If $a, b \in H$, then $b^{-1} \in H$ by condition (iii) of the proposition, and so condition (i) implies that $ab^{-1} \in H$.

Conversely, suppose that H is a nonempty subset of G that satisfies the given condition. Since H is nonempty, there is at least one element a that belongs to H. Then $e \in H$ since $e = aa^{-1}$, and this product belongs to H by assumption. Next, if $a \in H$, then a^{-1} can be expressed in the form $a^{-1} = ea^{-1}$, and this product must belong to H since e and a belong to H. Finally, we must show that H is closed under products: If $a, b \in H$, then we have already shown that $b^{-1} \in H$. We can express ab in the form $a(b^{-1})^{-1}$, and then the given condition shows that ab must belong to H. $\square$

In applying Corollary 3.2.3 it is crucial to show that the subset H is nonempty. Often the easiest way to do this is to show that H contains the identity element e.

3.2.4 Corollary.   Let G be a group, and let H be a finite, nonempty subset of G. Then H is a subgroup of G if and only if $ab \in H$ for all $a, b \in H$.

Proof. If H is a subgroup of G, then Proposition 3.2.2 implies that $ab \in H$ for all $a, b \in H$.

Conversely, assume that H is closed under the operation of G. We can use the previous corollary, provided we can show that $b^{-1} \in H$ whenever $b \in H$. Given $b \in H$, consider the powers $\{b, b^2, b^3, \ldots\}$ of b. These must all belong to H, by assumption, but since H is a finite set, they cannot all be distinct. There must be some repetition, say $b^n = b^m$ for positive integers $n > m$. The cancellation law then implies that $b^{n-m} = e$. Either $b = e$ or $n - m > 1$, and in the second case we then have $bb^{n-m-1} = e$, which shows that $b^{-1} = b^{n-m-1}$. Thus b^{-1} can be expressed as a positive power of b, which must belong to H.   □

Example 3.2.3

In S_3, the subset $\{(1),(1,2,3),(1,3,2)\}$ is closed under multiplication. The easiest way to see this is to look at the multiplication table given in Section 3.1. Since this subset is finite, Corollary 3.2.4 shows that it is a subgroup. The subsets $\{(1),(1,2)\}$, $\{(1),(1,3)\}$, and $\{(1),(2,3)\}$ can be shown in the same way to be subgroups of S_3.   □

Example 3.2.4

In the group $GL_2(\mathbf{R})$ of all invertible 2×2 matrices with real entries, let H be the following set of matrices:

$$\begin{bmatrix} 1 & 0 \\ 0 & 1 \end{bmatrix}, \quad \begin{bmatrix} -1 & 0 \\ 0 & 1 \end{bmatrix}, \quad \begin{bmatrix} 1 & 0 \\ 0 & -1 \end{bmatrix}, \quad \begin{bmatrix} -1 & 0 \\ 0 & -1 \end{bmatrix}.$$

It is easy to see that the product of any two of these matrices is a diagonal matrix with entries ± 1, which will again be in the set. Since the set is finite and closed under matrix multiplication, Corollary 3.2.4 implies that it is a subgroup of $GL_2(\mathbf{R})$.   □

Example 3.2.5

Let G be the group $GL_n(\mathbf{R})$ of all invertible $n \times n$ matrices with entries in the real numbers. Let H be the set of all diagonal matrices in G. That is, H consists of all matrices in G that have zeros in all entries except those along the main diagonal. Since the matrices in G are all invertible, the diagonal entries must all be nonzero. In showing that H is a subgroup of G, we can no longer apply Corollary 3.2.4 since H is not a finite set. It is probably easiest to just use Proposition 3.2.2. We only need to observe that the identity matrix belongs to H, that the product of two diagonal matrices with nonzero entries again has the same property, and that the inverse of a diagonal matrix with nonzero entries is again a diagonal matrix.   □

If G is a group, and a is any element of G, then Proposition 3.2.6 will show that the set of all powers of a is a subgroup of G, justifying the terminology of the next definition. In fact, this is the smallest subgroup of G that contains a, and turns out to have a particularly nice structure. Groups that consist of powers of a fixed element form an important class of groups and will be studied in depth in Section 3.5. We introduce a name for them at this point.

3.2.5 Definition.   The group G is called a *cyclic group* if there exists an element $a \in G$ such that $G = \{a^n \mid n \in \mathbf{Z}\}$. In this case a is called a *generator* of G.

More generally, for any group G and any element $a \in G$, the set $\{a^n \mid n \in \mathbf{Z}\}$ is called the *cyclic subgroup generated by a,* and is denoted by $\langle a \rangle$.

3.2.6 Proposition.   Let G be a group, and let $a \in G$.
(a) $\langle a \rangle$ is a subgroup of G.
(b) If K is any subgroup of G such that $a \in K$, then $\langle a \rangle \subseteq K$.

Proof. (a) The set $\langle a \rangle$ is closed under multiplication since if a^m, $a^n \in \langle a \rangle$, then $a^m a^n = a^{m+n} \in \langle a \rangle$. Furthermore, $\langle a \rangle$ includes the identity element and includes inverses, since by definition $a^0 = e$ and $(a^n)^{-1} = a^{-n}$.

(b) If K is any subgroup that contains a, then it must contain all positive powers of a since it is closed under multiplication. It also contains $a^0 = e$, and if $n < 0$, then $a^n \in K$ since $a^n = (a^{-n})^{-1}$. Thus $\langle a \rangle \subseteq K$. $\square$

The intersection of any collection of subgroups is again a subgroup (see Exercise 12). Given any subset S of a group G, the intersection of all subgroups of G that contain S is in fact the smallest subgroup that contains S. In the case $S = \{a\}$, by the previous proposition we obtain $\langle a \rangle$. In the case of two elements a, b of a nonabelian group G, it becomes much more complicated to describe the smallest subgroup of G that contains a and b. All products of powers of a and b must be included, and if $ab \neq ba$, it becomes very difficult to simplify terms. The general problem of listing all subgroups of a given group becomes difficult very quickly as the order of the group increases.

Example 3.2.6

In the multiplicative group $\mathbf{C}^\times$, consider the powers of i. We have $i^2 = -1$, $i^3 = -i$, and $i^4 = 1$. From this point on the powers repeat, since $i^5 = ii^4 = i$, $i^6 = i^2 i^4 = -1$, etc. For negative exponents we have $i^{-1} = -i$, $i^{-2} = -1$, and $i^{-3} = i$. Again, from this point on the powers repeat. Thus we have

$$\langle i \rangle = \{1, i, -1, -i\}.$$

The situation is quite different if we consider $\langle 2i \rangle$. In this case the powers of $2i$ are all distinct, and the subgroup generated by $2i$ is infinite:

$$\langle 2i \rangle = \left\{ \dots, \frac{1}{16}, \frac{1}{8}i, -\frac{1}{4}, -\frac{1}{2}i, 1, 2i, -4, -8i, 16, 32i, \dots \right\}.   \square$$

Example 3.2.7 (Z is Cyclic)

Consider the group $\mathbf{Z}$, using the standard operation of addition of integers. Since the operation is denoted additively rather than multiplicatively, we must consider multiples rather than powers. Thus $\mathbf{Z}$ is cyclic if and only if there exists an integer a such that $\mathbf{Z} = \{na \mid n \in \mathbf{Z}\}$. Either $a = 1$ or $a = -1$ will satisfy the condition, so $\mathbf{Z}$ is cyclic, with generators 1 and -1. $\square$

Example 3.2.8 (Z_n is Cyclic)

The additive group $\mathbf{Z}_n$ of integers modulo n is also cyclic, generated by $[1]_n$, since each congruence class can be expressed as a finite sum of $[1]_n$'s. To be precise, $[k]_n = k[1]_n$.

It is very interesting to determine all possible generators of $\mathbf{Z}_n$. If $[a]_n$ is a generator of $\mathbf{Z}_n$, then in particular $[1]_n$ must be a multiple of $[a]_n$. On the other hand, if $[1]_n$ is a multiple of $[a]_n$, then certainly every other congruence class modulo n is also a multiple of $[a]_n$. Thus to determine all of the generators of $\mathbf{Z}_n$ we only need to determine the integers a such that some multiple of a is congruent to 1. These are precisely the integers that are relatively prime to n.   □

Example 3.2.9

The multiplicative groups $\mathbf{Z}_n^\times$ provide many interesting examples. We first consider $\mathbf{Z}_5^\times$. We will omit the subscript on the congruence classes when it is clear from the context. We have $[2]^2 = [4]$, $[2]^3 = [3]$, and $[2]^4 = [1]$. Thus each element of $\mathbf{Z}_5^\times$ is a power of $[2]$, showing that the group is cyclic, which we write as $\mathbf{Z}_5^\times = \langle [2] \rangle$. We note that $[3]$ is also a generator, but $[4]$ is not, since $[4]^2 = [1]$, and so $\langle [4] \rangle = \{[1], [4]\} \neq \mathbf{Z}_5^\times$.

Next, consider $\mathbf{Z}_8^\times = \{[1], [3], [5], [7]\}$. The square of each element is the identity, so we have $\langle [3] \rangle = \{[1], [3]\}$, $\langle [5] \rangle = \{[1], [5]\}$, and $\langle [7] \rangle = \{[1], [7]\}$. Thus $\mathbf{Z}_8^\times$ is not cyclic.   □

Example 3.2.10

The group S_3 is not cyclic. We can list all cyclic subgroups as follows:

$\langle (1) \rangle = \{(1)\}$;

$\langle (1,2) \rangle = \{(1),(1,2)\}$, $\langle (1,3) \rangle = \{(1),(1,3)\}$, $\langle (2,3) \rangle = \{(1),(2,3)\}$;

$\langle (1,2,3) \rangle = \{(1),(1,2,3),(1,3,2)\} = \langle (1,3,2) \rangle$.

Since no cyclic subgroup is equal to all of S_3, it is not cyclic. The number of distinct powers of a given permutation is equal to its order, as defined in Section 2.3.   □

3.2.7 Definition.   Let a be an element of the group G. If there exists a positive integer n such that $a^n = e$, then a is said to have *finite order,* and the smallest such positive integer is called the *order* of a, denoted by $o(a)$.

If there does not exist a positive integer n such that $a^n = e$, then a is said to have *infinite order*.

We note that the proof of Corollary 3.2.4 shows that every element of a finite group must have finite order. We have already considered the order of a permutation in Section 2.3. You may recall that problems in Sections 1.3 and 1.4 dealt with the orders of elements in $\mathbf{Z}_n$ and $\mathbf{Z}_n^\times$.

3.2.8 Proposition.   Let a be an element of the group G.

(a) If a has infinite order, then $a^k \neq a^m$ for all integers $k \neq m$.

(b) If a has finite order and $k \in \mathbf{Z}$, then $a^k = e$ if and only if $o(a) | k$.

(c) If a has finite order $o(a) = n$, then for all integers k, m, we have $a^k = a^m$ if and only if $k \equiv m$ (mod n). Furthermore, $|\langle a \rangle| = o(a)$.

Proof. (a) Let a have infinite order. Suppose that $a^k = a^m$ for $k, m \in \mathbf{Z}$, with $k \geq m$. Then multiplying by $(a^m)^{-1}$ gives us $a^{k-m} = e$, and since a is not of finite order, we must have $k - m = 0$.

(b) Let $o(a) = n$, and suppose that $a^k = e$. Using the division algorithm we can write $k = nq + r$, where $0 \leq r < n$. Thus

$$a^r = a^{k-nq} = a^k a^{n(-q)} = a^k(a^n)^{-q} = e \cdot e^{-q} = e.$$

Since $0 \leq r < o(a)$ and $a^r = e$, by the definition of $o(a)$ we must have $r = 0$, and so $k = q \cdot o(a)$.

On the other hand, if $o(a) | k$, then $k = nq$ for some $q \in \mathbf{Z}$. But then $a^k = (a^n)^q = e^q = e$.

(c) We must show that if a has finite order, then $a^k = a^m$ if and only if $k \equiv m$ (mod n). We first observe that since we are working in a group, $a^k = a^m$ if and only if $a^{k-m} = e$. By part (b) this occurs if and only if $n | (k - m)$, which is equivalent to the statement that $k \equiv m$ (mod n).

Finally, using Corollary 3.2.4, it follows that the subset $S = \{e, a, \ldots, a^{n-1}\}$ is a subgroup that contains a. Hence $\langle a \rangle \subseteq S$. On the other hand, $a^j \in \langle a \rangle$ for all $j \in \mathbf{Z}$, and so $S \subseteq \langle a \rangle$. Thus $|\langle a \rangle| = |S| = o(a)$. $\square$

If you check all of the examples that we have given of subgroups of finite groups, you will see that in every case the number of elements in the subgroup is a divisor of the order of the group. This is always true, and is a very useful result. For example, in checking whether or not a subset of a finite group is in fact a subgroup, you can immediately answer no if the number of elements in the subset is not a divisor of the order of the group. We have separated out part of the proof as a lemma.

3.2.9 Lemma.   Let H be a subgroup of the group G. For $a, b \in G$ define $a \sim b$ if $ab^{-1} \in H$. Then $\sim$ is an equivalence relation.

Proof. The relation is reflexive since $aa^{-1} = e \in H$. It is symmetric since if $a \sim b$, then $ab^{-1} \in H$. Since H contains the inverse of each of its elements, we have $ba^{-1} = (ab^{-1})^{-1} \in H$, showing that $b \sim a$. Finally, the relation is transitive since if $a \sim b$ and $b \sim c$ for any $a, b, c \in G$, then both ab^{-1} and bc^{-1} belong to H, so since H is a subgroup, the product $ac^{-1} = (ab^{-1})(bc^{-1})$ also belongs to H, showing that $a \sim c$. $\square$

Example 3.2.11 (Congruence Modulo n)

The equivalence relation we have used to define congruence modulo n in $\mathbf{Z}$ is a special case of Lemma 3.2.9. If the operation of G is denoted additively, then the equivalence relation of Lemma 3.2.9 is defined by setting $a \sim b$ if $a + (-b)$ belongs to H. This is usually written as $a \sim b$ if $a - b \in H$. We consider the case when $G = \mathbf{Z}$ and H is the subgroup $n\mathbf{Z}$. Since $a \equiv b$ (mod n) if and only if $a - b$ is a multiple of n, we see that a is congruent to b modulo n if and only if $a - b$ is an element of the subgroup $n\mathbf{Z}$. This shows how to fit our earlier work on congruences into the new terminology of group theory. $\square$

3.2.10 Theorem (Lagrange).   If H is a subgroup of the finite group G, then the order of H is a divisor of the order of G.

Proof. Let H be a subgroup of the finite group G, with $|G| = n$ and $|H| = m$. Let $\sim$ denote the equivalence relation defined in the previous lemma, and for any $a \in G$ let $[a]$ denote the equivalence class of a.

We claim that the function $\rho_a : H \to [a]$ defined by $\rho_a(x) = xa$ for all $x \in H$ is a one-to-one correspondence between H and $[a]$. We first note that the stated range of ρ_a is correct since if $h \in H$, then $\rho_a(h) = ha \in [a]$ because $(ha)(a^{-1}) = h \in H$. To show that ρ_a is one-to-one, suppose that $h, k \in H$ with $\rho_a(h) = \rho_a(k)$. Then $ha = ka$, and since the cancellation property holds in any group, we can conclude that $h = k$. Finally, ρ_a is onto since if $y \in G$ with $y \sim a$, then we have $ya^{-1} = h$ for some $h \in H$, and thus the equation $\rho_a(x) = y$ has the solution $x = h$ since $ha = (ya^{-1})a = y$.

Since the equivalence classes of $\sim$ partition G, each element of G belongs to precisely one of the equivalence classes. We have shown that each equivalence class has m elements, since it is in one-to-one correspondence with H. Counting the elements of G according to the distinct equivalence classes, we simply get $n = mt$, where t is the number of distinct equivalence classes. $\square$

Example 3.2.12

In this example we will investigate equivalence relative to two different subgroups of S_3. The proof of Lagrange's theorem shows us how to proceed in a systematic way. First, let

$$H = \langle (1,2,3) \rangle = \{(1),(1,2,3),(1,3,2)\}.$$

By definition, the elements of H are equivalent to each other and form the first equivalence class. Any other equivalence class must be disjoint from the first one and have the same number of elements, so the only possibility is that the remaining elements of G must form a second equivalence class.

Therefore the equivalence relation defined by the subgroup H determines two equivalence classes:

$$\{(1),(1,2,3),(1,3,2)\}, \qquad \{(1,2),(1,3),(2,3)\}.$$

Next, let us consider the subgroup K generated by $(1,2)$. Then

$$K = \{(1),(1,2)\}$$

so the equivalence classes must each contain two elements. The proof of Lagrange's theorem shows that we can find the equivalence class of an element a by multiplying it on the left by all elements of K. If we let $a = (1,2,3)$, then we have two equivalent elements $(1)(1,2,3) = (1,2,3)$ and $(1,2)(1,2,3) = (2,3)$. There are two elements remaining, and they form the third equivalence class. Thus the equivalence relation defined by the subgroup K determines three equivalence classes:

$$\{(1),(1,2)\}, \qquad \{(1,2,3),(2,3)\}, \qquad \{(1,3,2),(1,3)\}. \quad \square$$

3.2.11 Corollary.   Let G be a finite group of order n.
(a) For any $a \in G$, $o(a)$ is a divisor of n.
(b) For any $a \in G$, $a^n = e$.

Proof. (a) The order of a is the same as the order of $\langle a \rangle$, which by Lagrange's theorem is a divisor of the order of G.

(b) If a has order m, then by part (a) we have $n = mq$ for some integer q. Thus $a^n = a^{mq} = (a^m)^q = e$. $\quad \square$

Example 3.2.13 (Euler's Theorem)

We can now give a short group theoretic proof of Euler's theorem (Theorem 1.4.9). We must show that if φ denotes the Euler φ function and a is any integer relatively prime to the positive integer n, then $a^{\varphi(n)} \equiv 1 \pmod{n}$. Let G be the multiplicative group of congruence classes modulo n. The order of G is given by $\varphi(n)$, and so by Corollary 3.2.11, raising any congruence class to the power $\varphi(n)$ must give the identity element. The statement $[a]^{\varphi(n)} = [1]$ is equivalent to $a^{\varphi(n)} \equiv 1 \pmod{n}$. $\quad \square$

3.2.12 Corollary.   Any group of prime order is cyclic.

Proof. Let G be a group of order p, where p is a prime number. Let a be an element of G different from e. Then the order of $\langle a \rangle$ is not 1, and so it must be p since it is a divisor of p. This implies that $\langle a \rangle = G$, and thus G is cyclic. $\quad \square$

EXERCISES: SECTION 3.2

1. In $GL_2(\mathbf{R})$, find the order of each of the following elements:

$$\begin{bmatrix} 1 & -1 \\ 1 & 0 \end{bmatrix}, \quad \begin{bmatrix} 1 & -1 \\ 0 & 1 \end{bmatrix}, \quad \begin{bmatrix} -1 & 1 \\ 0 & 1 \end{bmatrix}, \quad \begin{bmatrix} 1 & -1 \\ -1 & 0 \end{bmatrix}.$$

2. For each of the following groups, find all cyclic subgroups of the group:
   (a) $\mathbf{Z}_6$,   **(b)** $\mathbf{Z}_8$,   **(c)** S_4.

3. Show that $\{(1),(1,2)(3,4),(1,3)(2,4),(1,4)(2,3)\}$ is a subgroup of S_4.

4. Let $G = GL_2(\mathbf{R})$.

   (a) Show that $T = \left\{ \begin{bmatrix} a & b \\ 0 & d \end{bmatrix} \mid ad \neq 0 \right\}$ is a subgroup of G.

   (b) Show that $D = \left\{ \begin{bmatrix} a & 0 \\ 0 & d \end{bmatrix} \mid ad \neq 0 \right\}$ is a subgroup of G.

5. Let $G = GL_3(\mathbf{R})$. Show that $\left\{ \begin{bmatrix} 1 & a & b \\ 0 & 1 & c \\ 0 & 0 & 1 \end{bmatrix} \right\}$ is a subgroup of G.

6. Let S be a set, and let a be a fixed element of S. Show that $\{\sigma \in \mathrm{Sym}(S) \mid \sigma(a) = a\}$ is a subgroup of $\mathrm{Sym}(S)$.

7. (a) Find the set of elements of finite order of $\mathbf{R}^\times$.
   (b) Find the set of elements of finite order of $\mathbf{C}^\times$.

8. Let G be an abelian group, with its operation denoted additively. Show that $\{a \in G \mid 2a = 0\}$ is a subgroup of G. Compute this subgroup for $G = \mathbf{Z}_{12}$.

9. Let G be an abelian group. Show that the set of all elements of G of finite order forms a subgroup of G.

10. Prove that any cyclic group is abelian.

11. Prove or disprove this statement: If G is a group in which every proper subgroup is cyclic, then G is cyclic.

12. Prove that the intersection of any collection of subgroups of a group is again a subgroup.

13. Let G be a group, and let $a \in G$. The set $C(a) = \{x \in G \mid xa = ax\}$ of all elements of G that commute with a is called the *centralizer* of a.
   (a) Show that $C(a)$ is a subgroup of G.
   (b) Show that $\langle a \rangle \subseteq C(a)$.
   (c) Compute $C(a)$ if $G = S_3$ and $a = (1,2,3)$.
   (d) Compute $C(a)$ if $G = S_3$ and $a = (1,2)$.

14. Let G be a group. The set $Z(G) = \{x \in G \mid xg = gx \text{ for all } g \in G\}$ of all elements that commute with every other element of G is called the *center* of G.
   (a) Show that $Z(G)$ is a subgroup of G.
   (b) Show that $Z(G) = \cap_{a \in G} C(a)$.
   (c) Compute the center of S_3.
   (d) Compute the center of $GL_2(\mathbf{R})$.

15. Show that any finite group has an even number of elements of order three.

16. Let G be a group with $a, b \in G$.
    (a) Show that $o(a^{-1}) = o(a)$.
    (b) Show that $o(ab) = o(ba)$.
    (c) Show that $o(aba^{-1}) = o(b)$.

17. Let G be a group with $a, b \in G$. Assume that $o(a)$ and $o(b)$ are finite and relatively prime, and that $ab = ba$. Show that $o(ab) = o(a)o(b)$.

18. Find an example of a group G and elements $a, b \in G$ such that a and b each have finite order but ab does not.

3.3 CONSTRUCTING EXAMPLES

Being able to construct your own examples is very important. To do this you need to be familiar with a good variety of groups. In this section we first study groups with orders up to 6. Then we introduce the notion of the direct product of two groups, which can be used to construct new groups from known ones. Finally, we introduce some new matrix groups.

    As a corollary of Lagrange's theorem, we proved that any group of prime order must be cyclic. This shows that any group of order 2, 3, or 5 must be cyclic. It is then easy to write down what the multiplication table of the group must look like.

    Next, take the case of a group G of order 4. As another corollary of Lagrange's theorem, we proved that the order of any element of a group must be a divisor of the order of the group. Since $|G| = 4$, we can only have elements (different from the identity) of order 2 or 4. If there exists an element $a \in G$ of order 4, then its four powers e, a, a^2, and a^3 must be the only elements in G, and so G is cyclic. On the other hand, if there is no element of order 4, then every element not equal to e must have order 2. This means that in the multiplication table for G the element e must occur down the main diagonal, and then by using the fact that each element must occur exactly once in each row and column, it can be shown that there is only one possible pattern for the table.

    This analysis of the table does not imply that there exists a group with such a multiplication table. In particular, there is no guarantee that the associative law holds. However, a group that serves as a model for the table is not difficult to find. The group $\mathbf{Z}_8^\times$ of units modulo 8 is easily checked to have four elements, and

TABLES 3.3.1 and 3.3.2. Multiplication Tables for Groups of Order 4

	e	a	a^2	a^3
e	e	a	a^2	a^3
a	a	a^2	a^3	e
a^2	a^2	a^3	e	a
a^3	a^3	e	a	a^2

	e	a	b	c
e	e	a	b	c
a	a	e	c	b
b	b	c	e	a
c	c	b	a	e

TABLE 3.3.3. Multiplication Table for a Cyclic Group of Order 6

	e	a	a^2	a^3	a^4	a^5
e	e	a	a^2	a^3	a^4	a^5
a	a	a^2	a^3	a^4	a^5	e
a^2	a^2	a^3	a^4	a^5	e	a
a^3	a^3	a^4	a^5	e	a	a^2
a^4	a^4	a^5	e	a	a^2	a^3
a^5	a^5	e	a	a^2	a^3	a^4

the square of every element is the identity $[1]_8$. Thus the multiplication table for a group of four elements has one of only two possible patterns. These are listed in Tables 3.3.1 and 3.3.2.

We know of two basic examples of groups of order 6, the group $\mathbf{Z}_n$ of integers modulo n, which is cyclic, and the group S_3 of permutations on three elements. The multiplication table for a cyclic group of order 6 is given in Table 3.3.3.

We have described S_3 by explicitly listing the permutations that belong to it. It is possible to give another description, which is easier to work with in many cases. Let $e = (1)$, let $a = (1,2,3)$, and let $b = (1,2)$. Using the multiplication table given in Example 3.1.2, we see that $a^2 = (1,3,2)$ and then $a^3 = aa^2 = e$. For the element b we have $b^2 = e$. Using the convention $a^0 = e$ and $b^0 = e$, we can express each element of S_3 in a unique way in the form $a^i b^j$, for $i = 0, 1, 2$ and $j = 0, 1$. Specifically, we have $(1) = e$, $(1,2,3) = a$, $(1,3,2) = a^2$, $(1,2) = b$, $(1,3) = ab$, and $(2,3) = a^2b$. To multiply two elements in this form, we must be able to bring them back to the standard form, but to do so only requires the formula $ba = a^2b$. This allows us to give the following description of the symmetric group on three elements:

$$S_3 = \{e,a,a^2,b,ab,a^2b \mid a^3 = e,\ b^2 = e,\ ba = a^2b\}.$$

Using this notation we can rewrite the multiplication table as shown in Table 3.3.4.

TABLE 3.3.4. Multiplication Table for S_3

	e	a	a^2	b	ab	a^2b
e	e	a	a^2	b	ab	a^2b
a	a	a^2	e	ab	a^2b	b
a^2	a^2	e	a	a^2b	b	ab
b	b	a^2b	ab	e	a^2	a
ab	ab	b	a^2b	a	e	a^2
a^2b	a^2b	ab	b	a^2	a	e

The multiplication table for any group of order 6 must have the form of Table 3.3.3 or that of Table 3.3.4. We have indicated, in Exercises 12–14 at the end of this section, how this can be proved.

We now introduce a new method of combining two subgroups. We have already observed that the intersection of subgroups of a group is again a subgroup. If H and K are subgroups of a group G, then $H \cap K$ is the largest subgroup of G that is contained in both H and K. On the other hand, if we consider the smallest subgroup that contains both H and K, then it certainly must contain all products of the form hk, where $h \in H$ and $k \in K$. In certain cases, these products do form a subgroup. In general, though, it may be necessary to include more elements.

3.3.1 Definition.   Let G be a group, and let S and T be subsets of G. Then

$$ST = \{g \in G \mid g = st \text{ for some } s \in S, t \in T\}.$$

If H and K are subgroups of G, then we simply call HK the *product* of H and K. The next proposition shows that if G is abelian, then the product of any two subgroups is again a subgroup since the condition of the proposition is satisfied whenever the elements of H and K commute with each other. In this case, if the operation of G is denoted additively, then instead of HK we write $H + K$, and refer to the *sum* of H and K.

3.3.2 Proposition.   Let G be a group, and let H and K be subgroups of G. If $h^{-1}kh \in K$ for all $h \in H$ and $k \in K$, then HK is a subgroup of G.

Proof. To show that HK is closed, let $g_1, g_2 \in HK$. Then $g_1 = h_1 k_1$ and $g_2 = h_2 k_2$ for some $h_1, h_2 \in H$ and $k_1, k_2 \in K$. Applying the stated condition to h_2 and k_1, we have $h_2^{-1}k_1 h_2 \in K$, say $h_2^{-1}k_1 h_2 = k_3$. Thus $k_1 h_2 = h_2 k_3$, and so

$$g_1 g_2 = (h_1 k_1)(h_2 k_2) = h_1(k_1 h_2)k_2 = h_1(h_2 k_3)k_2 = (h_1 h_2)(k_3 k_2).$$

This shows that $g_1 g_2 \in HK$, since $h_1 h_2 \in H$ and $k_3 k_2 \in K$. Since the identity element belongs to both H and K, we have $e = e \cdot e \in HK$. Finally, if $g = hk$ for $h \in H$ and $k \in K$, then

$$g^{-1} = k^{-1}h^{-1} = (h^{-1}h)k^{-1}h^{-1} = (h^{-1})((h^{-1})^{-1}k^{-1}h^{-1}) \in HK$$

since $h^{-1} \in H$ and $k^{-1} \in K$. Therefore $(h^{-1})^{-1}k^{-1}h^{-1} \in K$ by the given condition.   □

Example 3.3.1

In the group $\mathbf{Z}_{15}^{\times}$, let $H = \{[1],[11]\}$ and $K = \{[1],[4]\}$. These are subgroups since $[11]$ and $[4]$ both have order two. The condition that $h^{-1}kh \in K$ for all $h \in H$ and $k \in K$ holds since $\mathbf{Z}_{15}^{\times}$ is abelian. Computing all possible products in HK gives us

$$[1][1] = [1], \qquad [1][4] = [4], \qquad [11][1] = [11], \qquad [11][4] = [14],$$

and so HK is a subgroup of order 4.

Let L be the cyclic subgroup $\{[1],[4],[7],[13]\}$ generated by $[7]$. Listing all of the distinct products shows that HL is all of $\mathbf{Z}_{15}^{\times}$.   □

Example 3.3.2

Let G be the additive group $\mathbf{Z}$ of integers, and let $H = a\mathbf{Z}$ and $K = b\mathbf{Z}$. Then $H + K$ is the subgroup generated by $d = (a,b)$. Any element of $H + K$ is a linear combination of a and b, and so it is divisible by d, which shows that $H + K \subseteq d\mathbf{Z}$. On the other hand, d is a linear combination of a and b, so $d \in H + K$, which implies that $d\mathbf{Z} \subseteq H + K$. Thus

$$a\mathbf{Z} + b\mathbf{Z} = (a,b)\mathbf{Z}.   □$$

Our basic examples thus far have come from ordinary sets of numbers, from the integers modulo n, and from groups of permutations and groups of matrices. We now study a construction that will allow us to give some additional examples that will prove to be very important.

3.3.3 Definition.   Let G_1 and G_2 be groups. The set of all ordered pairs (a_1,a_2) such that $a_1 \in G_1$ and $a_2 \in G_2$ is called the *direct product* of G_1 and G_2, denoted by $G_1 \times G_2$.

3.3.4 Proposition.   Let G_1 and G_2 be groups.
(a) The direct product $G_1 \times G_2$ is a group under the multiplication

$$(a_1,a_2)(b_1,b_2) = (a_1 b_1, a_2 b_2).$$

(b) If $a_1 \in G_1$ and $a_2 \in G_2$ have orders n and m, respectively, then in $G_1 \times G_2$ the element (a_1,a_2) has order $\mathrm{lcm}[n,m]$.

Proof. (a) The given multiplication defines a binary operation. The associative law holds since

$$(a_1,a_2)((b_1,b_2)(c_1,c_2)) = (a_1,a_2)(b_1 c_1, b_2 c_2) = (a_1(b_1 c_1), a_2(b_2 c_2))$$

$$= ((a_1 b_1)c_1, (a_2 b_2)c_2) = (a_1 b_1, a_2 b_2)(c_1,c_2)$$

$$= ((a_1,a_2)(b_1,b_2))(c_1,c_2).$$

If we use e_1 and e_2 to denote the identity elements in G_1 and G_2, respectively, then (e_1,e_2) is easily seen to be the identity element of the direct product. Finally, for any element $(a_1,a_2) \in G_1 \times G_2$ we have $(a_1,a_2)^{-1} = (a_1^{-1},a_2^{-1})$.

(b) Let $a_1 \in G_1$ and $a_2 \in G_2$ have orders n and m, respectively. Then in $G_1 \times G_2$, the order of (a_1,a_2) is the smallest positive power k such that $(a_1,a_2)^k = (e_1,e_2)$. Since $(a_1,a_2)^k = (a_1^k,a_2^k)$, this shows that the order is the smallest positive integer k such that $a_1^k = e_1$ and $a_2^k = e_2$. By Proposition 3.2.8(b) this must be the smallest positive integer divisible by both n and m, so $k = \mathrm{lcm}[n,m]$.   □

Example 3.3.3

   In this example we give the addition table for $\mathbf{Z}_2 \times \mathbf{Z}_2$. To simplify the notation we have omitted the brackets denoting congruence classes. The operation in the direct product uses the operations from the given groups in each component, so in this example we use addition modulo 2 in each component.

   This group is usually called the *Klein four-group*. The pattern in Table 3.3.5 is the same as that in Table 3.3.2. This group is characterized by the fact that it has order 4 and each element except the identity has order 2.   □

TABLE 3.3.5. Addition in $\mathbf{Z}_2 \times \mathbf{Z}_2$

	(0,0)	(1,0)	(0,1)	(1,1)
(0,0)	(0,0)	(1,0)	(0,1)	(1,1)
(1,0)	(1,0)	(0,0)	(1,1)	(0,1)
(0,1)	(0,1)	(1,1)	(0,0)	(1,0)
(1,1)	(1,1)	(0,1)	(1,0)	(0,0)

Example 3.3.4

   In the group $\mathbf{Z} \times \mathbf{Z}$, the subgroup generated by an element (m,n) consists of all multiples $k(m,n)$. This subgroup cannot contain both of the elements $(1,0)$ and $(0,1)$, and so no single element generates $\mathbf{Z} \times \mathbf{Z}$, showing that it is not cyclic. There are natural subgroups $\langle(1,0)\rangle$ and $\langle(0,1)\rangle$. The ''diagonal'' subgroup $\langle(1,1)\rangle$ is also interesting.   □

Example 3.3.5

   The group $\mathbf{Z}_2 \times \mathbf{Z}_3$ is cyclic, since the element $(1,1)$ must have order 6 by Proposition 3.3.4. (The order of 1 in $\mathbf{Z}_2$ is 2, while the order of 1 in $\mathbf{Z}_3$ is 3, and then we have $\mathrm{lcm}[2,3] = 6$.)

   On the other hand, the group $\mathbf{Z}_2 \times \mathbf{Z}_4$ is not cyclic, since in the first component the possible orders are 1 and 2, and in the second component the possible orders are 1, 2, and 4. The largest possible least common multiple we can have is 4, so there is no element of order 8 and the group is not cyclic.   □

   We now want to introduce some more general matrix groups. We first need the definition of a field, which we will phrase in the language of groups. The sets **Q**, **R**, and **C** that we have worked with form abelian groups under addition, and in each case the set of nonzero elements also forms an abelian group under multiplication. This is also the case for $\mathbf{Z}_p$, when p is a prime number, and we would like to be able to work with groups of matrices that have entries in any of these sets. Fields are fundamental objects in abstract algebra and will be studied at length in later chapters. Another definition is given in Chapter 4.

3.3.5 Definition.   Let F be a set with two binary operations $+$ and $\cdot$ with respective identity elements 0 and 1, where $0 \neq 1$. Then F is called a *field* if

(i) the set of all elements of F is an abelian group under $+$;

(ii) the set of all nonzero elements of F is an abelian group under $\cdot$;

(iii) $a(b + c) = ab + ac$ for all $a, b, c \in F$.

Axiom (iii) is the *distributive law,* which gives a connection between addition and multiplication. The properties of matrix multiplication depend heavily on this law.

We next want to consider matrices that have entries in a field F. (Even if **R** and $\mathbf{Z}_p$ are the only fields F with which you feel comfortable, using them in the following matrix groups will give you interesting and instructive examples.) If (a_{ij}) and (b_{ij}) are $n \times n$ matrices, then the product (c_{ij}) of the two matrices is defined as the matrix whose i,j-entry is

$$c_{ij} = \sum_{k=1}^{n} a_{ik} b_{kj}.$$

This product makes sense since in F the two operations of addition and multiplication are well-defined.

3.3.6 Definition.   Let F be a field. The set of all invertible $n \times n$ matrices with entries in F is called the *general linear group of degree n over F,* and is denoted by $GL_n(F)$.

3.3.7 Proposition.   Let F be a field. Then $GL_n(F)$ is a group under matrix multiplication.

Proof. If A and B are invertible matrices, then the formulas $(A^{-1})^{-1} = A$ and $(AB)^{-1} = B^{-1}A^{-1}$ show that $GL_n(F)$ has inverses for each element and is closed under matrix multiplication. The identity matrix (with 1 in each diagonal entry and 0 in every other entry) is an identity element. The proof that matrix multiplication is associative is left as an exercise. $\square$

Example 3.3.6

This example gives the multiplication table for $GL_2(\mathbf{Z}_2)$. The total number of 2×2 matrices over $\mathbf{Z}_2$ is $2^4 = 16$, but it can be checked that only six of the matrices are invertible, and these are listed in Table 3.3.6. We simply use 0 and 1 to denote the congruence classes $[0]_2$ and $[1]_2$. Note that the group $GL_2(\mathbf{Z}_2)$ is not abelian.

TABLE 3.3.6. Multiplication in $GL_2(\mathbf{Z}_2)$

$\begin{bmatrix}1&0\\0&1\end{bmatrix}$	$\begin{bmatrix}1&0\\0&1\end{bmatrix}$	$\begin{bmatrix}1&1\\1&0\end{bmatrix}$	$\begin{bmatrix}0&1\\1&1\end{bmatrix}$	$\begin{bmatrix}0&1\\1&0\end{bmatrix}$	$\begin{bmatrix}1&1\\0&1\end{bmatrix}$	$\begin{bmatrix}1&0\\1&1\end{bmatrix}$
$\begin{bmatrix}1&0\\0&1\end{bmatrix}$	$\begin{bmatrix}1&0\\0&1\end{bmatrix}$	$\begin{bmatrix}1&1\\1&0\end{bmatrix}$	$\begin{bmatrix}0&1\\1&1\end{bmatrix}$	$\begin{bmatrix}0&1\\1&0\end{bmatrix}$	$\begin{bmatrix}1&1\\0&1\end{bmatrix}$	$\begin{bmatrix}1&0\\1&1\end{bmatrix}$
$\begin{bmatrix}1&1\\1&0\end{bmatrix}$	$\begin{bmatrix}1&1\\1&0\end{bmatrix}$	$\begin{bmatrix}0&1\\1&1\end{bmatrix}$	$\begin{bmatrix}1&0\\0&1\end{bmatrix}$	$\begin{bmatrix}1&1\\0&1\end{bmatrix}$	$\begin{bmatrix}1&0\\1&1\end{bmatrix}$	$\begin{bmatrix}0&1\\1&0\end{bmatrix}$
$\begin{bmatrix}0&1\\1&1\end{bmatrix}$	$\begin{bmatrix}0&1\\1&1\end{bmatrix}$	$\begin{bmatrix}1&0\\0&1\end{bmatrix}$	$\begin{bmatrix}1&1\\1&0\end{bmatrix}$	$\begin{bmatrix}1&0\\1&1\end{bmatrix}$	$\begin{bmatrix}0&1\\1&0\end{bmatrix}$	$\begin{bmatrix}1&1\\0&1\end{bmatrix}$
$\begin{bmatrix}0&1\\1&0\end{bmatrix}$	$\begin{bmatrix}0&1\\1&0\end{bmatrix}$	$\begin{bmatrix}1&1\\0&1\end{bmatrix}$	$\begin{bmatrix}1&0\\1&1\end{bmatrix}$	$\begin{bmatrix}1&0\\0&1\end{bmatrix}$	$\begin{bmatrix}1&1\\1&0\end{bmatrix}$	$\begin{bmatrix}0&1\\1&1\end{bmatrix}$
$\begin{bmatrix}1&1\\0&1\end{bmatrix}$	$\begin{bmatrix}1&1\\0&1\end{bmatrix}$	$\begin{bmatrix}1&0\\1&1\end{bmatrix}$	$\begin{bmatrix}0&1\\1&0\end{bmatrix}$	$\begin{bmatrix}0&1\\1&1\end{bmatrix}$	$\begin{bmatrix}1&0\\0&1\end{bmatrix}$	$\begin{bmatrix}1&1\\1&0\end{bmatrix}$
$\begin{bmatrix}1&0\\1&1\end{bmatrix}$	$\begin{bmatrix}1&0\\1&1\end{bmatrix}$	$\begin{bmatrix}0&1\\1&0\end{bmatrix}$	$\begin{bmatrix}1&1\\0&1\end{bmatrix}$	$\begin{bmatrix}1&1\\1&0\end{bmatrix}$	$\begin{bmatrix}0&1\\1&1\end{bmatrix}$	$\begin{bmatrix}1&0\\0&1\end{bmatrix}$

Example 3.3.7 (Quaternion Group)

Let Q be the following set of matrices in $GL_2(\mathbf{C})$:

$$\pm\begin{bmatrix}1&0\\0&1\end{bmatrix}, \quad \pm\begin{bmatrix}i&0\\0&-i\end{bmatrix}, \quad \pm\begin{bmatrix}0&1\\-1&0\end{bmatrix}, \quad \pm\begin{bmatrix}0&i\\i&0\end{bmatrix}.$$

If we let

$$1=\begin{bmatrix}1&0\\0&1\end{bmatrix}, \quad \mathbf{i}=\begin{bmatrix}i&0\\0&-i\end{bmatrix}, \quad \mathbf{j}=\begin{bmatrix}0&1\\-1&0\end{bmatrix}, \quad \mathbf{k}=\begin{bmatrix}0&i\\i&0\end{bmatrix},$$

then computations show that we have the following identities:

$$\mathbf{i}^2=\mathbf{j}^2=\mathbf{k}^2=-1;$$

$$\mathbf{ij}=\mathbf{k}, \quad \mathbf{jk}=\mathbf{i}, \quad \mathbf{ki}=\mathbf{j}; \qquad \mathbf{ji}=-\mathbf{k}, \quad \mathbf{kj}=-\mathbf{i}, \quad \mathbf{ik}=-\mathbf{j},$$

which you may recall from your calculus course. Since they show that the set is closed under matrix multiplication, we have defined a subgroup of $GL_2(\mathbf{C})$. We can make a few observations: Q is not abelian and is not cyclic. In fact, -1 has order 2, while $\pm\mathbf{i}$, $\pm\mathbf{j}$, and $\pm\mathbf{k}$ have order 4.   □

EXERCISES: SECTION 3.3

1. Find HK in $\mathbf{Z}_{16}^{\times}$, if $H = \langle[3]\rangle$ and $K = \langle[5]\rangle$.

2. Find HK in $\mathbf{Z}_{21}^{\times}$, if $H = \{[1],[3]\}$ and $K = \{[1],[4],[10],[13],[16],[19]\}$.

3. Find the cyclic subgroup generated by $\begin{bmatrix} 2 & 1 \\ 0 & 2 \end{bmatrix}$ in $GL_2(\mathbf{Z}_3)$.

4. Find an example of two subgroups H and K of S_3 for which HK is not a subgroup.

5. Prove that if G_1 and G_2 are abelian groups, then the direct product $G_1 \times G_2$ is abelian.

6. Construct an abelian group of order 12 that is not cyclic.

7. Construct a group of order 12 that is not abelian.

8. Let G_1 and G_2 be groups, with subgroups H_1 and H_2, respectively. Show that $\{(g_1, g_2) \mid g_1 \in H_1, g_2 \in H_2\}$ is a subgroup of the direct product $G_1 \times G_2$.

9. Let G_1 and G_2 be groups, and let G be the direct product $G_1 \times G_2$. Let $H = \{(a,b) \in G_1 \times G_2 \mid b = e\}$ and let $K = \{(a,b) \in G_1 \times G_2 \mid a = e\}$.
   (a) Show that H and K are subgroups of G.
   (b) Show that $HK = KH = G$.
   (c) Show that $H \cap K = \{(e,e)\}$.

10. (a) Generalize Definition 3.3.3 to the case of the direct product of n groups.
    (b) Generalize Proposition 3.3.4 to the case of the direct product of n groups.

11. Let p, q be distinct prime numbers, and let $n = pq$. In $\mathbf{Z}_n^\times$, let $H = \{[a] \mid a \equiv 1 \pmod{p}\}$ and $K = \{[b] \mid b \equiv 1 \pmod{q}\}$. Show that $HK = \mathbf{Z}_n^\times$.
    Hint: Show that HK has $\varphi(n)$ elements.

12. Let G be a group of order 6. It must contain an element of order 2, since the order of G is even. Show that it cannot be true that every element different from e has order 2.
    Hint: Show that if every element had order 2, it would be possible to construct a subgroup of order 4.

13. Let G be a group of order 6, and suppose that a, $b \in G$ with a of order 3 and b of order 2. Show that either G is cyclic or $ab \neq ba$.

14. Let G be any group of order 6. Show that if G is not cyclic, then its multiplication table must look like that of S_3.
    Hint: If the group is not cyclic, use the previous exercises to produce elements $a, b \in G$ with $a^3 = e$, $b^2 = e$, and $ba = a^2b$.

3.4 ISOMORPHISMS

In studying groups we are interested in their algebraic properties, and not in the particular form in which they are presented. For example, if we construct the multiplication tables for two finite groups and find that they have the same patterns, although the elements might have different forms, then we would say that the groups have exactly the same algebraic properties.

Consider the subgroup $\{\pm 1\}$ of $\mathbf{Q}^\times$ and the group $\mathbf{Z}_2$. If you write out the group tables for these groups you will find precisely the same pattern, as shown in Tables 3.4.1 and 3.4.2.

Actually, if we have any group G with two elements, say the identity element e and one other element a, then there is only one possibility for the multiplication table for G. We have already observed that Propositions 3.1.6 and 3.1.7

TABLE 3.4.1.	Multiplication in $\{\pm 1\}$		TABLE 3.4.2.	Addition in $\mathbf{Z}_2$		TABLE 3.4.3.		
	1	-1		[0]	[1]		e	a
1	1	-1	[0]	[0]	[1]	e	e	a
-1	-1	1	[1]	[1]	[0]	a	a	e

imply that in each row and column of a group table, each element of the group must occur exactly once. Since e is the identity element, $e \cdot e = e$, $e \cdot a = a$, and $a \cdot e = a$. Since a cannot be repeated in the first row of the table, we must have $a \cdot a = e$. This gives us Table 3.4.3, and shows that all groups with two elements must have exactly the same algebraic properties.

We need a formal definition to describe when two groups have the same algebraic properties. To begin with, there should be a one-to-one correspondence between the elements of the groups. This means in essence that elements of one group could be renamed to correspond exactly to the elements of the second group. In addition, products of corresponding elements should correspond. If G_1 is a group with operation $*$ and G_2 is a group with operation $\cdot$, then any function $\phi : G_1 \rightarrow G_2$ that preserves products must have the property that $\phi(a * b) = \phi(a) \cdot \phi(b)$ for all $a, b \in G_1$. This expresses in a formula the fact that if we first multiply a and b to get $a * b$ and then find the corresponding element $\phi(a * b)$ of G_2, we should get exactly the same answer as if we find the corresponding elements $\phi(a)$ and $\phi(b)$ in G_2 and then compute their product $\phi(a) \cdot \phi(b)$ in G_2. It is rather cumbersome to write the two operations, so in our definition we will omit them, since it should be clear from the context which operation is to be used. A one-to-one correspondence that preserves products will be called an *isomorphism*. It is derived from the Greek words *isos* meaning "equal" and *morphe* meaning "form."

3.4.1 Definition. Let G_1 and G_2 be groups, and let $\phi : G_1 \rightarrow G_2$ be a function. Then ϕ is said to be a *group isomorphism* if ϕ is one-to-one and onto and

$$\phi(ab) = \phi(a)\phi(b)$$

for all $a, b \in G_1$. In this case, G_1 is said to be *isomorphic* to G_2, and this is denoted by $G_1 \cong G_2$.

There are several immediate consequences of the definition of an isomorphism. Using an induction argument it is possible to show that

$$\phi(a_1 a_2 \cdots a_n) = \phi(a_1)\phi(a_2) \cdots \phi(a_n)$$

for any isomorphism $\phi : G_1 \rightarrow G_2$. Let e_1 and e_2 be the identity elements of G_1 and G_2, respectively. Then

$$\phi(e_1) \cdot \phi(e_1) = \phi(e_1 \cdot e_1) = \phi(e_1) = \phi(e_1) \cdot e_2,$$

and then the cancellation law in G_2 implies that $\phi(e_1) = e_2$. Finally,

$$\phi(a^{-1}) \cdot \phi(a) = \phi(a^{-1} \cdot a) = \phi(e_1) = e_2,$$

and so $(\phi(a))^{-1} = \phi(a^{-1})$. We can summarize by saying that any group isomorphism preserves general products, the identity element, and inverses of elements. From now on we will generally use e to denote the identity element of a group, and we will not distinguish between the identity elements of various groups unless confusion would otherwise result.

We have emphasized that we really do not distinguish (algebraically) between groups that are isomorphic. Thus isomorphism should be almost like equality. In fact, it satisfies properties similar to those of an equivalence relation. There are several things to prove, since our definition of isomorphism of groups involves a function, say $\phi : G_1 \rightarrow G_2$, and this determines a direction to the isomorphism.

The reflexive property holds, since for any group G the identity mapping is an isomorphism, and so $G \cong G$. If $G_1 \cong G_2$, then there is an isomorphism $\phi : G_1 \rightarrow G_2$, and since ϕ is one-to-one and onto, there exists an inverse function ϕ^{-1}, which by part (a) of the next proposition is an isomorphism. Thus $G_2 \cong G_1$, and the symmetric property holds. Finally, the transitive property holds since if $G_1 \cong G_2$ and $G_2 \cong G_3$, then by part (b) of the next proposition the composition of the two given isomorphisms is an isomorphism, showing that $G_1 \cong G_3$.

3.4.2 Proposition.
(a) The inverse of a group isomorphism is a group isomorphism.
(b) The composition of two group isomorphisms is a group isomorphism.

Proof. (a) Let $\phi : G_1 \rightarrow G_2$ be a group isomorphism. Since ϕ is one-to-one and onto, by Proposition 2.1.8 there is an inverse function $\theta : G_2 \rightarrow G_1$ defined as follows. For each element $g_2 \in G_2$ there exists a unique element $g_1 \in G_1$ such that $\phi(g_1) = g_2$. Then we define $\theta(g_2) = g_1$. As a direct consequence of the definition of θ, we have $\theta\phi$ equal to the identity on G_1 and $\phi\theta$ equal to the identity on G_2. The definition also implies that θ is one-to-one and onto. All that remains is to show that θ preserves products. Let $a_2, b_2 \in G_2$, and let $\theta(a_2) = a_1$ and $\theta(b_2) = b_1$. Then $\phi(a_1) = a_2$ and $\phi(b_1) = b_2$, so

$$\phi(a_1 b_1) = \phi(a_1)\phi(b_1) = a_2 b_2,$$

which shows, by the definition of θ, that

$$\theta(a_2 b_2) = a_1 b_1 = \theta(a_2)\theta(b_2).$$

(b) Let $\phi : G_1 \rightarrow G_2$ and $\theta : G_2 \rightarrow G_3$ be group isomorphisms. By Proposition 2.1.6, the composition of two one-to-one and onto functions is again one-to-one and onto, so $\theta\phi$ is one-to-one and onto. If $a, b \in G_1$, then

$$\theta\phi(ab) = \theta(\phi(ab)) = \theta(\phi(a)\phi(b)) = \theta(\phi(a))\theta(\phi(b)) = \theta\phi(a)\theta\phi(b),$$

showing that $\theta\phi$ preserves products. $\square$

Example 3.4.1

Consider the subgroup $\langle i \rangle$ of $\mathbf{C}^{\times}$. In Table 3.4.4, each element is expressed as a power of i.

For the sake of comparison, in Table 3.4.5 we give the addition table for $\mathbf{Z}_4$. The elements of $\mathbf{Z}_4$ appear in precisely the same positions as the exponents of i did in the previous table. This illustrates in a concrete way the intuitive notion that multiplication of powers of i should correspond to addition of the exponents.

To make the isomorphism precise, define a function $\phi : \mathbf{Z}_4 \to \langle i \rangle$ by $\phi([n]) = i^n$. This formula depends on choosing a representative n of its equivalence class, so to show that the function is well-defined we must show that if $n \equiv m \pmod 4$, then $i^n = i^m$. This follows immediately from Proposition 3.2.8, since i has order 4. The function defines a one-to-one correspondence, so all that remains is to show that ϕ preserves the respective operations. We must first add elements $[n]$ and $[m]$ of $\mathbf{Z}_4$ (using the operation in $\mathbf{Z}_4$) and then apply ϕ, and compare this with the element we obtain by first applying ϕ and then multiplying (using the operation in $\langle i \rangle$):

$$\phi([n] + [m]) = \phi([n + m]) = i^{n+m} = i^n i^m = \phi([n])\phi([m]).$$

We conclude that ϕ is a group isomorphism. □

TABLE 3.4.4. Multiplication in $\langle i \rangle$

	i^0	i^1	i^2	i^3
i^0	i^0	i^1	i^2	i^3
i^1	i^1	i^2	i^3	i^0
i^2	i^2	i^3	i^0	i^1
i^3	i^3	i^0	i^1	i^2

Example 3.4.2 (Exponential and Logarithmic Functions)

The groups $\mathbf{R}$ (under addition) and $\mathbf{R}^+$ (under multiplication) are isomorphic. Define $\phi : \mathbf{R} \to \mathbf{R}^+$ by $\phi(x) = e^x$. Then ϕ preserves the respective operations since $\phi(x + y) = e^{x+y} = e^x e^y = \phi(x)\phi(y)$. To show that ϕ is one-to-one, suppose that $\phi(x) = \phi(y)$. Then taking the natural logarithm of each side of the equation $e^x = e^y$ gives $x = y$. Finally, ϕ is an onto mapping since for any $y \in \mathbf{R}^+$ we have $y = e^{\ln(y)} = \phi(\ln(y))$. □

TABLE 3.4.5. Addition in $\mathbf{Z}_4$

	[0]	[1]	[2]	[3]
[0]	[0]	[1]	[2]	[3]
[1]	[1]	[2]	[3]	[0]
[2]	[2]	[3]	[0]	[1]
[3]	[3]	[0]	[1]	[2]

To show that two groups G_1 and G_2 are isomorphic, it is often necessary to actually define the function that gives the isomorphism. In practice, there is usually some natural correspondence between elements which suggests how to define the necessary function.

On the other hand, to show that the groups are not isomorphic, it is not practical to try to check all one-to-one correspondences between the groups to see that none of them preserve products. (Of course, if there is no one-to-one correspondence between the groups, they are not isomorphic. For example, no group with four elements can be isomorphic to a group with five elements.) What we need to do is to find a property of the first group which (i) the second group does not have and (ii) would be preserved by any isomorphism. The next proposition identifies several structural properties that are preserved by group isomorphisms.

3.4.3 Proposition.   Let $\phi : G_1 \rightarrow G_2$ be an isomorphism of groups.
(a) If a has order n in G_1, then $\phi(a)$ has order n in G_2.
(b) If G_1 is abelian, then so is G_2.
(c) If G_1 is cyclic, then so is G_2.

Proof. (a) Suppose that $a \in G_1$ with $a^n = e$. Then we must have $(\phi(a))^n = \phi(a^n) = \phi(e) = e$. This shows that the order of $\phi(a)$ is a divisor of the order of a. Since ϕ is an isomorphism, then there exists an inverse isomorphism that maps $\phi(a)$ to a, and a similar argument shows that the order of a is a divisor of the order of $\phi(a)$. It follows that a and $\phi(a)$ must have the same order.

(b) Assume that G_1 is abelian, and let $a_2, b_2 \in G_2$. Since ϕ is an onto mapping, there exist $a_1, b_1 \in G_1$ with $\phi(a_1) = a_2$ and $\phi(b_1) = b_2$. Then

$$a_2 b_2 = \phi(a_1)\phi(b_1) = \phi(a_1 b_1) = \phi(b_1 a_1) = \phi(b_1)\phi(a_1) = b_2 a_2,$$

showing that G_2 is abelian.

(c) Suppose that G_1 is cyclic, with $G_1 = \langle a \rangle$. For any element $y \in G_2$ we have $y = \phi(x)$ for some $x \in G_1$, since ϕ is onto. Using the assumption that G_1 is cyclic, we can write $x = a^n$ for some $n \in \mathbf{Z}$. Then $y = \phi(a^n) = (\phi(a))^n$, which shows that each element of G_2 can be expressed as a power of $\phi(a)$. Thus G_2 is cyclic, generated by $\phi(a)$. $\square$

Example 3.4.3

The additive group $\mathbf{R}$ of real numbers is not isomorphic to the multiplicative group $\mathbf{R}^{\times}$ of nonzero real numbers. One way to see this is to observe that $\mathbf{R}^{\times}$ has an element of order 2, namely, -1. On the other hand, an element of order 2 in $\mathbf{R}$ must satisfy the equation $2x = 0$ (in additive notation). The only solution is $x = 0$, and so this shows that $\mathbf{R}$ has no element of order 2. Thus there cannot be an isomorphism between the two groups, since by Proposition 3.4.3 it would preserve the orders of all elements.   □

Example 3.4.4

The cyclic group $\mathbf{Z}_4$ and the Klein four-group $\mathbf{Z}_2 \times \mathbf{Z}_2$ are not isomorphic. In $\mathbf{Z}_4$ there is an element of order 4, namely, [1]. On the other hand, the order of an element in a direct product is the least common multiple of the orders of its components, and so any element of $\mathbf{Z}_2 \times \mathbf{Z}_2$ not equal to the identity must have order 2.   □

To motivate some further work with isomorphisms, let us ask the following question. Which of the groups S_3, $GL_2(\mathbf{Z}_2)$, $\mathbf{Z}_6$, and $\mathbf{Z}_2 \times \mathbf{Z}_3$ are isomorphic? The first two groups we know to be nonabelian. On the other hand, any cyclic group is abelian, since powers of a fixed element will commute with each other. The element ([1],[1]) of $\mathbf{Z}_2 \times \mathbf{Z}_3$ has order 6 (the least common multiple of 2 and 3), and so $\mathbf{Z}_2 \times \mathbf{Z}_3$ is cyclic, as well as $\mathbf{Z}_6$. Thus the four groups represent at least two different isomorphism classes. Proposition 3.4.5 will show that $\mathbf{Z}_2 \times \mathbf{Z}_3$ is isomorphic to $\mathbf{Z}_6$, and the next example shows that the two nonabelian groups we are considering are also isomorphic.

Example 3.4.5

Refer to Table 3.3.6 for a multiplication table for $GL_2(\mathbf{Z}_2)$. To establish the connection between S_3 and $GL_2(\mathbf{Z}_2)$, let

$$e = \begin{bmatrix} 1 & 0 \\ 0 & 1 \end{bmatrix}, \qquad a = \begin{bmatrix} 1 & 1 \\ 1 & 0 \end{bmatrix}, \quad \text{and} \quad b = \begin{bmatrix} 0 & 1 \\ 1 & 0 \end{bmatrix}.$$

Then direct computations show that $a^3 = e$, $b^2 = e$, and $ba = a^2b$. Furthermore, each element of $GL_2(\mathbf{Z}_2)$ can be expressed uniquely in one of the following forms: e, a, a^2, b, ab, a^2b. If we make these substitutions in the multiplication table for $GL_2(\mathbf{Z}_2)$, we obtain Table 3.4.6.

In Section 3.3 we described S_3 by letting $a = (1,2,3)$ and $b = (1,2)$, which allowed us to write

$$S_3 = \{e, a, a^2, b, ab, a^2b \mid a^3 = e, b^2 = e, ba = a^2b\}$$

TABLE 3.4.6. Multiplication in $GL_2(\mathbf{Z}_2)$

	e	a	a^2	b	ab	a^2b
e	e	a	a^2	b	ab	a^2b
a	a	a^2	e	ab	a^2b	b
a^2	a^2	e	a	a^2b	b	ab
b	b	a^2b	ab	e	a^2	a
ab	ab	b	a^2b	a	e	a^2
a^2b	a^2b	ab	b	a^2	a	e

without using permutations. This indicates how to define an isomorphism from S_3 to $GL_2(\mathbf{Z}_2)$. Let

$$\phi((1,2,3)) = \begin{bmatrix} 1 & 1 \\ 1 & 0 \end{bmatrix} \quad \text{and} \quad \phi((1,2)) = \begin{bmatrix} 0 & 1 \\ 1 & 0 \end{bmatrix}$$

and then extend this to all elements by letting

$$\phi((1,2,3)^i(1,2)^j) = \begin{bmatrix} 1 & 1 \\ 1 & 0 \end{bmatrix}^i \begin{bmatrix} 0 & 1 \\ 1 & 0 \end{bmatrix}^j$$

for $i = 0, 1, 2$ and $j = 0, 1$. Our remarks about the unique forms of the respective elements show that ϕ is a one-to-one correspondence. The fact that the multiplication tables are identical shows that ϕ respects the two operations. This verifies that ϕ is an isomorphism.   □

Example 3.4.6

Using the idea of the previous example, to show that $\mathbf{Z}_6$ and $\mathbf{Z}_2 \times \mathbf{Z}_3$ are isomorphic, we can look for elements that can be used to describe each group. Since we have already observed that both groups are cyclic, we can let a be a generator for $\mathbf{Z}_6$ and b be a generator for $\mathbf{Z}_2 \times \mathbf{Z}_3$. Then the function $\phi(na) = nb$ can be shown to define an isomorphism. (Remember that we are using additive notation.)   □

The next proposition gives an easier way to check that a function which preserves products is one-to-one. In additive notation, it depends on the fact that for any mapping which preserves sums we have $\phi(x_1) = \phi(x_2)$ if and only if $\phi(x_1 - x_2) = 0$. Any vector space is an abelian group under vector addition, and any linear transformation preserves sums. Thus the result that a linear transformation is one-to-one if and only if its null space is trivial is a special case of our next proposition.

3.4.4 Proposition.   Let G_1 and G_2 be groups, and let $\phi : G_1 \to G_2$ be a function such that $\phi(ab) = \phi(a)\phi(b)$ for all $a, b \in G_1$. Then ϕ is one-to-one if and only if $\phi(x) = e$ implies $x = e$, for all $x \in G_1$.

Proof. Let $\phi : G_1 \to G_2$ satisfy the hypothesis of the proposition. If ϕ is one-to-one, then the only element that can map to the identity of G_2 is the identity of G_1. On the other hand, suppose that $\phi(x) = e$ implies $x = e$, for all $x \in G_1$. If $\phi(x_1) = \phi(x_2)$ for some $x_1, x_2 \in G_1$, then multiplying both sides of this equation by $(\phi(x_2))^{-1}$ gives us $\phi(x_1 x_2^{-1}) = \phi(x_1)(\phi(x_2))^{-1} = e$, which shows by assumption that $x_1 x_2^{-1} = e$, and thus $x_1 = x_2$. This shows that ϕ is one-to-one.   □

3.4.5 Proposition.   If m, n are positive integers such that $\gcd(m,n) = 1$, then $\mathbf{Z}_m \times \mathbf{Z}_n$ is isomorphic to $\mathbf{Z}_{mn}$.

Proof. Define $\phi : \mathbf{Z}_{mn} \to \mathbf{Z}_m \times \mathbf{Z}_n$ by $\phi([x]_{mn}) = ([x]_m, [x]_n)$. If $a \equiv b$ (mod mn), then $a \equiv b$ (mod m) and $a \equiv b$ (mod n), and so ϕ is well-defined. It is easy to check that ϕ preserves sums. To show that ϕ is one-to-one we can use the previous proposition. If $\phi([x]_{mn}) = ([0]_m, [0]_n)$, then both m and n must be divisors of x. Since gcd $(m,n) = 1$, it follows that mn must be a divisor of x, which shows that $[x]_{mn} = [0]_{mn}$. Since the two groups have the same number of elements, any one-to-one mapping must be onto, and thus ϕ is an isomorphism. $\square$

In most of our exercises, to show that two groups G_1 and G_2 are isomorphic, you must define a one-to-one function from one group to the other. Sometimes it is easier to define the function in one direction than in the other, so when you are working on a problem, it may be worth checking both ways. The inverse of the function we used in proving Proposition 3.4.5 is also interesting. Given $\phi : \mathbf{Z}_{mn} \to \mathbf{Z}_m \times \mathbf{Z}_n$ defined by $\phi([x]_{mn}) = ([x]_m, [x]_n)$, the inverse must assign to each $([a]_m, [b]_n) \in \mathbf{Z}_m \times \mathbf{Z}_n$ an element $[x]_{mn} \in \mathbf{Z}_{mn}$ such that $x \equiv a$ (mod m) and $x \equiv b$ (mod n). We could have applied the Chinese remainder theorem (Theorem 1.3.6) to define our function in this direction, but the other way seemed more natural.

EXERCISES: SECTION 3.4

1. Show that the multiplicative group $\mathbf{Z}_{10}^\times$ is isomorphic to the additive group $\mathbf{Z}_4$.
   Hint: Find a generator $[a]_{10}$ of $\mathbf{Z}_{10}^\times$ and define $\phi : \mathbf{Z}_4 \to \mathbf{Z}_{10}^\times$ by $\phi([n]_4) = [a]_{10}^n$.

2. Show that the multiplicative group $\mathbf{Z}_7^\times$ is isomorphic to the additive group $\mathbf{Z}_6$.

3. Show that the multiplicative group $\mathbf{Z}_8^\times$ is isomorphic to the group $\mathbf{Z}_2 \times \mathbf{Z}_2$.

4. Show that $\mathbf{Z}_5^\times$ is not isomorphic to $\mathbf{Z}_8^\times$ by showing that the first group has an element of order 4 but the second group does not.

5. Is the additive group $\mathbf{C}$ of complex numbers isomorphic to the multiplicative group $\mathbf{C}^\times$ of nonzero complex numbers?

6. Prove that any group with three elements must be isomorphic to $\mathbf{Z}_3$.

7. Find two abelian groups of order 8 that are not isomorphic.

8. Show that the group $\{f : \mathbf{R} \to \mathbf{R} \mid f(x) = mx + b, m \neq 0\}$ of affine functions from $\mathbf{R}$ to $\mathbf{R}$ is isomorphic to the group of all 2×2 matrices over $\mathbf{R}$ of the form $\begin{bmatrix} m & b \\ 0 & 1 \end{bmatrix}$ with $m \neq 0$. (See Exercises 8 and 9 of Section 3.1.)

9. Let G be the following set of matrices over $\mathbf{R}$:
$$\begin{bmatrix} 1 & 0 \\ 0 & 1 \end{bmatrix}, \quad \begin{bmatrix} -1 & 0 \\ 0 & 1 \end{bmatrix}, \quad \begin{bmatrix} 1 & 0 \\ 0 & -1 \end{bmatrix}, \quad \begin{bmatrix} -1 & 0 \\ 0 & -1 \end{bmatrix}.$$
   Show that G is isomorphic to $\mathbf{Z}_2 \times \mathbf{Z}_2$. (See Example 3.2.4.)

10. Let $G = \mathbf{R} - \{-1\}$. Define $*$ on G by $a * b = a + b + ab$. Show that G is isomorphic to the multiplicative group $\mathbf{R}^\times$. (See Exercise 13 in Section 3.1.)

Hint: Remember that an isomorphism maps identity to identity. Use this fact to help find the necessary mapping.

11. Let G be any group, and let a be a fixed element of G. Define a function $\phi_a : G \to G$ by $\phi_a(x) = axa^{-1}$, for all $x \in G$. Show that ϕ_a is an isomorphism.

12. Let G be any group. Define $\phi : G \to G$ by $\phi(x) = x^{-1}$, for all $x \in G$.
   (a) Prove that ϕ is one-to-one and onto.
   (b) Prove that ϕ is an isomorphism if and only if G is abelian.

13. Let $\phi : G_1 \to G_2$ be a group isomorphism. Prove that if H is a subgroup of G_1, then $\phi(H) = \{\phi(x) \mid x \in H\}$ is a subgroup of G_2.

14. Define $\phi : \mathbf{C}^\times \to \mathbf{C}^\times$ by $\phi(a + bi) = a - bi$, for all nonzero complex numbers $a + bi$. Show that ϕ is an isomorphism.

15. Show that $\mathbf{C}^\times$ is isomorphic to the subgroup $GL_2(\mathbf{R})$ consisting of all matrices of the form $\begin{bmatrix} a & b \\ -b & a \end{bmatrix}$ such that $a^2 + b^2 \neq 0$.

16. Let G_1 and G_2 be groups. Show that G_1 is isomorphic to the subgroup of the direct product $G_1 \times G_2$ defined by $\{(x,y) \mid y = e\}$.

17. Let G_1 and G_2 be groups. A function from G_1 into G_2 that preserves products but is not necessarily a one-to-one correspondence will be called a *group homomorphism,* from the Greek word *homos* meaning ''same.'' Show that $\phi : GL_2(\mathbf{R}) \to \mathbf{R}^\times$ defined by $\phi(A) = \det(A)$ for all matrices $A \in GL_2(\mathbf{R})$ is a group homomorphism.

18. Using the definition in the previous problem, let $\phi : G_1 \to G_2$ be a group homomorphism. We define the *kernel* of ϕ to be

$$\ker(\phi) = \{x \in G_1 \mid \phi(x) = e\}.$$

Prove that $\ker(\phi)$ is a subgroup of G_1.

3.5 CYCLIC GROUPS

The class of cyclic groups will turn out to play a crucial role in studying the solution of equations by radicals. Yet this class can be characterized very simply, since we will show that a cyclic group must be isomorphic either to $\mathbf{Z}$ or $\mathbf{Z}_n$ for some n. This allows us to apply some elementary number theory to describe all subgroups of a cyclic group and to find all possible generators.

3.5.1 Theorem.   Every subgroup of a cyclic group is cyclic.

Proof. Let G be a cyclic group with generator a, so that $G = \langle a \rangle$, and let H be any subgroup of G. If H is the trival subgroup consisting only of e, then we are done since $H = \langle e \rangle$. If H is nontrivial, then it contains some element different from the identity, which can then be written in the form a^n for some integer $n \neq 0$. Since $a^{-n} = (a^n)^{-1}$ must also belong to H, we can assume that H contains some power a^k with $k > 0$.

Let m be the smallest positive integer such that $a^m \in H$. We claim that $H = \langle a^m \rangle$. Since $a^m \in H$, we have $\langle a^m \rangle \subseteq H$, and so the main point is to show that each element of H can be expressed as some power of a^m. Let $x \in H$. Then since $G = \langle a \rangle$, we have $x = a^k$ for some $k \in \mathbf{Z}$. By the division algorithm, $k = mq + r$ for $q, r \in \mathbf{Z}$ with $0 \leq r < m$. Then $x = a^k = a^{mq+r} = (a^m)^q a^r$. This shows that $a^r = (a^m)^{-q}x$ belongs to H (since a^m and x belong to H). This contradicts the definition of a^m as the smallest positive power of a in H unless $r = 0$. Therefore $k = mq$ and $x = (a^m)^q \in \langle a^m \rangle$. We conclude that $H = \langle a^m \rangle$ and so H is cyclic. $\square$

In our current terminology, Theorem 1.1.4 shows that every subgroup of $\mathbf{Z}$ is cyclic. This result can actually be used to give a very short proof of Theorem 3.5.1. Let $G = \langle a \rangle$ and let H be a subgroup of G. Let $I = \{n \in \mathbf{Z} \mid a^n \in H\}$. It follows from the rules for exponents that I is closed under addition and subtraction, so Theorem 1.1.4 implies that $I = m\mathbf{Z}$ for some integer m. We conclude that $H = \langle a^m \rangle$, and so H is cyclic.

The next theorem shows that any cyclic group is isomorphic either to $\mathbf{Z}$ or to $\mathbf{Z}_n$. Thus any two infinite cyclic groups are isomorphic to each other. Furthermore, two finite cyclic groups are isomorphic if and only if they have the same order.

3.5.2 Theorem. Let G be a cyclic group.
(a) If G is infinite, then $G \cong \mathbf{Z}$.
(b) If $|G| = n$, then $G \cong \mathbf{Z}_n$.

Proof. (a) Let $G = \langle a \rangle$ be an infinite cyclic group. Define $\phi : \mathbf{Z} \to G$ by $\phi(m) = a^m$, for all $m \in \mathbf{Z}$. The mapping ϕ is onto since $G = \langle a \rangle$, and Proposition 3.2.8 (a) shows that $\phi(m) \neq \phi(k)$ for $m \neq k$, so ϕ is also one-to-one. Finally, ϕ preserves the respective operations since

$$\phi(m + k) = a^{m+k} = a^m a^k = \phi(m)\phi(k).$$

This shows that ϕ is an isomorphism.

(b) Let $G = \langle a \rangle$ be a finite cyclic group with n elements. Define $\phi : \mathbf{Z}_n \to G$ by $\phi([m]) = a^m$, for all $[m] \in \mathbf{Z}_n$. In order to show that ϕ is a function, we must check that the formula we have given is well-defined. That is, we must show that if $k \equiv m \pmod{n}$, then $a^k = a^m$. This follows from Proposition 3.2.8 (c). Furthermore, if $\phi([k]) = \phi([m])$, then the same proposition shows that $[k] = [m]$, and so ϕ is one-to-one. It is clear that ϕ is onto, since $G = \langle a \rangle$. Finally, ϕ preserves the respective operations since

$$\phi([m] + [k]) = a^{m+k} = a^m a^k = \phi([m])\phi([k]).$$

This shows that ϕ is an isomorphism. $\square$

The subgroups of $\mathbf{Z}$ have the form $m\mathbf{Z}$, for $m \in \mathbf{Z}$. In addition, $m\mathbf{Z} \subseteq n\mathbf{Z}$ if and only if $n \mid m$. Thus $m\mathbf{Z} = n\mathbf{Z}$ if and only if $m = \pm n$.

The subgroups of $\mathbf{Z}_n$ take more work to describe. Given $m \in \mathbf{Z}$, we wish to find the multiples of $[m]$ in $\mathbf{Z}_n$. That is, we need to determine the integers b such that $[b] = k[m]$ for some $k \in \mathbf{Z}$. Equivalently, we need to know when $mx \equiv b$ (mod n) has a solution. By Theorem 1.3.5, the values of b are precisely the multiples of $\gcd(m,n)$.

In the next proposition, we have chosen to describe the subgroups of a cyclic group with n elements using multiplicative notation. We also give a proof appropriate to this context, which is independent of earlier results from Chapter 1.

3.5.3 Proposition.   Let $G = \langle a \rangle$ be a cyclic group with $|G| = n$.
(a) If $m \in \mathbf{Z}$, then $\langle a^m \rangle = \langle a^d \rangle$, where $d = \gcd(m,n)$, and a^m has order n/d.
(b) The element a^k generates G if and only if $\gcd(k,n) = 1$.
(c) The subgroups of G are in one-to-one correspondence with the positive divisors of n.

Proof. Since $d \,|\, m$, we have $a^m \in \langle a^d \rangle$, and so $\langle a^m \rangle \subseteq \langle a^d \rangle$. On the other hand, there exist integers s, t such that $d = sm + tn$, and so

$$a^d = a^{sm+tn} = (a^m)^s(a^n)^t = (a^m)^s$$

since $a^n = e$. Thus $a^d \in \langle a^m \rangle$, and so $\langle a^d \rangle \subseteq \langle a^m \rangle$. The order of a^d is n/d, and so a^m has order n/d. This proves part (a), and the remaining statements follow immediately. $\square$

We will use the notation $m\mathbf{Z}_n$ for the subgroup $\langle [m] \rangle$ consisting of all multiples of $[m]$. If m and k are divisors of n, then we have $m\mathbf{Z}_n \subseteq k\mathbf{Z}_n$ if and only if $k \,|\, m$. For small values of n, we can easily give a diagram showing all subgroups of $\mathbf{Z}_n$ and the inclusion relations between them. This is called a *lattice diagram* for the subgroups. Since subgroups correspond to divisors of n and inclusions are the opposite of divisibility relations, we can find the lattice diagram of divisors of n and simply turn it upside down.

Example 3.5.1

In Example 1.2.2 we gave the lattice diagram of all divisors of 12. This leads to the lattice diagram given in Figure 3.5.1. $\square$

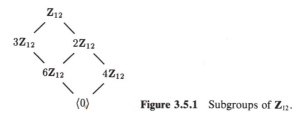

Figure 3.5.1 Subgroups of $\mathbf{Z}_{12}$.

Example 3.5.2

If n is a prime power, then the lattice of subgroups of $\mathbf{Z}_n$ is particularly simple, since the subgroups can be linearly ordered. We give the lattice diagram for $\mathbf{Z}_{125}$ in Figure 3.5.2. $\square$

$$\mathbf{Z}_{125}$$
$$|$$
$$5\mathbf{Z}_{125}$$
$$|$$
$$25\mathbf{Z}_{125}$$
$$|$$
$$\langle 0 \rangle$$

Figure 3.5.2 Subgroups of $\mathbf{Z}_{125}$.

In Definition 3.3.3 we introduced the direct product of two groups. This definition can easily be extended to the direct product $G_1 \times \cdots \times G_n$ of n groups $G_1, \ldots, G_n$ by considering n-tuples in which the ith entry is an element of G_i, with componentwise multiplication. As with the direct product of two groups, the order of an element is the least common multiple of the orders of each component.

The following proposition implies that every finite cyclic group is isomorphic to a direct product of cyclic groups of prime power order. We could call this a structure theorem for finite cyclic groups, in the sense that we can show how they are built up from combinations of simpler cyclic groups. This is a special case of the general structure theorem for finite abelian groups, proved in Section 7.5, which states that any finite abelian group is isomorphic to a direct product of cyclic groups of prime power order.

3.5.4 Theorem. Let n be a positive integer with prime decomposition $n = p_1^{\alpha_1} p_2^{\alpha_2} \cdots p_m^{\alpha_m}$, where $p_1 < p_2 < \ldots < p_n$. Then

$$\mathbf{Z}_n \cong \mathbf{Z}_{p_1^{\alpha_1}} \times \mathbf{Z}_{p_2^{\alpha_2}} \times \cdots \times \mathbf{Z}_{p_m^{\alpha_m}}.$$

Proof. In the direct product of the given groups, the element with [1] in each component has order n, since the least common multiple of the given prime powers is n. Thus the direct product is cyclic of order n, so by Theorem 3.5.2, it must be isomorphic to $\mathbf{Z}_n$. $\quad\square$

For a positive integer n, the Euler φ-function $\varphi(n)$ is defined to be the number of positive integers less than or equal to n and relatively prime to n. Thus $\varphi(n)$ gives the number of elements of $\mathbf{Z}_n$ that are generators of $\mathbf{Z}_n$. In Section 1.4 we gave a formula for $\varphi(n)$, without proof. The proof is an easy consequence of our description of $\mathbf{Z}_n$.

3.5.5 Corollary. Let n be a positive integer with prime decomposition $n = p_1^{\alpha_1} p_2^{\alpha_2} \cdots p_m^{\alpha_m}$, where $p_1 < p_2 < \ldots < p_n$. Then

$$\varphi(n) = n \left(1 - \frac{1}{p_1}\right)\left(1 - \frac{1}{p_2}\right) \cdots \left(1 - \frac{1}{p_m}\right).$$

Proof. To count the generators of $\mathbf{Z}_n$, it is easier to use the isomorphic direct product

$$\mathbf{Z}_{p_1^{\alpha_1}} \times \mathbf{Z}_{p_2^{\alpha_2}} \times \cdots \times \mathbf{Z}_{p_m^{\alpha_m}} \cong \mathbf{Z}_n$$

obtained in Theorem 3.5.4 than it is to use $\mathbf{Z}_n$, since an isomorphism preserves generators. An element of this direct product is a generator if and only if it has order n, and so this means that the least common multiple of the orders of its components in their respective groups must be n. An element of the direct product has order $p_1^{\beta_1} p_2^{\beta_2} \cdots p_m^{\beta_m}$ with $\beta_i \leq \alpha_i$ for each i. For this order to be equal to n we must have $\beta_i = \alpha_i$ for each i. It follows that an element of the direct product is a generator if and only if each of its components is a generator in its respective cyclic group of prime power order. Thus the total number of possible generators is equal to the product of the number of generators in each component.

We have reduced the problem to counting the number of generators in $\mathbf{Z}_{p^\alpha}$, for any prime p. The elements that are *not* generators are the multiples of p, and among the p^α elements of $\mathbf{Z}_{p^\alpha}$ there are $p^{\alpha-1}$ such multiples. Thus

$$\varphi(p^\alpha) = p^\alpha - p^{\alpha-1} = p^\alpha \left(1 - \frac{1}{p} \right).$$

Taking the product of these values for each of the primes in the decomposition of n gives the formula we want. $\square$

If G is a finite group, then as we noted following the definition of order, each element of G must have finite order. Thus for each $a \in G$, we have $a^{o(a)} = e$. If N is the least common multiple of the integers $o(a)$, for all $a \in G$, then $a^N = e$ for all $a \in G$. Since $o(a)$ is a divisor of $|G|$ for any $a \in G$, it follows that N is a divisor of $|G|$. Using this concept, we are able to characterize cyclic groups among all finite abelian groups.

3.5.6 Definition.   Let G be a group. If there exists a positive integer N such that $a^N = e$ for all $a \in G$, then the smallest such positive integer is called the *exponent* of G.

Example 3.5.3

The exponent of any finite group is the least common multiple of the orders of its elements. Thus the exponent of S_3 is 6, since S_3 has elements of order 1, 2, and 3. The exponent of $\mathbf{Z}_2 \times \mathbf{Z}_2$ is 2. Since we are using additive notation, remember to use multiples instead of powers. The exponent of $\mathbf{Z}_2 \times \mathbf{Z}_3$ is 6, since there are elements of order 1, 2, 3, and 6. $\square$

3.5.7 Lemma.   Let G be a group, and let $a, b \in G$ be elements such that $ab = ba$. If the orders of a and b are relatively prime, then $o(ab) = o(a)o(b)$.

Proof. Let $o(a) = n$ and $o(b) = m$. Then since $ab = ba$, we must have $(ab)^{mn} = (a^n)^m (b^m)^n = e$, which shows that ab has finite order, say $o(ab) = k$. Furthermore, $(ab)^{mn} = e$ implies that $k \mid mn$. On the other hand, $(ab)^k = e$, which shows that $a^k = b^{-k}$. Therefore $a^{km} = (a^k)^m = (b^{-k})^m = (b^m)^{-k} = e$, showing that $n \mid km$. Since $(n,m) = 1$, we must have $n \mid k$. A similar argument shows that $m \mid k$, and then $mn|k$ since $(n,m) = 1$. Since m, n, k are positive integers with $mn|k$ and $k|mn$, we have $k = mn$.   □

   3.5.8 Proposition.   Let G be a finite abelian group.
 (a) The exponent of G is equal to the order of any element of G of maximal order.
 (b) The group G is cyclic if and only if its exponent is equal to its order.

   Proof. (a) Let a be an element of G that has maximal order. Let $b \in G$ and suppose that $o(b)$ is not a divisor of $o(a)$. Then in the prime factorizations of $o(a)$ and $o(b)$, there exists a prime p with $o(a) = p^{\alpha}n$, $o(b) = p^{\beta}m$, where p is relatively prime to both n and m, and $\beta > \alpha \geq 0$. Then $o(a^{p^{\alpha}}) = n$ and $o(b^{m}) = p^{\beta}$, so these orders are relatively prime. It follows from Lemma 3.5.7 that $o(a^{p^{\alpha}}b^{m}) = np^{\beta}$, and this is greater than $o(a)$, a contradiction. Thus $o(b)|o(a)$ for all $b \in G$, and $o(a)$ is therefore the exponent of G.
   (b) Part (b) follows immediately from part (a), since G is cyclic if and only if there exists an element of order $|G|$.   □

EXERCISES: SECTION 3.5

 1. Let G be a group and let $a \in G$ be an element of order 12. What is the order of a^{j} for $j = 2, \ldots, 11$?
 2. Let G be a group and let $a \in G$ be an element of order 30. List the powers of a that have order 2, order 3, or order 5.
 3. Give the lattice diagram of subgroups of $\mathbf{Z}_{24}$ and $\mathbf{Z}_{36}$.
 4. Give the lattice diagram of subgroups of $\mathbf{Z}_{60}$.
 5. Find the cyclic subgroup of $\mathbf{C}^{\times}$ generated by $1/\sqrt{2} + i/\sqrt{2}$.
 6. Find the cyclic subgroup of $\mathbf{C}^{\times}$ generated by $1 + i$.
 7. Find $\langle \pi \rangle$ in $\mathbf{R}^{\times}$.
 8. Which of $\mathbf{Z}_{15}^{\times}$, $\mathbf{Z}_{18}^{\times}$, $\mathbf{Z}_{20}^{\times}$, $\mathbf{Z}_{27}^{\times}$ are cyclic?
 9. Which of $\mathbf{Z}_{7}^{\times}$, $\mathbf{Z}_{10}^{\times}$, $\mathbf{Z}_{12}^{\times}$, $\mathbf{Z}_{14}^{\times}$ are isomorphic?
 10. Find all cyclic subgroups of $\mathbf{Z}_{4} \times \mathbf{Z}_{2}$.
 11. Find all cyclic subgroups of $\mathbf{Z}_{6} \times \mathbf{Z}_{3}$.
 12. Show that in a finite cyclic group of order n, the equation $x^{m} = e$ has exactly m solutions, for each positive integer m that is a divisor of n.
 13. Prove that any cyclic group with more than two elements has at least two different generators.
 14. Prove that any finite cyclic group with more than two elements has an even number of distinct generators.
 15. Let G be any group with no proper, nontrivial subgroups, and assume that $|G| > 1$. Prove that G must be isomorphic to $\mathbf{Z}_{p}$ for some prime p.

16. Let G be the set of all 3×3 matrices of the form

$$\begin{bmatrix} 1 & a & b \\ 0 & 1 & c \\ 0 & 0 & 1 \end{bmatrix}.$$

   (a) Show that if $a, b, c \in \mathbf{Z}_3$, then G is a group with exponent 3.
   (b) Show that if $a, b, c \in \mathbf{Z}_2$, then G is a group with exponent 4.

17. Prove that $\Sigma_{d|n} \varphi(d) = n$ for any positive integer n.
   Hint: Interpret the equation in the cyclic group $\mathbf{Z}_n$, by considering all of its subgroups.

18. Let $n = 2^k$ for $k > 2$. Prove that $\mathbf{Z}_n^{\times}$ is not cyclic.
   Hint: Show that ± 1 and $(n/2) \pm 1$ satisfy the equation $x^2 = 1$, and that this is impossible in any cyclic group.

19. Let G be a group with p^k elements, where p is a prime number. Prove that G has a subgroup of order p.

3.6 PERMUTATION GROUPS

When groups were first studied, they were thought of as sets of permutations closed under products and including the identity, together with inverses of all elements. The abstract definition that we now use was not given until later. The content of Cayley's theorem, which we are about to prove, is the surprising result that this abstract definition is not any more general than the original concrete definition.

In the case of a finite group, a little thought about the group multiplication table may convince the reader that the theorem is not so surprising after all. As we have observed, each row in the multiplication table represents a permutation of the group elements. Furthermore, each row corresponds to multiplication by a given element, and so there is a natural way to assign a permutation to each element of the group.

3.6.1 Definition.   Any subgroup of the symmetric group Sym(S) on a set S is called a *permutation group*.

In the following proof of Cayley's theorem, we must show that any group G is isomorphic to a subgroup of Sym(S) for some set S, so the first problem is to find an appropriate set S. Our choice is to let S be G itself. Next we must assign to each element a of G some permutation of G. The natural one is the function $\lambda_a : G \to G$ defined by $\lambda_a(x) = ax$ for all $x \in G$. (We use the notation λ_a to indicate multiplication on the left by a.) The values $\lambda_a(x)$ are the entries in the group table that occur in the row corresponding to multiplication by a, and this makes λ_a a permutation of G. Finally we must show that assigning λ_a to a respects the two operations and gives a one-to-one correspondence.

3.6.2 Theorem (Cayley).   Every group is isomorphic to a permutation group.

Proof. Let G be any group. Given $a \in G$, define $\lambda_a : G \to G$ by $\lambda_a(x) = ax$, for all $x \in G$. Then λ_a is onto since the equation $ax = b$ has a solution for each $b \in G$, and it is one-to-one since the solution is unique, so we conclude that λ_a is a permutation of G. This shows that the function $\phi : G \to \mathrm{Sym}(G)$ defined by $\phi(a) = \lambda_a$ is well-defined.

We next want to show that $G_\lambda = \phi(G)$ is a subgroup of $\mathrm{Sym}(G)$, and to do so we need several facts. The formula $\lambda_a \lambda_b = \lambda_{ab}$ holds since for all $x \in G$ we have $\lambda_a(\lambda_b(x)) = a(bx) = (ab)x = \lambda_{ab}(x)$. Because λ_e is the identity function, this formula also implies that $(\lambda_a)^{-1} = \lambda_{a^{-1}}$. This shows that G_λ is closed, contains the identity, and contains inverses for its elements, so it is a subgroup.

To show that ϕ preserves products, we must show that $\phi(ab) = \phi(a)\phi(b)$. This follows from the formula $\lambda_{ab} = \lambda_a \lambda_b$. To complete the proof that $\phi : G \to G_\lambda$ is an isomorphism, it is only necessary to show that ϕ is one-to-one, since it is onto by the definition of G_λ. If $\phi(a) = \phi(b)$ for $a, b \in G$, then we have $\lambda_a(x) = \lambda_b(x)$ for all $x \in G$. In particular, $ae = \lambda_a(e) = \lambda_b(e) = be$, and so $a = b$.

In summary, we have found a subgroup G_λ of $\mathrm{Sym}(G)$ and an isomorphism $\phi : G \to G_\lambda$ defined by assigning to each $a \in G$ the permutation λ_a.   □

In this chapter we have assumed as a matter of course that all permutations in S_n are expressed in the natural way as a product of disjoint cycles. A formal proof that this can be done is given in Section 2.2. Let $\sigma \in S_n$. Recall that if σ is written as a product of disjoint cycles, then the order of σ is the least common multiple of the lengths of its cycles.

Example 3.6.1

Groups of symmetries are very useful in geometry. We now look at the group of rigid motions of a square. (See Figure 3.6.1.) Imagine a square of cardboard, placed in a box just large enough to contain it. Picking up the square and then replacing it in the box, in what may be a new position, gives what is called a *rigid motion* of the square. Each of the rigid motions determines a permutation of the vertices of the square, and the permutation notation gives a convenient way to describe these motions. To count the number of rigid motions, fix a vertex and label it A. Label one of the adjacent vertices B. We have a total of eight rigid motions, since we have four choices of a position in which to place vertex A, and then two choices for vertex B because it must be adjacent to A.

We have used $(1,2,3,4)$ to describe the rigid motion in which the corner of the square currently occupying position 1 is placed in position 2, while the corner currently in position 2 is moved to position 3, the one in position 3 is moved to position 4, and the one in position 4 is moved to position 1. In the rigid motion $(2,4)$ the square is replaced so that the corners originally in positions 2 and 4 are interchanged, while the corners originally in positions 1

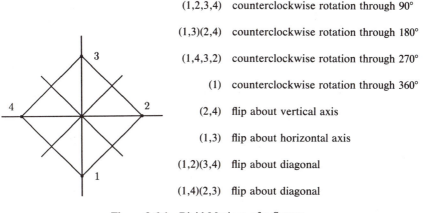

(1,2,3,4) counterclockwise rotation through 90°

(1,3)(2,4) counterclockwise rotation through 180°

(1,4,3,2) counterclockwise rotation through 270°

(1) counterclockwise rotation through 360°

(2,4) flip about vertical axis

(1,3) flip about horizontal axis

(1,2)(3,4) flip about diagonal

(1,4)(2,3) flip about diagonal

Figure 3.6.1 Rigid Motions of a Square.

and 3 remain in the same positions. Note that we do not obtain all elements of S_4 as rigid motions, since, for example, (1,2) would represent an impossible configuration.

It is possible to introduce an operation on the rigid motions, by simply saying that the "product" of two rigid motions will be given by first performing one and then the other. This defines a group, since following one rigid motion by another gives a third rigid motion, the identity is a rigid motion, and any rigid motion can be reversed, providing inverses. In terms of permutations of the vertices of the square, this operation just corresponds to ordinary multiplication of permutations. We give the multiplication table for this subgroup of S_4 in Table 3.6.1. In order to make the table smaller, we have found it necessary to use a more compact notation that omits commas.

In our notation, the motion σ carries the vertex currently in position i to position $\sigma(i)$. Thus we may think of a motion as a function from the set of position numbers into itself. As dictated by our convention for functions, the motion $\sigma\tau$ is the motion obtained by first performing τ and then σ. The

TABLE 3.6.1. Rigid Motions of a Square

	(1)	(1234)	(13)(24)	(1432)	(24)	(12)(34)	(13)	(14)(23)
(1)	(1)	(1234)	(13)(24)	(1432)	(24)	(12)(34)	(13)	(14)(23)
(1234)	(1234)	(13)(24)	(1432)	(1)	(12)(34)	(13)	(14)(23)	(24)
(13)(24)	(13)(24)	(1432)	(1)	(1234)	(13)	(14)(23)	(24)	(12)(34)
(1432)	(1432)	(1)	(1234)	(13)(24)	(14)(23)	(24)	(12)(34)	(13)
(24)	(24)	(14)(23)	(13)	(12)(34)	(1)	(1432)	(13)(24)	(1234)
(12)(34)	(12)(34)	(24)	(14)(23)	(13)	(1234)	(1)	(1432)	(13)(24)
(13)	(13)	(12)(34)	(24)	(14)(23)	(13)(24)	(1234)	(1)	(1432)
(14)(23)	(14)(23)	(13)	(12)(34)	(24)	(1432)	(13)(24)	(1234)	(1)

reader should be warned that our convention of labeling positions is not followed by all authors; some prefer to follow the convention of labeling vertices. □

Example 3.6.2

The rigid motions of an equilateral triangle yield the group S_3. With the vertices labeled as in Figure 3.6.2, the counterclockwise rotations are given by the permutations (1,2,3), (1,3,2), and (1). Flipping the triangle about one of the angle bisectors gives one of the permutations (1,2), (1,3) or (2,3). The multiplication table for S_3 has already been given in Example 3.1.2. □

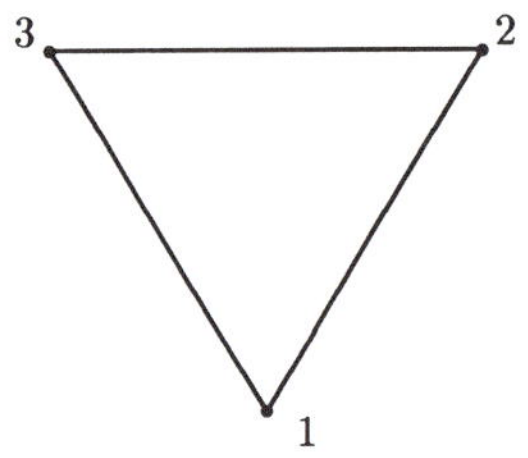

(1,2,3)	counterclockwise rotation through 120°
(1,3,2)	counterclockwise rotation through 240°
(1)	counterclockwise rotation through 360°
(2,3)	flip about vertical axis
(1,3)	flip about angle bisector
(1,2)	flip about angle bisector

Figure 3.6.2   Rigid Motions of an Equilateral Triangle.

Example 3.6.3

In this example we will determine the group of all rigid motions of a regular n-gon. In Section 3.3 we have seen that S_3, the group of rigid motions of an equilateral triangle, can be described using elements a (of order 3) and b (of order 2) which satisfy the identity $ba = a^2b$. The elements of S_3 can then be written (uniquely) as e, a, a^2, b, ab, and a^2b.

In Example 3.6.1, letting $a = (1,2,3,4)$ and $b = (2,4)$, we have elements of order 4 and 2, respectively, which can be shown to satisfy the identity $ba = a^3b$. Furthermore, using these elements the group can then be described as the set $\{e,a,a^2,a^3,b,ab,a^2b,a^3b\}$. The identity $ba = a^3b$ shows us how to multiply two elements in this form and then bring them back to the "standard form."

Now let us consider the general case of the rigid motions of a regular n-gon. Since a rigid motion followed by another rigid motion is again a rigid motion, and since any rigid motion can be reversed, the set of all rigid motions of a regular n-gon forms a group. To see that there are $2n$ rigid motions, fix two adjacent vertices. There are n places to send the first of these vertices, and then there are two choices for the adjacent vertex, giving a total of $2n$ motions. (The relationship between the vertices determines whether or not the n-gon has been flipped over, that is, whether or not the orientation has been reversed.)

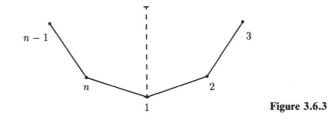

Figure 3.6.3

Figure 3.6.3 represents part of a regular n-gon. Let a be a counter-clockwise rotation about the center, through $360/n$ degrees. Thus a is the cycle $(1,2,3,\ldots,n)$ of length n and has order n. Let b be a flip about the line of symmetry through position number 1. Thus b has order 2 and is given by the product of transpositions $(2,n)(3,n-1)\cdots$.

Consider the set $S = \{a^k, a^k b \mid 0 \le k < n\}$ of rigid motions. It is easy to see that the elements a^k for $0 \le k < n$ are all distinct, and that the elements $a^k b$ for $0 \le k < n$ are also distinct. Since the rigid motion represented by a^k does not change orientation, while the motion represented by $a^j b$ does, it is never the case that $a^k = a^j b$. Thus $|S| = 2n$, and so $G = S$. Since we have listed (uniquely) all the elements of G, it only remains to show how they can be multiplied.

Clearly, $a^n = e$ and $b^2 = e$. After multiplying two elements, to bring the product into one of the standard forms listed above, we only need to know how to move b past a. That is, we must compute ba, and to do so it turns out to be easiest to compute bab. From Figure 3.6.4, it is easy to see that we obtain $bab = a^{-1}$, and then multiplying on the right by $b^{-1} = b$, we obtain the identity $ba = a^{n-1}b$.

Thus, for example, if we want to multiply ab by a^2, we use the identity $ba = a^{n-1}b$ as follows:

$$ab \cdot a^2 = aa^{n-1}ba = a^n ba = ba = a^{n-1}b.$$

We have obtained a complete description of the group of rigid motions of a regular n-gon in terms of elements a and b and the identities they satisfy. $\square$

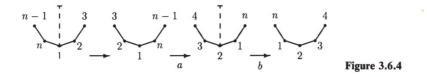

Figure 3.6.4

3.6.3 Definition.   Let $n \ge 3$ be an integer. The group of rigid motions of a regular n-gon is called the nth *dihedral* group, denoted by D_n.

Example 3.6.4

In Figure 3.6.5 we give the lattice of subgroups of S_3, using the notation of Example 3.6.3. By Lagrange's theorem, the only possible orders of proper subgroups are 1, 2, or 3. Since subgroups of order 2 or 3 must be cyclic, it is relatively simple to find all subgroups. □

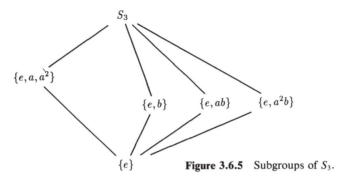

Figure 3.6.5 Subgroups of S_3.

Example 3.6.5

In Figure 3.6.6 we will give the lattice of all subgroups of D_4, again using the notation of Example 3.6.3. The possible orders of proper subgroups are 1, 2, or 4. We first find all cyclic subgroups: a has order 4, while each of the elements a^2, b, ab, a^2b, a^3b has order 2. Any subgroup of order 4 that is not cyclic must be isomorphic to the Klein four-group, so it must contain two elements of order 2 and their product. By considering all possible pairs of elements of order 2 it is possible to find the remaining two subgroups of order 4. Just as a cyclic subgroup is the smallest subgroup containing the generator, these subgroups are the smallest ones containing the two elements used to construct it. In general, to find all subgroups, one would need to consider all possible combinations of elements, a difficult task in a large group. □

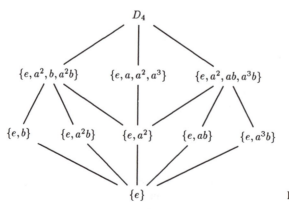

Figure 3.6.6 Subgroups of D_4.

In Section 2.2 we proved that any permutation in S_n can be written as a product of transpositions (cycles of length two) and then proved that the number of transpositions in such a decomposition of a given permutation must either be always even or always odd. Thus we can call a permutation *even* if it can be expressed as an even number of transpositions, and *odd* otherwise.

3.6.4 Proposition.   The set of all even permutations of S_n is a subgroup of S_n.

Proof. If σ and τ are even permutations, then each can be expressed as a product of an even number of transpositions. It follows that $\tau\sigma$ can be expressed as a product of an even number of transpositions, and so the set of all even permutations of S_n is closed under multiplication of permutations. Furthermore, the identity permutation is even. Since S_n is a finite set, this is enough to imply that we have a subgroup. $\square$

3.6.5 Definition.   The set of all even permutations of S_n is called the *alternating group* on n elements, and will be denoted by A_n.

When we considered even and odd permutations in Chapter 2, our proof of Theorem 2.3.11 (justifying the definition of even and odd permutations) was different from the one usually given. We now give the standard approach to parity of elements of S_n. Let Δ_n be the polynomial in n variables $x_1, x_2, \ldots, x_n$ defined by

$$\Delta_n = \prod_{1 \le i < j \le n} (x_i - x_j).$$

Any permutation $\sigma \in S_n$ acts on Δ_n by permuting the subscripts. If $i < j$ and $\sigma(i) < \sigma(j)$, then the sign of the factor $x_i - x_j$ of Δ_n remains unchanged, but if $\sigma(i) > \sigma(j)$ then the sign of the factor is changed.

For example, we have $\Delta_3 = (x_1 - x_2)(x_1 - x_3)(x_2 - x_3)$. Letting the permutation $(1,2,3)$ act on Δ_3 gives the new polynomial $(x_2 - x_3)(x_2 - x_1)(x_3 - x_1)$, in which the signs of two factors have been changed. On the other hand, the transposition $(1,2)$ applied to Δ_3 gives the new polynomial $(x_2 - x_1)(x_2 - x_3)(x_1 - x_3)$, in which the sign of only one factor has changed. This suggests that an even permutation should leave the sign unchanged, while an odd permutation should change the sign.

3.6.6 Lemma.   Any transposition in S_n changes the sign of Δ_n.

Proof. Let (h, k) be a transposition in S_n ($n > 1$). We must discover what its action is on Δ_n. The factor $x_i - x_j$ is affected by the action of (h, k) only if i or j is equal to h or k. Table 3.6.2 lists the seven cases in which this occurs, and lists a $-$ or $+$ depending on whether or not there is a sign change.

TABLE 3.6.2

	Case	Old Factor	New Factor	Sign
$h < i$	$i = k < j$	$x_k - x_j$	$x_h - x_j$	$+$
$h < i$	$i < k = j$	$x_i - x_k$	$x_i - x_h$	$-$
$h = i$	$i < k < j$	$x_h - x_j$	$x_k - x_j$	$+$
$h = i$	$i < k = j$	$x_h - x_k$	$x_k - x_h$	$-$
$h = i$	$i < j < k$	$x_h - x_j$	$x_k - x_j$	$-$
$i < h < j$	$k = j$	$x_i - x_k$	$x_i - x_h$	$+$
$i < h = j$	$j < k$	$x_i - x_h$	$x_i - x_k$	$+$

The $k - h - 1$ factors $x_h - x_j$ with $h < j < k$ can be matched with the $k - h - 1$ factors $x_i - x_k$ with $h < i < k$, and so their associated sign changes cancel each other out. Thus the sign change of the single factor $x_h - x_k$ determines that there is a net change of sign. This shows that the transposition (h, k) changes the sign of $\Delta_n(x)$. □

3.6.7 Theorem.   A permutation in S_n is even if and only if it leaves the sign of Δ_n unchanged.

Proof. We first note that if $\sigma, \tau \in S_n$, then $\sigma\tau(\Delta_n) = \sigma(\tau(\Delta_n))$. Thus the product of an even number of transpositions will leave the sign of Δ_n unchanged, whereas the product of an odd number of transpositions will change the sign. □

EXERCISES: SECTION 3.6

1. Find the orders of each of these permutations:
   (a) $(1,2)(2,3)(3,4)$
   (b) $(1,2,5)(2,3,4)(5,6)$
   (c) $(1,3)(2,6)(1,4,5)$
   (d) $(1,2,3)(2,4,3,5)(1,3,2)$

2. Write out the addition tables for $\mathbf{Z}_4$ and for $\mathbf{Z}_2 \times \mathbf{Z}_2$. Write out the permutation (in cyclic notation) determined by each row of each of the addition tables.

3. Answer the previous question for $\mathbf{Z}_4 \times \mathbf{Z}_2$.

4. Find the permutations that correspond to the rigid motions of a rectangle that is not a square. Do the same for the rigid motions of a rhombus (diamond) that is not a square.

5. Find the largest possible order of an element in S_4. Answer the same question for S_5, S_6, S_7, and S_8.

6. List the elements of A_4.

7. Without writing down all 60 elements of A_5, describe what their cycle structures would be and how many of each type there are.

8. Show that D_n is isomorphic to a subgroup of S_n, for $n \geq 3$.

9. Show that in any group of permutations, the set of all even permutations forms a subgroup.

10. For any elements $\sigma, \tau \in S_n$, show that $\sigma\tau\sigma^{-1}\tau^{-1} \in A_n$.

11. Find the order of the group of rigid motions of a cube.

    Hint: Think of how many ways you can put the cube into a box just large enough to hold it.

12. A rigid motion of a cube can be thought of either as a permutation of its eight vertices or as a permutation of its six sides. Find a rigid motion of the cube that has order 3, and express the permutation that represents it in both ways, as a permutation on eight elements and as a permutation on six elements.

13. Let S be an infinite set. Let H be the set of all elements $\sigma \in \text{Sym}(S)$ such that $\sigma(x) = x$ for all but finitely many $x \in S$. Prove that H is a subgroup of $\text{Sym}(S)$.

14. The center of a group is the set of all elements that commute with every other element of the group. That is,

$$Z(G) = \{z \in G \mid za = az \text{ for all } a \in G\}.$$

    Show that if $n \geq 3$, then the center of S_n is trivial.

15. Find the center of D_n.

    Hint: Consider two cases, depending on whether n is odd or even.

16. Show that in S_n the only elements which commute with the cycle $(1,2,\dots,n)$ are its powers.

17. Let $\tau = (a,b,c)$ and let σ be any permutation. Show that $\sigma\tau\sigma^{-1} = (\sigma(a),\sigma(b),\sigma(c))$.

18. Show that the product of two transpositions is one of (i) the identity; (ii) a 3-cycle; (iii) a product of two 3-cycles.

19. Show that S_n is isomorphic to a subgroup of A_{n+2}.

3.7 HOMOMORPHISMS

In Section 3.4 we studied the notion of a group isomorphism. If G_1 and G_2 are groups, then in addition to the one-to-one correspondence between them, the crucial property of an isomorphism $\phi : G_1 \to G_2$ is that $\phi(ab) = \phi(a)\phi(b)$ for all a, $b \in G$. In words, we can describe this by saying that an isomorphism respects the operations of the two groups. The fact that an isomorphism preserves the operations provides the connection between the algebraic properties of the two groups. In Definition 3.7.1 (which follows shortly), a function satisfying this condition will be called a *group homomorphism*.

Functions that preserve the algebraic properties of groups are important in many situations far more general than the setting of an isomorphism. From linear algebra you should recall the fact that the determinant of a product is the product of the determinants. This can be put into a group theoretic context (see Example 3.7.2) and provides a good example of a function that is not one-to-one, but still respects the essential algebraic structure.

If the operations in both G_1 and G_2 are denoted additively, then the formula defining a homomorphism becomes $\phi(a + b) = \phi(a) + \phi(b)$. Again, a familiar operation in calculus can be put into this context: The derivative of a sum is the sum of the derivatives.

One of the most important examples of a group homomorphism is provided by the rule for exponents: $a^{n+m} = a^n a^m$. Our first example considers the appropriate function that relates integers to powers of a group element a.

Example 3.7.1 (Exponential Function for Groups)

Let G be a group, and let a be any element of G. Define $\phi : \mathbf{Z} \to G$ by $\phi(n) = a^n$, for all $n \in \mathbf{Z}$. The rules we have developed for exponents show that for all $n, m \in \mathbf{Z}$,

$$\phi(n + m) = a^{n+m} = a^n a^m = \phi(n) \cdot \phi(m).$$

Thus ϕ is consistent with the operations in the respective groups.

If G is abelian, with its operation denoted additively, then we define $\phi : \mathbf{Z} \to G$ by $\phi(n) = na$. The fact that ϕ is a homomorphism is expressed by the formula $(n + m)a = na + ma$, which holds for all $n, m \in \mathbf{Z}$. After we have studied homomorphisms in more detail we will return to these examples to show how the ideas we have developed can be applied to help understand the order of an element and the cyclic subgroup generated by an element. $\quad\square$

Example 3.7.2

Many concepts from linear algebra provide examples of the general group theoretic concepts we are studying. Let V and W be vector spaces. Recall that a function $L : V \to W$ is called a *linear transformation* if $L(v_1 + v_2) = L(v_1) + L(v_2)$ and $L(av_1) = aL(v_1)$ for all vectors $v_1, v_2 \in V$ and all scalars a. Since any vector space is an abelian group under vector addition, any linear transformation between vector spaces is actually a homomorphism of the underlying abelian groups. (The condition involving scalar multiplication is not involved in the group theory setting.) Linear differential equations fit into this context and thus provide examples of homomorphisms of abelian groups.

The formula $\det(AB) = \det(A)\det(B)$ for $n \times n$ matrices shows that the function $\phi : GL_n(\mathbf{R}) \to \mathbf{R}^\times$ defined by $\phi(A) = \det(A)$ is a group homomorphism. Note that this function does not define a homomorphism on the abelian group of all $n \times n$ matrices under addition since it is possible to find matrices A, B for which $\det(A + B) \neq \det(A) + \det(B)$. $\quad\square$

Example 3.7.3

For a fixed integer m, define $\phi : \mathbf{Z}_n \to \mathbf{Z}_n$ by $\phi([x]) = [mx]$, for all $[x] \in \mathbf{Z}_n$. This is a function since if $a \equiv b \pmod{n}$, then $ma \equiv mb \pmod{n}$. It is a homomorphism since

$$\phi([a] + [b]) = \phi([a + b]) = [m(a + b)] = [ma] + [mb] = \phi([a]) + \phi([b]). \quad\square$$

We now formally record our definition of a homomorphism between groups. It follows immediately from the definition that an isomorphism is simply a homomorphism that is one-to-one and onto.

3.7.1 Definition.   Let G_1 and G_2 be groups, and let $\phi : G_1 \rightarrow G_2$ be a function. Then ϕ is said to be a *group homomorphism* if

$$\phi(ab) = \phi(a)\phi(b)$$

for all $a, b \in G_1$.

3.7.2 Proposition.   Let $\phi : G_1 \rightarrow G_2$ be a group homomorphism.
  (a) $\phi(e) = e$;
  (b) $(\phi(a))^{-1} = \phi(a^{-1})$ for all $a \in G_1$;
  (c) for any integer n and any $a \in G_1$, $\phi(a^n) = (\phi(a))^n$;
  (d) if $a \in G_1$ and a has order n, then the order of $\phi(a)$ in G_2 is a divisor of n.

Proof. (a) Since $\phi(e)\phi(e) = \phi(e^2) = \phi(e)$, cancellation gives $\phi(e) = e$.
(b) This follows since $\phi(a)\phi(a^{-1}) = \phi(aa^{-1}) = e$.
(c) This can be proved using a simple induction argument.
(d) Suppose that $a \in G_1$ with $a^m = e$. Since $\phi : G_1 \rightarrow G_2$ is a homomorphism, we must have $(\phi(a))^m = \phi(a^m) = \phi(e) = e$. Taking $m = o(a)$ shows that the order of $\phi(a)$ is a divisor of the order of a. $\square$

Example 3.7.4 (Homomorphisms Defined on Cyclic Groups)

In this example we will completely describe all homomorphisms defined on any cyclic group. Let C be a cyclic group, denoted multiplicatively, with generator a. If $\phi : C \rightarrow G$ is any group homomorphism, and $\phi(a) = g$, then the formula $\phi(a^m) = g^m$ must hold. Since every element of C is of the form a^m for some $m \in \mathbf{Z}$, this means that ϕ is completely determined by its value on a. Note that if a has finite order, then by the previous proposition the order of g must be a divisor of the order of a.

We next consider how to define homomorphisms on C. If C is infinite, then for an element g of any group G, the formula $\phi(a^m) = g^m$ defines a homomorphism since

$$\phi(a^m a^k) = \phi(a^{m+k}) = g^{m+k} = g^m g^k = \phi(a^m)\phi(a^k).$$

If $|C| = n$ and g is any element of G whose order is a divisor of n, then the formula $\phi(a^m) = g^m$ defines a homomorphism. We must first show that the formula defines a function, since the formula depends on the choice of an exponent in writing an element $x \in C$ as a power of the generator a. If $x = a^m$ and $x = a^k$, then $m \equiv k \pmod{n}$, since a has order n. Thus we can write $m = k + qn$ for some integer q, and then

$$g^m = g^{k+qn} = g^k(g^n)^q = g^k$$

since $g^n = e$. This depends on the crucial assumption that the order of g is a divisor of n. Now the previous argument can be used to show that ϕ is a homomorphism.   □

Example 3.7.5

As a particular case of the previous example, we now give explicit formulas for all homomorphisms $\phi : \mathbf{Z}_n \to \mathbf{Z}_k$. Any such homomorphism is completely determined by $\phi([1]_n)$, and this must be an element m of $\mathbf{Z}_k$ whose order is a divisor of n. Then the formula $\phi([x]_n) = [mx]_k$, for all $[x]_n \in \mathbf{Z}_n$, defines a homomorphism. Furthermore, every homomorphism from $\mathbf{Z}_n$ into $\mathbf{Z}_k$ must be of this form. Note that $\phi(\mathbf{Z}_n)$ is the cyclic subgroup generated by m, and so ϕ will map $\mathbf{Z}_n$ onto $\mathbf{Z}_k$ if and only if m is a generator of $\mathbf{Z}_k$.   □

Let $\phi : G_1 \to G_2$ be a group homomorphism. Recall the statement of Proposition 3.4.4: ϕ is one-to-one if and only if $\phi(x) = e$ implies $x = e$. The set $\{x \in G_1 \mid \phi(x) = e\}$ plays an important role in studying group homomorphisms. It should already be familiar to the student in the setting of linear algebra, where the kernel (or null space) of a linear transformation is studied.

3.7.3 Definition.   Let $\phi : G_1 \to G_2$ be a group homomorphism. Then $\{x \in G \mid \phi(x) = e\}$ is called the *kernel* of ϕ, and is denoted by $\ker(\phi)$.

3.7.4 Proposition.   Let $\phi : G_1 \to G_2$ be a group homomorphism, with $K = \ker(\phi)$.
   (a) K is a subgroup of G_1 such that $gkg^{-1} \in K$ for all $k \in K$ and $g \in G_1$.
   (b) ϕ is one-to-one if and only if $K = \{e\}$.

Proof. (a) The kernel of ϕ is nonempty since it contains e. If $a, b \in K$, then

$$\phi(ab^{-1}) = \phi(a)(\phi(b))^{-1} = e \cdot e = e$$

and this implies that K is a subgroup of G_1. Furthermore, if $k \in K$ and $g \in G_1$, then

$$\phi(gkg^{-1}) = \phi(g)\phi(k)(\phi(g))^{-1} = \phi(g)e(\phi(g))^{-1} = e.$$

Thus $gkg^{-1} \in K$.
   (b) If ϕ is one-to-one, then the only element that can map to the identity of G_2 is the identity of G_1. On the other hand, suppose that $K = \{e\}$ and $\phi(a) = \phi(b)$ for some $a, b \in G_1$. Multiplying both sides of this equation by $(\phi(b))^{-1}$ gives us $\phi(ab^{-1}) = \phi(a)(\phi(b))^{-1} = e$, which shows that $ab^{-1} \in \ker(\phi)$. But then by assumption, $ab^{-1} = e$, and thus $a = b$. This shows that ϕ is one-to-one.   □

The previous proposition shows that the kernel of a group homomorphism is a special type of subgroup, which we define below. We will study these subgroups in much greater detail in Section 3.8, so in this section our interest is only in their behavior with respect to group homomorphisms.

3.7.5 Definition.   A subgroup H of the group G is called a *normal* subgroup if $ghg^{-1} \in H$ for all $h \in H$ and $g \in G$.

It is obvious from the definition that if $H = G$ or $H = \{e\}$, then H is normal. It is also clear that any subgroup of an abelian group is normal. As one of the exercises at the end of the section asks you to show, the only proper nontrivial normal subgroup of S_3 is its three element subgroup. The next proposition investigates how subgroups are related via a homomorphism.

3.7.6 Proposition.   Let $\phi : G_1 \to G_2$ be a group homomorphism.
 (a) If H_1 is a subgroup of G_1, then $\phi(H_1)$ is a subgroup of G_2. If ϕ is onto and H_1 is normal in G_1, then $\phi(H_1)$ is normal in G_2.
 (b) If H_2 is a subgroup of G_2, then $\phi^{-1}(H_2) = \{x \in G_1 \mid \phi(x) \in H_2\}$ is a subgroup of G_1. If H_2 is a normal in G_2, then $\phi^{-1}(H_2)$ is normal in G_1.

Proof. (a) Let H_1 be a subgroup of G_1, and let $y, z \in \phi(H_1)$. Then there exist $a, b \in H_1$ with $\phi(a) = y$ and $\phi(b) = z$, and

$$yz^{-1} = \phi(a)(\phi(b))^{-1} = \phi(a)\phi(b^{-1}) = \phi(ab^{-1}) \in \phi(H_1).$$

Since $e \in \phi(H_1)$, this shows that $\phi(H_1)$ is a subgroup of G_2.

If ϕ is onto and H_1 is normal in G_1, let $y \in G_2$ and $z \in \phi(H_1)$. There exist $a \in G_1$ and $b \in H_1$ with $\phi(a) = y$ and $\phi(b) = z$. Then

$$yzy^{-1} = \phi(a)\phi(b)\phi(a^{-1}) = \phi(aba^{-1}) \in \phi(H_1)$$

since H_1 is normal and therefore $aba^{-1} \in H_1$.

(b) Let H_2 be a subgroup of G_2, and let

$$H_1 = \phi^{-1}(H_2) = \{x \in G_1 \mid \phi(x) \in H_2\}.$$

Then $e \in H_1$ since $\phi(e) = e \in H_2$. If $a, b \in H_1$, then $ab^{-1} \in H_1$ since $\phi(ab^{-1}) = \phi(a)(\phi(b))^{-1} \in H_2$ because H_2 is a subgroup. Thus H_1 is a subgroup.

If H_2 is a normal subgroup, then to show that H_1 is also normal, let g be any element of G_1, and let $h \in H_1$. Then $ghg^{-1} \in H_1$ because

$$\phi(ghg^{-1}) = \phi(g)\phi(h)(\phi(g))^{-1} \in H_2$$

since H_2 is normal. Thus H_1 is a normal subgroup. $\square$

If $\phi : G_1 \to G_2$ is a group homomorphism, then there is a natural equivalence relation on G_1 associated with the function ϕ by defining $a \sim b$ if $\phi(a) = \phi(b)$, where $a, b \in G_1$. For arbitrary functions, this equivalence relation is studied in detail in Section 2.2, where the notation G_1/ϕ is used for the set of equivalence classes of the relation. We will use $[a]_\phi$ to denote the equivalence class of $a \in G_1$.

It may be useful to review the proof that we have in fact defined an equivalence relation. We have $a \sim a$ since $\phi(a) = \phi(a)$. If $a \sim b$, then $\phi(a) = \phi(b)$ implies $\phi(b) = \phi(a)$, which shows that $b \sim a$. Finally, if $a \sim b$ and $b \sim c$, then $\phi(a) = \phi(b)$ and $\phi(b) = \phi(c)$ implies $\phi(a) = \phi(c)$, so $a \sim c$.

The formula $[a]_n[b]_n = [ab]_n$ for multiplication of congruence classes in $\mathbf{Z}_n$ suggests that we might try a similar formula in G_1/ϕ, since we have a multiplication defined in G_1. Part of the next proposition shows that this natural multiplication is in fact well-defined.

3.7.7 Proposition.   Let $\phi : G_1 \to G_2$ be a group homomorphism. Then multiplication of equivalence classes in G_1/ϕ is well-defined, and G_1/ϕ is a group under this multiplication. The natural mapping $\pi : G_1 \to G_1/\phi$ defined by $\pi(x) = [x]_\phi$ is a homomorphism.

Proof. To show that multiplication is well-defined, we must show that if $a \sim b$ and $c \sim d$, then $ac \sim bd$. If $\phi(a) = \phi(b)$ and $\phi(c) = \phi(d)$, then

$$\phi(ac) = \phi(a)\phi(c) = \phi(b)\phi(d) = \phi(bd).$$

The associative law for G_1/ϕ follows from that of G_1, since

$$[a]_\phi([b]_\phi[c]_\phi) = [a]_\phi[bc]_\phi = [a(bc)]_\phi = [(ab)c]_\phi = [ab]_\phi[c]_\phi = ([a]_\phi[b]_\phi)[c]_\phi$$

for all a, b, c, $\in G_1$. The class $[e]_\phi$ is an identity element since

$$[e]_\phi[a]_\phi = [ea]_\phi = [a]_\phi \qquad \text{and} \qquad [a]_\phi[e]_\phi = [ae]_\phi = [a]_\phi$$

for all $a \in G_1$. Finally, for any equivalence class $[a]_\phi$, there exists an inverse $[a^{-1}]_\phi$ since $[a^{-1}]_\phi[a]_\phi = [a^{-1}a]_\phi = [e]_\phi$ and $[a]_\phi[a^{-1}]_\phi = [e]_\phi$. Thus $([a]_\phi)^{-1} = [a^{-1}]_\phi$.

Since multiplication is well-defined, we have

$$\pi(ab) = [ab]_\phi = [a]_\phi[b]_\phi = \pi(a)\pi(b)$$

for all $a, b \in G_1$, and so π is a homomorphism. $\square$

The following theorem is extremely important, and we will return to it in the next section, where we give another proof. Theorem 2.2.6 shows that if $f : S \to T$ is any function, then there is a one-to-one correspondence between the elements of $f(S)$ and the set S/f. Thus the basic one-to-one correspondence that we will give in Theorem 3.7.8 comes from set theory, and not from the algebraic structure of either group or the fact that ϕ is a homomorphism. (We choose to reprove this fact in Theorem 3.7.8 to make its proof self-contained.)

Now suppose that $\phi : G_1 \to G_2$ is a homomorphism. Using the results in Section 2.2 we can factor ϕ into a composition of functions $\iota\bar{\phi}\pi$, where π is the function of Proposition 3.7.7 and ι is the inclusion mapping.

$$\begin{array}{ccccc} & \pi & & \bar{\phi} & & \iota \\ G_1 & \to & G_1/\phi & \to & \phi(G_1) & \to & G_2 \end{array}$$

In the diagram both π and ι are homomorphisms. We now show that the function $\overline{\phi}$ is an isomorphism of groups.

3.7.8 Theorem.   Let $\phi : G_1 \to G_2$ be a group homomorphism. Then G_1/ϕ is isomorphic to $\phi(G_1)$.

Proof. Define $\overline{\phi} : G_1/\phi \to \phi(G_1)$ by $\overline{\phi}([a]_\phi) = \phi(a)$, for all equivalence classes $[a]_\phi \in G_1/\phi$. This is a well-defined function since if $[a]_\phi = [b]_\phi$, then by definition $\phi(a) = \phi(b)$, and so $\overline{\phi}([a]_\phi) = \overline{\phi}([b]_\phi)$. If $\overline{\phi}([a]_\phi) = \overline{\phi}([b]_\phi)$, then $\phi(a) = \phi(b)$, and so $[a]_\phi = [b]_\phi$, which shows that $\overline{\phi}$ is one-to-one. The image of G_1/ϕ is

$$\{\overline{\phi}([a]_\phi) \mid a \in G_1\} = \{\phi(a) \mid a \in G_1\} = \phi(G_1)$$

so $\overline{\phi}$ maps G_1/ϕ onto $\phi(G_1)$. Finally, $\overline{\phi}$ is a homomorphism since

$$\overline{\phi}([a]_\phi)\overline{\phi}([b]_\phi) = \phi(a)\phi(b) = \phi(ab) = \overline{\phi}([ab]_\phi) = \overline{\phi}([a]_\phi[b]_\phi)$$

for all equivalence classes $[a]_\phi, [b]_\phi \in G_1/\phi$. $\square$

Example 3.7.6 (Characterization of Cyclic Groups)

The power of Theorem 3.7.8 can be illustrated by giving another proof that every cyclic group is isomorphic to either $\mathbf{Z}$ or $\mathbf{Z}_n$, for some n. Given $G = \langle a \rangle$, define $\phi : \mathbf{Z} \to G$ by $\phi(m) = a^m$, as in Example 3.7.1. If a has infinite order, then ϕ is one-to-one, so in this case, $\mathbf{Z}$ is isomorphic to $\phi(\mathbf{Z}) = G$. If a has order n, then $a^m = a^k$ if and only if $m \equiv k \pmod{n}$. Thus $\phi(m) = \phi(k)$ if and only if $m \equiv k \pmod{n}$, which shows that $\mathbf{Z}/\phi$ is the additive group of congruence classes modulo n. Thus if a has order n, then $G \cong \mathbf{Z}_n$. $\square$

Example 3.7.7 (Cayley's Theorem)

Theorem 3.7.8 is also useful in giving a more concise proof of Cayley's theorem. Given any group G, define $\phi : G \to \text{Sym}(G)$ by $\phi(a) = \lambda_a$, for any $a \in G$, where λ_a is the function defined by $\lambda_a(x) = ax$ for all $x \in G$. (It is necessary to check that λ_a is one-to-one and onto.) Then ϕ is a homomorphism since $\lambda_a \lambda_b = \lambda_{ab}$ for all $a, b \in G$. Because λ_a is the identity permutation only if $a = e$, we have $\ker(\phi) = \{e\}$. Since ϕ is one-to-one, the congruence classes of G/ϕ are just the individual elements of G, and thus G itself is isomorphic to $\phi(G)$, which is a permutation group. $\square$

Example 3.7.8

Define $\phi : \mathbf{R} \to \mathbf{C}^\times$ by $\phi(\theta) = \cos\theta + i\sin\theta$, for all $\theta \in \mathbf{R}$. The trigonometric formulas for the cosine and sine of the sum of two angles can be used (see Appendix E on the complex numbers) to show that $\phi(\alpha + \beta) = \phi(\alpha) \cdot \phi(\beta)$, and so ϕ is a group homomorphism. Geometrically, the function ϕ can be visualized as wrapping the real line around the unit circle. In this process, numbers that differ by a multiple of 2π are identified. It follows that

$$[\theta]_\phi = \{x \in \mathbf{R} \mid x = \theta + 2k\pi, k \in \mathbf{Z}\}.   \square$$

We conclude the section with a proposition that gives a more complete description of the equivalence classes of the equivalence relation defined by a homomorphism. It shows that the equivalence relation defined by ϕ is the same as the one defined by $\ker(\phi)$. Thus in Section 3.8 we will switch from the notation G/ϕ to the more standard notation $G/\ker(\phi)$.

3.7.9 Proposition.   Let $\phi : G_1 \to G_2$ be a group homomorphism, and $a, b \in G_1$. The following conditions are equivalent:

(1) $\phi(a) = \phi(b)$;

(2) $ab^{-1} \in \ker(\phi)$;

(3) $a = kb$ for some $k \in \ker(\phi)$;

(4) $b^{-1}a \in \ker(\phi)$;

(5) $a = bk$ for some $k \in \ker(\phi)$.

Proof. (1) implies (2): If $\phi(a) = \phi(b)$, then multiplying both sides of the equation by $(\phi(b))^{-1}$ we have

$$e = \phi(a)(\phi(b))^{-1} = \phi(a)\phi(b^{-1}) = \phi(ab^{-1}).$$

(2) implies (3): If $ab^{-1} = k \in \ker(\phi)$, then $a = kb$.
(3) implies (1): If $a = kb$ for some $k \in \ker(\phi)$, then

$$\phi(a) = \phi(kb) = \phi(k)\phi(b) = e\phi(b) = \phi(b).$$

Similarly it can be shown that (1) implies (4) implies (5) implies (1).   □

If $\phi : G_1 \to G_2$ is a homomorphism of abelian groups, with operations denoted additively, then Proposition 3.7.9 has the following form: For $a, b \in G_1$, the following conditions are equivalent: (1) $\phi(a) = \phi(b)$; (2) $a - b \in \ker(\phi)$; and (3) $a = b + k$ for some $k \in \ker(\phi)$. The following examples use additive notation.

Example 3.7.9

Let A be an $m \times n$ matrix, and consider the nonhomogeneous equation $Ax = \mathbf{b}$. We may view the matrix A as defining a homomorphism $\phi : \mathbf{R}^n \to \mathbf{R}^m$, where $\phi(\mathbf{x}) = A\mathbf{x}$. By Proposition 3.7.9, if we find a particular solution $\mathbf{x}_0$ with $A\mathbf{x}_0 = \mathbf{b}$, then the set of solutions to the nonhomogeneous equation $A\mathbf{x} = \mathbf{b}$ consists of all vectors of the form $\mathbf{x}_0 + \mathbf{k}$, where $\mathbf{k}$ is any solution of the homogeneous equation $A\mathbf{x} = \mathbf{0}$, since $\ker(\phi)$ is the solution space of the homogeneous equation $A\mathbf{x} = \mathbf{0}$.   □

Example 3.7.10

An analysis similar to the previous example shows that the standard theorem stating that the general solution to a nonhomogeneous linear differential equation is obtained by finding any particular solution and adding to it all solutions of the associated homogeneous equation is really just a consequence of the fact that linear differential operators preserve sums of functions.   □

EXERCISES: SECTION 3.7

1. Write down the formulas for all homomorphisms from $\mathbf{Z}_{24}$ into $\mathbf{Z}_{18}$.

2. Write down the formulas for all homomorphisms from $\mathbf{Z}$ onto $\mathbf{Z}_{12}$.

3. Show that $\phi_3 : \mathbf{Z}_3 \rightarrow \mathbf{Z}_3$ defined by $\phi_3([x]) = [x]^3$ and $\phi_5 : \mathbf{Z}_5 \rightarrow \mathbf{Z}_5$ defined by $\phi_5([x]) = [x]^5$ are homomorphisms but $\phi_4 : \mathbf{Z}_4 \rightarrow \mathbf{Z}_4$ defined by $\phi_4([x]) = [x]^4$ is not.

4. Let G be an abelian group, and let n be any positive integer. Show that the function $\phi : G \rightarrow G$ defined by $\phi(x) = x^n$ is a homomorphism.

5. Let G be the multiplicative group $\{1, 2, 4, 7, 8, 11, 13, 14\}$ modulo 15, and let $n = 2$. Compute the values of the function defined in the previous problem, and find its kernel and the image of G.

6. Define $\phi : \mathbf{C}^\times \rightarrow \mathbf{R}^\times$ by $\phi(a + bi) = a^2 + b^2$, for all $a + bi \in \mathbf{C}^\times$. Show that ϕ is a homomorphism.

7. Which of the following functions are homomorphisms?

    (a) $\phi : \mathbf{R}^\times \rightarrow GL_2(\mathbf{R})$ defined by $\phi(a) = \begin{bmatrix} a & 0 \\ 0 & 1 \end{bmatrix}$

    (b) $\phi : \mathbf{R} \rightarrow GL_2(\mathbf{R})$ defined by $\phi(a) = \begin{bmatrix} 1 & a \\ 0 & 1 \end{bmatrix}$

    (c) $\phi : M_2(\mathbf{R}) \rightarrow \mathbf{R}$ defined by $\phi\left(\begin{bmatrix} a & b \\ c & d \end{bmatrix}\right) = a$

    (d) $\phi : GL_2(\mathbf{R}) \rightarrow \mathbf{R}$ defined by $\phi\left(\begin{bmatrix} a & b \\ c & d \end{bmatrix}\right) = a + d$

    (e) $\phi : GL_2(\mathbf{R}) \rightarrow \mathbf{R}^\times$ defined by $\phi\left(\begin{bmatrix} a & b \\ c & d \end{bmatrix}\right) = ad - bc$

    (f) $\phi : GL_2(\mathbf{R}) \rightarrow \mathbf{R}^\times$ defined by $\phi\left(\begin{bmatrix} a & b \\ c & d \end{bmatrix}\right) = ab$

8. Let $\phi : G_1 \rightarrow G_2$ and $\theta : G_2 \rightarrow G_3$ be homomorphisms. Prove that $\theta\phi : G_1 \rightarrow G_3$ is a homomorphism. Prove that $\ker(\phi) \subseteq \ker(\theta\phi)$.

9. Let ϕ be a homomorphism of G_1 onto G_2. Prove that if G_1 is abelian, then so is G_2; prove that if G_1 is cyclic, then so is G_2. In each case, give a counterexample to the converse of the statement.

10. Let G be a finite group of even order, with n elements, and let H be a subgroup with $n/2$ elements. Prove that H must be normal.

    Hint: Define $\phi : G \rightarrow \mathbf{R}^\times$ by $\phi(x) = 1$ if $x \in H$ and $\phi(x) = -1$ if $x \notin H$ and show that ϕ is a homomorphism with kernel H. To show that ϕ preserves products, show that if $g \notin H$, then $\{x \mid gx \in H\} = G - H$.

11. Show that the only proper nontrivial normal subgroup of S_3 is the one with three elements.

12. Determine which subgroups of D_4 are normal.

13. Recall that the center of a group G is $\{g \in G \mid ga = ag$ for all $a \in G\}$. Prove that the center of any group is a normal subgroup.

14. Prove that the intersection of two normal subgroups is a normal subgroup.

15. Given an example to show that the assumption that ϕ is onto is needed in part (a) of Proposition 3.7.6.

3.8 COSETS, NORMAL SUBGROUPS, AND FACTOR GROUPS

In this section we introduce the important notion of a coset of a subgroup, motivated by the results in the previous section on the equivalence relation defined by a homomorphism. We also show that the set of all cosets of any normal subgroup can be given a natural group structure, just as we did in the previous section for the equivalence classes determined by a homomorphism.

The congruence classes of the integers modulo 2 are the sets of even and odd integers. We have denoted the set of even integers by $2\mathbf{Z}$, and so we could denote the set of odd integers by $1 + 2\mathbf{Z}$, to show that each odd integer can be expressed as 1 plus an even integer. Of course, we could use any odd integer in place of 1, say $3 + 2\mathbf{Z}$ or $5 + 2\mathbf{Z}$, still giving the set of all odd integers. More generally, for any integer k we can express its congruence class in $\mathbf{Z}_n$ in the form $[k]_n = k + n\mathbf{Z}$.

Let $\phi : G_1 \rightarrow G_2$ be a group homomorphism, and let a be a fixed element in G_1. By Proposition 3.7.9, for $b \in G_1$ we have $\phi(b) = \phi(a)$ if and only if b can be written in the form $b = ak$ for some element $k \in \ker(\phi)$. We can express this by writing $b \in aK$, where $K = \ker(\phi)$, and aK consists of all elements that can be written in the form ak for some $k \in K$. (This is the product of the sets $\{a\}$ and K, as defined in Section 3.3.) If $\sim$ is the equivalence relation defined by letting $a \sim b$ if $\phi(a) = \phi(b)$, then the equivalence class $[a]_\phi$ defined by $\sim$ is precisely the set aK. Proposition 3.7.9 also shows that $aK = Ka$.

In the proof of Lagrange's theorem (Theorem 3.2.10), for a subgroup H of the finite group G, we introduced the equivalence relation $a \sim b$ on G by letting $a \sim b$ if $ab^{-1} \in H$. In the course of the proof we showed that the elements of the congruence class $[a]$ are precisely the elements of the form ha for $h \in H$. Thus we can write $[a] = Ha$.

We will now develop this idea further, for sets of the form aH. Let G be a group and let H be a subgroup of G. To review the work in Section 3.2 without retracing precisely the same steps, we will consider the relation $a \sim b$ if $a^{-1}b \in H$, for $a, b \in G$. This defines an equivalence relation: $a \sim a$ since $a^{-1}a \in H$; if $a \sim b$, then $a^{-1}b$ is in H, so its inverse $b^{-1}a$ is in H, and thus $b \sim a$; if $a \sim b$ and $b \sim c$, then $a^{-1}b$ and $b^{-1}c$ are in H, so their product is in H, and thus $a \sim c$. Our first proposition identifies the equivalence classes of this equivalence relation as sets of the form aH, for $a \in G$.

3.8.1 Proposition.   Let H be a subgroup of the group G, and let $a, b \in G$. Then the following conditions are equivalent:

(1) $bH = aH$;

(2) $bH \subseteq aH$;

(3) $b \in aH$;

(4) $a^{-1}b \in H$.

Proof. It is obvious that (1) implies (2). Furthermore, (2) implies (3) since $b = be \in bH$.

(3) implies (4): If $b = ah$ for some $h \in H$, then $a^{-1}b = h \in H$.

(4) implies (1): Suppose that $a^{-1}b = h$ for some $h \in H$, so that $b = ah$ and $a = bh^{-1}$. To show that $bH \subseteq aH$, let $x \in bH$. Then $x = bh'$ for some $h' \in H$, and substituting for b gives $x = ahh'$, which shows that $x \in aH$. On the other hand, to show that $aH \subseteq bH$, let $x \in aH$. Then $x = ah''$ for some $h'' \in H$, and so $x = bh^{-1}h'' \in bH$. Thus we have shown that $bH = aH$. □

In Proposition 3.8.1, the symmetry in the condition $bH = aH$ shows that the roles of a and b can be reversed. Thus $a^{-1}b \in H$ if and only if $b^{-1}a \in H$. Since we are working with an equivalence relation, the equivalence classes must partition G, and so we know that if $aH \cap bH \neq \varnothing$, then $aH = bH$.

3.8.2 Definition.   Let H be a subgroup of the group G, and let $a \in G$. The set

$$aH = \{g \in G \mid g = ah \text{ for some } h \in H\}$$

is called the *left coset* of H in G determined by a. Similarly, the *right coset* of H in G determined by a is the set

$$Ha = \{g \in G \mid g = ha \text{ for some } h \in H\}.$$

The number of left cosets of H in G is called the *index* of H in G, and is denoted by $[G : H]$.

A result similar to Proposition 3.8.1 holds for right cosets. Let H be a subgroup of the group G, and let $a, b \in G$. Then the following conditions are equivalent: (1) $Ha = Hb$; (2) $Ha \subseteq Hb$; (3) $a \in Hb$; (4) $ab^{-1} \in H$; (5) $ba^{-1} \in H$; (6) $b \in Ha$; (7) $Hb \subseteq Ha$. The index of H in G could also be defined as the number of right cosets of H in G, since there is a one-to-one correspondence between left cosets and right cosets. (See Exercise 5.)

We next show that all left cosets of H have the same number of elements. Given any left coset aH of H, define the function $f : H \to aH$ by $f(h) = ah$, for all $h \in H$. Then f is one-to-one since if $f(h_1) = f(h_2)$, then $ah_1 = ah_2$ and so $h_1 = h_2$ by the cancellation law. It is obvious that f is onto, and so the one-to-one correspondence $f : H \to aH$ shows that aH has the same number of elements as H.

In the next example we list the left cosets of a given subgroup H of a finite group. For any $a \in H$ we have $aH = H$, so we begin by choosing any element a not in H. Then aH is found by listing all products of the form ah for $h \in H$. Now any element in aH determines the same coset, so for the next coset we choose any element not in H or aH (if possible). Continuing in this way provides a method for listing all cosets.

Example 3.8.1

Let G be the multiplicative group $\mathbf{Z}_{11}^{\times}$ of nonzero elements of $\mathbf{Z}_{11}$. Let H be the subgroup $\{[1], [10]\}$ generated by $[10]$. The first coset we can identify is H itself. Choosing an element not in H, say $[2]$, we form the products $[2][1]$ and $[2][10] = [9]$, to obtain the coset $[2]H = \{[2], [9]\}$. Next we choose any element not in the first two cosets, say $[3]$, which gives us $[3]H = \{[3], [8]\}$, since $[3][1] = [3]$ and $[3][10] = [8]$. Continuing in this fashion, we obtain $[4]H = \{[4], [7]\}$ and $[5]H = \{[5], [6]\}$. Thus the cosets of H are the following sets:

$$H = \{[1], [10]\}, \qquad [2]H = \{[2], [9]\}, \qquad [3]H = \{[3], [8]\},$$

$$[4]H = \{[4], [7]\}, \qquad [5]H = \{[5], [6]\}.$$

As another example, let $K = \{[1], [3], [9], [5], [4]\}$ be the subgroup generated by $[3]$. Since the left cosets all have the same number of elements and we already have a coset with half of the total number of elements, there must be only one other coset, containing the rest of the elements. Thus the left cosets of K are the following sets:

$$K = \{[1], [3], [9], [5], [4]\}, \qquad [2]K = \{[2], [6], [7], [10], [8]\}. \quad \square$$

Example 3.8.2

Let $G = S_3$, the group of all permutations on a set with three elements, and let $G = \{e, a, a^2, b, ab, a^2b\}$, where $a^3 = e$, $b^2 = e$, and $ba = a^2b$. Let $H = \{e, b\}$. We must be careful since this is the first example in a non-abelian group, so we must distinguish between left and right cosets.

The left cosets of H are easily computed to be

$$H = \{e, b\}, \qquad aH = \{a, ab\}, \qquad a^2H = \{a^2, a^2b\}.$$

Since $ba = a^2b$ and $ba^2 = ab$, the right cosets of H are

$$H = \{e, b\}, \qquad Ha = \{a, a^2b\}, \qquad Ha^2 = \{a^2, ab\}. \quad \square$$

If G is an abelian group with the operation denoted by $+$, then the cosets of a subgroup H have the form

$$a + H = \{g \in G \mid g = a + h \text{ for some } h \in H\}.$$

Proposition 3.8.1 shows that in this case, $a + H = b + H$ if and only if $a - b \in H$.

Example 3.8.3

Let $G = \mathbf{Z}_{12}$, and let H be the subgroup $4\mathbf{Z}_{12} = \{[0], [4], [8]\}$. To find all cosets of H, we begin by noting that $[4] + H = [8] + H = H$, so we start with an element not in H, say $[1]$. Then

$$[1] + H = \{[1] + [0], [1] + [4], [1] + [8]\} = \{[1], [5], [9]\}.$$

Next we choose an element not in H or $[1] + H$, say $[2]$. Then $[2] + H = \{[2], [6], [10]\}$, and we have only one coset remaining, $[3] + H = \{[3], [7], [11]\}$. Since we have used all elements of the group, we have found the following cosets:

$$H = \{[0], [4], [8]\}, \qquad [1] + H = \{[1], [5], [9]\},$$
$$[2] + H = \{[2], [6], [10]\}, \qquad [3] + H = \{[3], [7], [11]\}.$$

Since G is abelian, the right cosets are precisely the same as the left cosets. $\square$

Example 3.8.4

Let m be a real number, and consider the line $L = \{(x, y) \mid y = mx\}$ in the plane $\mathbf{R}^2$. Vector addition gives $\mathbf{R}^2$ a group structure, and L is then a subgroup of $\mathbf{R}^2$. The coset determined by a vector $(0, b)$ consists of all vectors of the form $(0, b) + (x, mx)$, or just $(x, mx + b)$. This is a line parallel to L, since it is the set $\{(x, y) \mid y = mx + b\}$. In fact, any line parallel to $y = mx$ is a coset of L. It is true in general that the cosets of a line through the origin are the lines parallel to the given line. $\square$

For a homomorphism $\phi : G_1 \rightarrow G_2$, we saw in Section 3.7 that the set G_1/ϕ of equivalence classes defined by ϕ forms a group. Since $\phi(a) = \phi(b)$ if and only if $\phi(ab^{-1}) = e$, we have $a \sim b$ if and only if $ab^{-1} \in \ker(\phi)$. The equivalence classes of this equivalence relation are the right cosets of $\ker(\phi)$. Proposition 3.7.9 shows that the equivalence classes are also the left cosets of $\ker(\phi)$, since $a \sim b$ if and only if $a^{-1}b \in \ker(\phi)$.

Recall that a subgroup H of G is said to be normal if $aha^{-1} \in H$ for all $a \in G$ and $h \in H$. (See Definition 3.7.5.) Since $(a^{-1})^{-1} = a$, we can interchange the roles of a and a^{-1}, showing that H is normal if and only if $a^{-1}ha \in H$ for all $a \in G$ and $h \in H$. We will show that a subgroup is normal if and only if its left and right cosets coincide, that is, if and only if $aH = Ha$ for all $a \in G$. This provides an additional way to determine whether a subgroup is normal. In this case we will see that multiplication of cosets is compatible with the structure of G and that the cosets themselves define a group.

Since $\mathbf{Z}$ is abelian, the left and right cosets of any subgroup coincide. The equivalence relation defined by the subgroup $n\mathbf{Z}$, when stated in additive notation, just says that $a - b \in n\mathbf{Z}$, and so we have the well-known equivalence relation of congruence modulo n. We are already familiar with the resulting group $\mathbf{Z}_n$ defined by the corresponding equivalence classes.

Let N be a normal subgroup of the group G. We now want to introduce a multiplication for the cosets of N in G. We will do this by analogy with the multiplication in $\mathbf{Z}_n$ and in G/ϕ, where $\phi : G_1 \rightarrow G_2$ is a homomorphism. From a given coset of N we may choose any element a to use as a representative, so that we can write aN for the coset. The formula $aNbN = abN$ can then be interpreted in the following way: To multiply two cosets, choose representatives of the cosets, multiply them, and define the product to be the new coset in which the product of the representatives lies. Of course, there is a problem with the definition. We must make sure that it is independent of the choice of the representatives by which we have named the cosets. The following proposition takes care of the difficulty.

3.8.3 Proposition.   Let N be a normal subgroup of G, and let a, b, c, $d \in G$. If $aN = cN$ and $bN = dN$, then $abN = cdN$.

Proof. If $aN = cN$ and $bN = dN$, then by Proposition 3.8.1 we have $a^{-1}c \in N$ and $b^{-1}d \in N$. Since N is normal, $d^{-1}(a^{-1}c)d \in N$. But then since $b^{-1}d \in N$, we have $(ab)^{-1}cd = (b^{-1}d)(d^{-1}a^{-1}cd) \in N$, and so $abN = cdN$. $\square$

3.8.4 Theorem.   If N is a normal subgroup of G, then the set of left cosets of N forms a group under the coset multiplication given by

$$aNbN = abN$$

for all a, $b \in G$.

Proof. Proposition 3.8.3 shows that the given multiplication of left cosets is in fact well-defined. The subgroup N itself serves as an identity element, since $N = eN$ and therefore $eNaN = aN$ and $aNeN = aN$ for all $a \in G$. Furthermore, the inverse of aN is $a^{-1}N$ because $aNa^{-1}N = eN$ and $a^{-1}NaN = eN$. Finally, to show the associative law, let a, b, $c \in G$. Then

$$(aNbN)cN = abNcN = (ab)cN = a(bc)N = aNbcN = aN(bNcN).$$

This completes the proof. $\square$

3.8.5 Definition.   If N is a normal subgroup of G, then the group of left cosets of N in G is called the *factor group* of G determined by N. It will be denoted by G/N.

Example 3.8.5 (Order of an Element in G/N)

Let N be a normal subgroup of G. It is interesting to compute the order of an element of G/N. If $a \in G$, then the order of aN is the smallest positive integer n such that $(aN)^n = a^nN = N$. That is, the order of aN is the smallest positive integer n such that $a^n \in N$. $\square$

3.8.6 Proposition.   Let N be a normal subgroup of G.

(a) The natural projection mapping $\pi : G \to G/N$ defined by $\pi(a) = aN$ is a homomorphism, and $\ker(\pi) = N$.

(b) There is a one-to-one correspondence between subgroups of G/N and subgroups H of G with $H \supseteq N$. Under this correspondence, normal subgroups correspond to normal subgroups.

Proof. (a) Let $a, b \in G$. Then $\pi(ab) = abN = aNbN = \pi(a)\pi(b)$, showing that π is a homomorphism. Furthermore, we have $a \in \ker(\pi)$ if and only if $aN = N$, and this is equivalent to the statement that $a \in N$.

(b) Since π is a homomorphism, we can apply Proposition 3.7.6. If K is a subgroup of G/N, then $\pi^{-1}(K)$ is a subgroup of G that contains N, and if K is normal, then so is $\pi^{-1}(K)$. Since π is onto, it is clear that assigning to each subgroup of G/N its inverse image in G is a one-to-one mapping. To show that this mapping is onto, let H be a subgroup of G with $H \supseteq N$. We claim that $H = \pi^{-1}(\pi(H))$. By definition,

$$\pi^{-1}(\pi(H)) = \{a \in G \mid \pi(a) \in \pi(H)\},$$

and so it is clear that $H \subseteq \pi^{-1}(\pi(H))$. To show the reverse inclusion, let $a \in \pi^{-1}(\pi(H))$. Then $aN = hN$ for some $h \in H$, so we have $ah^{-1} \in N$. But since $N \subseteq H$, this implies $ah^{-1} \in H$, and so $a = (ah^{-1})h \in H$. Finally, it follows directly from Proposition 3.7.6 that if H is normal, then so is its image H/N under the projection map. $\square$

Example 3.8.6

Let $G = \mathbf{Z}_{12}$ and let $N = \{[0], [3], [6], [9]\}$, the cyclic subgroup generated by the congruence class of 3. Then there are three elements of G/N, found by adding [1] and [2] to each element of N:

$$\{[0], [3], [6], [9]\}, \qquad \{[1], [4], [7], [10]\}, \qquad \{[2], [5], [8], [11]\}.$$

Since we only have three elements, the factor group G/N must be isomorphic to $\mathbf{Z}_3$. This can also be seen by considering the order of the equivalence class of [1]. Its order is the smallest positive multiple that gives the identity element of G/N, and so that is the smallest positive multiple of [1] that belongs to N. Thus [1] has order 3. $\square$

Let N be a normal subgroup of G. When we introduced the multiplication of cosets we did so by choosing representatives and multiplying them. It is very useful to have another way of viewing the product of cosets as products of sets.

For subsets S_1, S_2 of G, in Definition 3.3.1 we used $S_1 S_2$ to denote all elements $g \in G$ that have the form $g = s_1 s_2$ for some $s_1 \in S_1$ and $s_2 \in S_2$. Using our notation for multiplicative cosets, the elements of G/N have the form aN for elements $a \in G$. We can then consider the product $(aN)(bN)$, for $a, b \in G$, as the product of two subsets of G. To show that this product is equivalent to the one used in Theorem 3.8.4, we need to show that $(aN)(bN) = abN$.

3.8.7 Proposition. Let H be a subgroup of the group G. The following conditions are equivalent:

(1) H is a normal subgroup of G;

(2) $aH = Ha$ for all $a \in G$;

(3) for all $a, b \in G$, abH is the set theoretic product $(aH)(bH)$;

(4) for all $a, b \in G$, $ab^{-1} \in H$ if and only if $a^{-1}b \in H$.

Proof. (1) implies (2): Let $a \in G$. To show that $aH \subseteq Ha$, let $h \in H$. Then $aha^{-1} \in H$ since H is normal, and therefore $aha^{-1} = h'$ for some $h' \in H$. Thus $ah = h'a \in Ha$. The proof of the reverse inclusion is similar, so $aH = Ha$.

(2) implies (3): Assume that $Hb = bH$ for all $b \in G$. It is always true that $abH \subseteq (aH)(bH)$, since any element of the form abh, with $h \in H$, can be re-written as $abh = (ae)(bh)$, and the latter form shows it to be in $(aH)(bH)$. To show the reverse inclusion, let $(ah_1)(bh_2) \in (aH)(bH)$, for $h_1, h_2 \in H$. Then $h_1b \in Hb$ and $Hb = bH$, so $h_1b = bh_3$ for some $h_3 \in H$. Thus $(ah_1)(bh_2) = ab(h_3h_2) \in abH$.

(3) implies (1): If $(aH)(bH) = abH$ for all $a, b \in G$, then in particular we have $(aH)(a^{-1}H) = H$. Thus for any element $h \in H$, we have $aha^{-1} = aha^{-1}e \in H$, showing that H is normal.

(2) if and only if (4): Condition (2) holds if and only if the left and right cosets of H coincide. The left cosets of H are the equivalence classes of the equivalence relation determined by setting $a \sim b$ if $a^{-1}b \in H$, for all $a, b \in G$. The right cosets are the equivalence classes determined by the symmetric condition $ab^{-1} \in H$. Since the two equivalence relations coincide if and only if their equivalence classes are identical, condition (4) holds if and only if condition (2) holds. $\square$

Example 3.8.7 (Subgroups of Index 2 are Normal)

Let H be a normal subgroup of G, and assume that H has only two left cosets. Then these must be H and $G - H$, and since these must also be the right cosets, it follows from Proposition 3.8.7 that H is normal. This gives a much simpler proof than the one outlined in Exercise 10 of Section 3.7. $\square$

In Section 3.7, for any group homomorphism $\phi : G_1 \to G_2$ we constructed a group, which we denoted by G_1/ϕ. It consisted of the congruence classes of the relation $a \sim b$ if $\phi(a) = \phi(b)$. By Proposition 3.7.9 this is precisely the relation of congruence modulo $\ker(\phi)$. Thus if we let $K = \ker(\phi)$, we have $[a]_\phi = aK$ for all $a \in G_1$. We now have in hand the notation in which the fundamental homomorphism theorem is usually stated. We include an outline of the proof using coset notation.

3.8.8 Theorem (Fundamental Homomorphism Theorem).   If $\phi : G_1 \rightarrow G_2$ is a group homomorphism, with $K = \ker(\phi)$, then $G_1/K \cong \phi(G_1)$.

Proof. For each coset aK of G_1/K define $\overline{\phi}(aK) = \phi(a)$. The function $\overline{\phi}$ is well-defined since if $aK = bK$ for $a, b \in G_1$, then $a = bk$ for some $k \in \ker(\phi)$, and so

$$\phi(a) = \phi(bk) = \phi(b)\phi(k) = \phi(b).$$

It is a homomorphism since

$$\overline{\phi}(aKbK) = \overline{\phi}(abK) = \phi(ab) = \phi(a)\phi(b) = \overline{\phi}(aK)\overline{\phi}(bK)$$

for all $a, b \in G_1$. If $\phi(a) = \phi(b)$ for $a, b \in G_1$, then we have $\phi(ab^{-1}) = \phi(a)(\phi(b))^{-1} = e$ and so $ab^{-1} \in K$, showing that $aK = bK$. Thus $\overline{\phi}$ is one-to-one, and it is clearly onto.   □

Let $\phi : G_1 \rightarrow G_2$ be a group homomorphism. Then we know that ϕ is one-to-one if and only if its kernel is the trivial subgroup of G_1, and in this case G_1 is isomorphic to $\phi(G_1)$. At the other extreme, if $\ker(\phi) = G_1$, then ϕ is the trivial mapping, which sends every element to the identity. The group G is called a *simple* group if it has no proper nontrivial normal subgroups. Thus if G_1 is a simple group and $\phi : G_1 \rightarrow G_2$ is a group homomorphism, then ϕ is either one-to-one or trivial.

For any prime p, $\mathbf{Z}_p$ is simple, since it has no proper nontrivial subgroups of any kind (every nonzero element is a generator). We will study simple groups in Chapter 7 because they play an important role in determining the structure of groups. We will show that A_n is simple, if $n \geq 5$, and this fact will play an important role in the proof that the general equation of degree 5 is not solvable by radicals.

Example 3.8.8 ($\mathbf{Z}_n/m\mathbf{Z}_n \cong \mathbf{Z}_m$ if $m|n$)

In Example 3.8.6 we computed a particular factor group of $\mathbf{Z}_{12}$. It is not difficult to do this computation for any factor group of $\mathbf{Z}_n$. Any homomorphic image of a cyclic group is again cyclic, and so all factor groups of $\mathbf{Z}_n$ must be cyclic, and hence isomorphic to $\mathbf{Z}_m$ for some m. However, we prefer to give a direct proof of the isomorphism. Let n be any positive integer. Then the subgroups of $\mathbf{Z}_n$ correspond to divisors of n, and so to describe all factor groups of $\mathbf{Z}_n$ we only need to describe $\mathbf{Z}_n/m\mathbf{Z}_n$ for all positive divisors m of n.

Define $\phi : \mathbf{Z}_n \rightarrow \mathbf{Z}_m$ by $\phi([x]_n) = [x]_m$. The mapping is well-defined since $m|n$. It is a homomorphism by Example 3.7.4, which describes homomorphisms of cyclic groups, and it is clearly onto. Finally,

$$\ker(\phi) = \{[x]_n \mid [x]_m = [0]_m\} = \{[x]_n \mid x \text{ is a multiple of } m\}$$

and so $\ker(\phi)$ is equal to $m\mathbf{Z}_n$. It follows from the fundamental homomorphism theorem that $\mathbf{Z}_n/m\mathbf{Z}_n \cong \mathbf{Z}_m$.   □

Example 3.8.9

Let G be the dihedral group D_4, described by elements a of order 4 and b or order 2, satisfying the identity $ba = a^3b$. Let N be the subgroup $\{e, a^2\}$. The computation $b \cdot a^2 = a^3ba = a^3a^3b = a^2b$ shows that a^2 commutes with b, and since a^2 commutes with powers of a, it must commute with every element of G. This implies, in particular, that N is normal.

The factor group G/N consists of the four cosets N, $aN = \{a, a^3\}$, $bN = \{b, a^2b\}$, and $abN = \{ab, a^3b\}$. We have $aNaN = N$, $bNbN = N$ and $abNabN = N$, which shows that every nontrivial element of G/N has order 2. Therefore G/N is isomorphic to $\mathbf{Z}_2 \times \mathbf{Z}_2$ rather than $\mathbf{Z}_4$.   □

Example 3.8.10

Let $G = \mathbf{Z}_4 \times \mathbf{Z}_4$ and let $N = \{(0, 0), (2, 0), (0, 2), (2, 2)\}$. (We have omitted the brackets denoting congruence classes because that makes the notation too cumbersome.) There are four cosets of this subgroup, which we can denote as follows:

$$N, \qquad (1, 0) + N, \qquad (0, 1) + N, \qquad (1, 1) + N.$$

The representatives of the cosets have been carefully chosen to show that each nontrivial element of the factor group has order 2, making the factor group G/N isomorphic to $\mathbf{Z}_2 \times \mathbf{Z}_2$.

Let K be the subgroup $\{(0, 0), (1, 0), (2, 0), (3, 0)\}$. It can be checked that the cosets of K have the following form:

$$K, \qquad (0, 1) + K, \qquad (0, 2) + K, \qquad (0, 3) + K.$$

Having chosen these representatives, it is clear that the factor group must be cyclic, generated by $(0, 1) + K$, and so G/K is isomorphic to $\mathbf{Z}_4$.   □

The previous example provides the motivation for a general comment concerning factor groups of direct products. One way to define a subgroup of a direct product $G_1 \times G_2$ is to use normal subgroups $N_1 \subseteq G_1$ and $N_2 \subseteq G_2$ to construct the following subgroup:

$$N_1 \times N_2 = \{(g_1, g_2) \mid g_1 \in N_1, g_2 \in N_2\} \subseteq G_1 \times G_2.$$

In the previous example we computed the factor groups for the subgroups $2\mathbf{Z}_4 \times 2\mathbf{Z}_4$ and $\mathbf{Z}_4 \times \langle 0 \rangle$, respectively. The factor group is easy to describe in this case. (See Example 3.8.11 for a subgroup of a direct product that cannot be described in this manner.)

Let N_1 and N_2 be normal subgroups of the groups G_1 and G_2, respectively. Then $N_1 \times N_2$ is a normal subgroup of the direct product $G_1 \times G_2$ and

$$(G_1 \times G_2)/(N_1 \times N_2) \cong (G_1/N_1) \times (G_2/N_2).$$

To prove this statement, define $\phi : G_1 \times G_2 \rightarrow G_1/N_1 \times G_2/N_2$ by $\phi(g_1, g_2) = (g_1N_1, g_2N_2)$ for all $g_1 \in G_1$, $g_2 \in G_2$. It is easy to verify the following: ϕ is a homomorphism, ϕ is onto, and ker(ϕ) = $N_1 \times N_2$. The desired conclusions follow from the fundamental homomorphism theorem.

Example 3.8.11

Let $G = \mathbf{Z}_4 \times \mathbf{Z}_4$ and let N be the "diagonal" subgroup $\{(0, 0), (1, 1), (2, 2), (3, 3)\}$ generated by $(1, 1)$. The factor group G/N will have four elements, and so it must be isomorphic to either $\mathbf{Z}_4$ or $\mathbf{Z}_2 \times \mathbf{Z}_2$. The smallest positive multiple of $(1, 0)$ that belongs to N is $4 \cdot (1, 0) = (0, 0)$, showing that the coset $(1, 0) + N$ has order 4. Thus G/N is cyclic, and hence isomorphic to $\mathbf{Z}_4$. $\square$

Example 3.8.12

To give an example involving a group of matrices, consider $GL_n(\mathbf{R})$ and the subgroup $SL_n(\mathbf{R})$ of all $n \times n$ matrices with determinant 1. Define $\phi : GL_n(\mathbf{R}) \to \mathbf{R}^{\times}$ by letting $\phi(A) = \det(A)$, for any matrix $A \in GL_n(\mathbf{R})$. The formula $\det(AB) = \det(A) \det(B)$ shows that ϕ is a homomorphism. It is easy to see that ϕ is onto, and ker(ϕ) is precisely $SL_n(\mathbf{R})$. Applying the fundamental homomorphism theorem, we see that $SL_n(\mathbf{R})$ is a normal subgroup, and we obtain $GL_n(\mathbf{R})/SL_n(\mathbf{R}) \cong \mathbf{R}^{\times}$. $\square$

EXERCISES: SECTION 3.8

1. In $\mathbf{Z}_{24}$, list all cosets of $\langle [3] \rangle$ and of $\langle [16] \rangle$.
2. Let $G = \mathbf{Z}_3 \times \mathbf{Z}_6$, let $H = \langle (1, 2) \rangle$, and let $K = \langle (1, 3) \rangle$. List all cosets of H and K.
3. For the subgroup $\{e, ab\}$ of S_3, list all left and right cosets.
4. For the subgroups $\{e, a^2\}$ and $\{e, b\}$ of D_4, list all left and right cosets.
5. Let G be a group with subgroup H. Prove that there is a one-to-one correspondence between the left and right cosets of H. (Your proof must include the case in which G is infinite.)
6. Prove that if N is a normal subgroup of G, and H is any subgroup of G, then $H \cap N$ is a normal subgroup of H.
7. Let H be a subgroup of G, and let $a \in G$. Show that aHa^{-1} is a subgroup of G with $|aHa^{-1}| = |H|$.
8. Let H be a subgroup of G. Show that H is normal in G if and only if $aHa^{-1} = H$ for all $a \in G$.
9. Let G be a finite group, and let n be a divisor of $|G|$. Show that if H is the only subgroup of G of order n, then H must be normal in G.
10. Let N be a normal subgroup of index m in G. Show that $a^m \in N$ for all $a \in G$.
11. Let N be a normal subgroup of G. Show that the order of any coset aN in G/N is a divisor of $o(a)$, when $o(a)$ is finite.
12. Let H and K be normal subgroups of G such that $H \cap K = \langle e \rangle$. Show that $hk = kh$ for all $h \in H$ and $k \in K$.

13. Let N be a normal subgroup of G. Prove that G/N is abelian if and only if N contains all elements of the form $aba^{-1}b^{-1}$ for $a, b \in G$.

14. Let N be a subgroup of the center of G. Show that if G/N is a cyclic group, then G must be abelian.

15. Find all factor groups of the dihedral group D_4.

16. Let H and K be normal subgroups of G such that $H \cap K = \langle e \rangle$ and $HK = G$. Prove that $G \cong H \times K$.

17. Show that $\mathbf{R}^{\times}/\langle -1 \rangle$ is isomorphic to the group of positive real numbers under multiplication.

18. Compute the factor group $(\mathbf{Z}_6 \times \mathbf{Z}_4)/\langle (3, 2) \rangle$.

19. Show that $(\mathbf{Z} \times \mathbf{Z})/\langle (0, 1) \rangle$ is an infinite cyclic group.

20. Show that $(\mathbf{Z} \times \mathbf{Z})/\langle (1, 1) \rangle$ is an infinite cyclic group.

21. Show that $(\mathbf{Z} \times \mathbf{Z})/\langle (2, 2) \rangle$ is not a cyclic group.

22. Let A be an infinite set. Let H be the set of all elements $\sigma \in \text{Sym}(A)$ such that $\sigma(x) = x$ for all but finitely many $x \in A$. Prove that H is a normal subgroup of $\text{Sym}(A)$.

23. Let H and N be subgroups of a group G, and assume that N is a normal subgroup of G. Prove the following statements:
   (a) N is a normal subgroup of HN.
   (b) Each element of HN/N has the form hN, for some $h \in H$.
   (c) $\phi : H \to HN/N$ defined by $\phi(h) = hN$ for all $h \in H$ is an onto homomorphism.
   (d) $HN/N \cong H/K$, where $K = H \cap N$.

24. Let H and N be normal subgroups of a group G, with $N \subseteq H$. Define $\phi : G/N \to G/H$ by $\phi(aN) = aH$.
   (a) Show that ϕ is a well-defined onto homomorphism.
   (b) Show that $(G/N)/(H/N) \cong G/H$.

4

Polynomials

In this chapter we will study polynomial equations from a very concrete point of view. We will find it convenient, though, to introduce one abstract concept. We will study polynomials in which the coefficients can come from any set in which addition, subtraction, multiplication, and division are possible. In this setting we will prove numerous results analogous to those for the set of integers: a prime factorization theorem, a division algorithm, and an analog of the Euclidean algorithm. Congruences for polynomials will be used to show that roots of polynomials can always be found, by enlarging (if necessary) the set over which the polynomials are considered.

In an appendix at the end of the chapter we give solutions by radicals to the general cubic and quartic equations. We now give a brief historical note on the developments leading up to the formulation of these solutions.

Leonardo of Pisa (1170–1250) published *Liber abbaci* in 1202. (He was called Fibonacci, as a member of the Bonacci family.) His book contained, among other topics, an introduction to Hindu-Arabic numerals, a variety of problems of interest in trade, a chapter on calculations with square roots and cube roots, and a chapter containing a systematic treatment of linear and quadratic equations. Leonardo's book built on work from the Arab world and set the stage for further development.

Arab mathematicians, including Omar Khayyam (about 1100), had studied cubic equations. Their solutions were found as intersections of conic sections. The Italians were also interested in solving cubic equations. For example, in a book published in the middle of the fourteenth century there are problems involving interest rates which lead to cubic equations. At that date, solutions to certain very special cases of cubic equations were already known.

The general cubic equation

$$ax^3 + bx^2 + cx + d = 0$$

can be reduced to the form

$$x^3 + px + q = 0$$

by dividing through by a and then introducing a new variable $y = x + b/3a$. If only positive coefficients and positive values of x are allowed, then there are three cases:

$$x^3 + px = q$$
$$x^3 = px + q$$
$$x^3 + q = px.$$

The first case was solved by Scipione del Ferro, a professor at the University of Bologna, who died in 1526. His solution was never published, although he told it to a few friends.

In 1535 there was a mathematical contest between one of del Ferro's students and a Venetian mathematics teacher named Niccolò Fontana (1499–1557). (He is usually known as Tartaglia, which means the "stammerer.") Tartaglia found the general solution to the equations (of the first type above) posed by del Ferro's student and won the contest. Then four years later Tartaglia was approached by Gerolamo Cardano, who lived in Milan, with a request to share his method of solution. Tartaglia finally did so, but only after obtaining a sworn oath from Cardano that he would never publish the solution.

Shortly afterward, Cardano succeeded in extending the method of solution of cubics of the first type to the remaining two types. In doing so, he was probably the first person to use complex numbers in the form $a + \sqrt{-b}$. Cardano's younger friend and secretary, Lodovico Ferrari (1522–1565), discovered that the general fourth degree equation can be reduced to a cubic equation, and hence could be solved. This put Cardano and Ferrari in a difficult position, being in possession of important new results which they could not publish because of the oath Cardano had sworn. In 1543 they were able to examine del Ferro's papers in Bologna, and found out that he had indeed discovered a solution to the one case of the cubic.

Cardano decided to publish the solution of the cubic and fourth-degree equation in a book entitled *Ars Magna* (1545), in which he stated that del Ferro deserved credit for the solution of cubic equations of the first type, although he had learned it from Tartaglia, who had rediscovered it. He stated that he had extended the solution to the two remaining cases, and that Ferrari had given him the solution to the fourth-degree equation. This marked the beginning of a dispute between Cardano and Tartaglia, and a year later Tartaglia published the story of the oath, including its full text.

In the appendix to this chapter we also include the solution of the general fourth-degree equation that was given by René Descartes (1596–1650). Much of our modern notation is due to Descartes, who used the last letters of the alphabet to denote unknown quantities, and the first letters to denote known quantities. He used the term "imaginary numbers" for expressions of the form $a \pm \sqrt{-b}$, and was the first to systematically write powers with exponents in the modern form. Analytic geometry, whose primary aim is to solve geometric problems using algebraic methods, was invented independently (and very nearly simultaneously) by Descartes and Fermat.

4.1 FIELDS; ROOTS OF POLYNOMIALS

We introduced the concept of a field in Section 3.3 in order to work with the group $GL_n(F)$ of invertible $n \times n$ matrices with entries in a field F. We now want to work with polynomials with coefficients in an arbitrary field, and so we need to develop this concept further. In order to be able to add, subtract, multiply, and divide the polynomials under discussion, we need to be able to assume that the same operations can be performed on the set of coefficients. Of course, all four operations are possible in the familiar fields $\mathbf{Q}$, $\mathbf{R}$, $\mathbf{C}$, and $\mathbf{Z}_p$, where p is prime. However, we also need to be able to consider other sets as coefficients.

For example, a typical procedure in our later work is the following: To find all solutions of a polynomial equation (with rational coefficients), we proceed in stages. After we have found one solution, we let F be the smallest set of complex numbers that is a field and contains both $\mathbf{Q}$ and the solution. We then treat the original polynomial as if its coefficients come from F. Since the polynomial has a root in F, it can be factored into polynomials of lower degree (with coefficients in F). Then each of the factors can be investigated in a similar manner, and the process is continued until a field E is obtained that is the smallest field inside $\mathbf{C}$ containing $\mathbf{Q}$ and all solutions of the original equation.

Example 4.1.1

Given the equation

$$x^4 - 2x^3 - 3x^2 + 4x + 2 = 0,$$

we can factor to obtain

$$(x^2 - 2)(x^2 - 2x - 1) = 0,$$

and so $\sqrt{2}$ is a solution. To find the smallest field that contains all coefficients and roots of the equation, we first find the smallest set of complex numbers that contains $\mathbf{Q}$ and $\sqrt{2}$ and is closed under addition, subtraction, multiplication, and division. We denote this set by $\mathbf{Q}(\sqrt{2})$.

It is obvious that $\mathbf{Q}(\sqrt{2}) \supseteq \{a + b\sqrt{2} \mid a, b \in \mathbf{Q}\}$, since we must have closure under addition and subtraction. In fact, this is as large a set as is necessary. It is easy to see that $\{a + b\sqrt{2} \mid a, b \in \mathbf{Q}\}$ is closed under addition and subtraction. It is only slightly more difficult to check that the same is true for multiplication and division:

$$(a + b\sqrt{2})(c + d\sqrt{2}) = ac + ad\sqrt{2} + bc\sqrt{2} + 2bd$$

$$= (ac + 2bd) + (ad + bc)\sqrt{2},$$

$$\frac{a + b\sqrt{2}}{c + d\sqrt{2}} = \frac{(a + b\sqrt{2})(c - d\sqrt{2})}{(c + d\sqrt{2})(c - d\sqrt{2})}$$

$$= \left(\frac{ac - 2bd}{c^2 - 2d^2}\right) + \left(\frac{bc - ad}{c^2 - 2d^2}\right)\sqrt{2}.$$

In both cases the answer has the correct form: a rational number plus a rational number times the square root of 2. In dividing, we must of course assume that either $c \neq 0$ or $d \neq 0$, and then $c^2 - 2d^2 \neq 0$ since $\sqrt{2}$ is not a rational number, which also shows that $c - d\sqrt{2} \neq 0$.

Thus $\mathbf{Q}(\sqrt{2})$ is the smallest field that contains $\mathbf{Q}$ and the one root $\sqrt{2}$. The remaining roots are $-\sqrt{2}$, $1 + \sqrt{2}$, and $1 - \sqrt{2}$, which are all contained in $\mathbf{Q}(\sqrt{2})$, so we have found the smallest field that contains the coefficients and roots of the equation

$$x^4 - 2x^3 - 3x^2 + 4x + 2 = 0. \quad \square$$

Example 4.1.2

The polynomial $x^4 - x^2 - 2$ factors as $(x^2 - 2)(x^2 + 1)$. In this case the field $\mathbf{Q}(\sqrt{2})$ contains two roots $\pm\sqrt{2}$, but not the two remaining roots $\pm i$, and so we need a larger field to be certain of including all roots. Such a field will be constructed in Example 4.4.3. $\square$

In Definition 3.3.5 we stated the definition of a field in group theoretic terms: the elements of a field form an abelian group under addition, the nonzero elements form an abelian group under multiplication, and addition and multiplication are connected by the distributive law. Roughly speaking, then, a field is a set in which the operations of addition, subtraction, multiplication, and division can be performed. We now repeat the definition of a field, with somewhat more detail, and without reference to groups.

4.1.1 Definition. Let F be a set on which two binary operations are defined, called addition and multiplication, and denoted by $+$ and $\cdot$ respectively. Then F is called a *field* with respect to these operations if the following properties hold:

(i) *Closure.* For all $a, b \in F$ the sum $a + b$ and the product $a \cdot b$ are uniquely defined and belong to F.

(ii) *Associative laws.* For all $a, b, c \in F$,

$$a + (b + c) = (a + b) + c \quad \text{and} \quad a \cdot (b \cdot c) = (a \cdot b) \cdot c.$$

(iii) *Commutative laws.* For all $a, b, \in F$,

$$a + b = b + a \quad \text{and} \quad a \cdot b = b \cdot a.$$

(iv) *Distributive laws.* For all $a, b, c \in F$,

$$a \cdot (b + c) = (a \cdot b) + (a \cdot c) \quad \text{and} \quad (a + b) \cdot c = (a \cdot c) + (b \cdot c).$$

(v) *Identity elements.* The set F contains an *additive identity* element, denoted by 0, such that for all $a \in F$,

$$a + 0 = a \quad \text{and} \quad 0 + a = a.$$

The set F also contains a *multiplicative identity* element, denoted by 1 (and assumed to be different from 0) such that for all $a \in F$,

$$a \cdot 1 = a \quad \text{and} \quad 1 \cdot a = a.$$

(vi) *Inverse elements.* For each $a \in F$, the equations

$$a + x = 0 \quad \text{and} \quad x + a = 0$$

have a solution $x \in F$, called an *additive inverse* of a, and denoted by $-a$.

For each nonzero element $a \in F$, the equations

$$a \cdot x = 1 \quad \text{and} \quad x \cdot a = 1$$

have a solution $x \in F$, called a *multiplicative inverse* of a, and denoted by a^{-1}.

The basic examples you should keep in mind in this section are **Q, R, C,** and $\mathbf{Z}_p$, where p is prime. In Section 4.4 we will show how to construct other important examples of fields, by generalizing the construction of $\mathbf{Q}(\sqrt{2})$ given in Example 4.1.1. This will also allow us to construct finite fields, which have many important applications. We now provide the first example of a finite field different from $\mathbf{Z}_p$, for p prime.

Example 4.1.3

The following set of matrices, with entries from $\mathbf{Z}_2$, forms a field:

$$F = \left\{ \begin{bmatrix} 0 & 0 \\ 0 & 0 \end{bmatrix}, \begin{bmatrix} 1 & 0 \\ 0 & 1 \end{bmatrix}, \begin{bmatrix} 1 & 1 \\ 1 & 0 \end{bmatrix}, \begin{bmatrix} 0 & 1 \\ 1 & 1 \end{bmatrix} \right\}.$$

Here we have omitted the brackets from the congruence classes [0] and [1], so that we simply have $1 + 1 = 0$, etc. You should check that F is closed under addition and multiplication. The associative and distributive laws hold for all matrices. You can check that these particular matrices commute under multiplication. The additive identity is $\begin{bmatrix} 0 & 0 \\ 0 & 0 \end{bmatrix}$, and the multiplicative identity is $\begin{bmatrix} 1 & 0 \\ 0 & 1 \end{bmatrix}$. You should check that F has both additive and multiplicative inverses. ☐

Each nonzero element of a field F is invertible; we will use $F^\times$ to denote the set of all nonzero elements of F. When there is no risk of confusion, we will often write ab instead of $a \cdot b$. If we assume that multiplication is done before addition, we can often eliminate parentheses. For example, the distributive laws can be written as $a(b + c) = ab + ac$ and $(a + b)c = ac + bc$. You should also note that the list of properties used to define a field is redundant in a number of places. For example, since multiplication is commutative, we really only need to state one of the distributive laws.

To simplify matters, we have dealt only with the two operations of addition and multiplication. To define subtraction we use the additive inverse of an element: $a - b = a + (-b)$. Similarly, we use the multiplicative inverse of an element to define division, as follows: If $b \neq 0$, then $a \div b = ab^{-1}$. We will sometimes write a/b in place of ab^{-1}. There is a problem with these definitions, since $a - b$ and $a \div b$ should be unique, whereas we do not as yet even know that $-b$ and b^{-1} are unique. The next proposition takes care of this problem.

4.1.2 Proposition. Let F be a field, with $a, b, c \in F$.
  (a) *Cancellation laws.* If $a + c = b + c$, then $a = b$. If $c \neq 0$ and $a \cdot c = b \cdot c$, then $a = b$.
  (b) *Uniqueness of identity elements.* If $a + b = a$, then $b = 0$. If $a \cdot c = a$ and $a \neq 0$, then $c = 1$.
  (c) *Uniqueness of inverses.* If $a + b = 0$, then $b = -a$. If $a \neq 0$, and $ab = 1$, then $b = a^{-1}$.

Proof. Proposition 3.1.2 shows that identity elements and inverses are unique for any binary operation. The axioms that hold for the field F show that F is an abelian group under addition, and so the additive cancellation law follows from Proposition 3.1.6 of Chapter 3.

Suppose that $a \cdot c = b \cdot c$, with $c \neq 0$. Since $c \neq 0$, c^{-1} exists, so the proof can be given in the following steps:

$$(ac)c^{-1} = (bc)c^{-1}$$
$$a(cc^{-1}) = b(cc^{-1})$$
$$a \cdot 1 = b \cdot 1$$
$$a = b. \quad ☐$$

4.1.3 Proposition.   Let F be a field.
(a) For all $a \in F$, $a \cdot 0 = 0$.
(b) If $a, b \in F$ with $a \neq 0$ and $b \neq 0$, then $ab \neq 0$.
(c) For all $a \in F$, $-(-a) = a$.
(d) For all $a, b \in F$, $(a)(-b) = (-a)(b) = -ab$.
(e) For all $a, b \in F$, $(-a)(-b) = ab$.

Proof. (a) We will use the fact that $0 + 0 = 0$. By the distributive law,

$$a \cdot 0 + a \cdot 0 = a \cdot (0 + 0) = a \cdot 0 = a \cdot 0 + 0,$$

so the cancellation law for addition shows that $a \cdot 0 = 0$.
   (b) If $a \neq 0$ and $a \cdot b = 0$, then $a \cdot b = a \cdot 0$ and the cancellation law for multiplication shows that $b = 0$.
   (c) In words, the equation $-(-a) = a$ states that the additive inverse of $-a$ is a, and this follows from the equation $-a + a = 0$ which defines $-a$.
   (d) Using the distributive law,

$$a \cdot b + a \cdot (-b) = a \cdot (b + (-b)) = a \cdot 0 = 0,$$

which shows that $(a)(-b)$ is the additive inverse of ab, and so $(a)(-b) = -(ab)$. Similarly, $(-a)(b) = -ab$
   (e) Now consider $(-a)(-b)$. By what we have just shown,

$$(-a)(-b) = -((-a)(b)) = -(-ab) = ab. \quad \square$$

Having proved some elementary results on fields, we are now ready to discuss polynomials with coefficients in a field.

4.1.4 Definition.   Let F be a field. If $a_m, a_{m-1}, \ldots, a_1, a_0 \in F$, then any expression of the form

$$a_m x^m + a_{m-1} x^{m-1} + \ldots + a_1 x + a_0$$

is called a *polynomial over F* in the *indeterminate x* with coefficients $a_m, a_{m-1}, \ldots, a_0$. The set of all polynomials with coefficients in F is denoted by $F[x]$.
   If n is the largest nonnegative integer such that $a_n \neq 0$, then we say that the polynomial $f(x) = a_n x^n + \ldots + a_0$ has *degree n*, written $\deg(f(x)) = n$, and a_n is called the *leading coefficient* of $f(x)$. If the leading coefficient is 1, then $f(x)$ is said to be *monic*.

According to this definition, the zero polynomial (each of whose coefficients is zero) has no degree. For convenience, it is often assigned $-\infty$ as a degree. A constant polynomial a_0 has degree 0 when $a_0 \neq 0$. Thus a polynomial belongs to the coefficient field if and only if it has degree 0 or $-\infty$.
   Two polynomials are equal by definition if they have the same degree and all corresponding coefficients are equal. It is important to distinguish between the

polynomial $f(x)$ as an element of $F[x]$ and the corresponding polynomial function from F into F defined by substituting elements of F in place of x. If $f(x) = a_m x^m + \ldots + a_0$ and $c \in F$, then $f(c) = a_m c^m + \ldots + a_0$. In fact, if F is a finite field, it is possible to have two different polynomials that define the same polynomial function. For example, let F be the field $\mathbf{Z}_5$ and consider the polynomials $x^5 - 2x + 1$ and $4x + 1$. For any $c \in \mathbf{Z}_5$, by Fermat's theorem (Theorem 1.4.10) we have $c^5 \equiv c \pmod 5$, and so

$$c^5 - 2c + 1 \equiv -c + 1 \equiv 4c + 1 \pmod 5,$$

which shows that as polynomial functions $x^5 - 2x + 1$ and $4x + 1$ are identical.

For the polynomials

$$f(x) = a_m x^m + a_{m-1} x^{m-1} + \ldots + a_1 x + a_0$$

and

$$g(x) = b_n x^n + b_{n-1} x^{n-1} + \ldots + b_1 x + b_0,$$

the sum of $f(x)$ and $g(x)$ is defined by just adding corresponding coefficients. The product $f(x) g(x)$ is defined to be

$$a_m b_n x^{n+m} + \ldots + (a_2 b_0 + a_1 b_1 + a_2 b_0) x^2 + (a_1 b_0 + a_0 b_1) x + a_0 b_0.$$

The coefficient c_k of $f(x) g(x)$ can be described by the formula

$$c_k = \sum_{i+j=k} a_i b_j = \sum_{i=0}^{k} a_i b_{k-i}.$$

This definition of the product is consistent with what we would expect to obtain using a naive approach: Expand the product using the distributive law repeatedly (this amounts to multiplying each term by every other) and then collect similar terms.

With this addition and multiplication, F[x] has properties similar to those of the integers. Checking that the following properties hold is tedious, though not difficult. The necessary proofs use the definitions, and depend on the properties of the coefficient field. We have omitted all details.

(i) Associative laws. For any polynomials $f(x)$, $g(x)$, $h(x)$ over F,

$$f(x) + (g(x) + h(x)) = (f(x) + g(x)) + h(x),$$
$$f(x) \cdot (g(x) \cdot h(x)) = (f(x) \cdot g(x)) \cdot h(x).$$

(ii) Commutative laws. For any polynomials $f(x)$, $g(x)$ over F,

$$f(x) + g(x) = g(x) + f(x),$$
$$f(x) \cdot g(x) = g(x) \cdot f(x).$$

(iii) Distributive laws. For any polynomials $f(x)$, $g(x)$, $h(x)$ over F,

$$f(x) \cdot (g(x) + h(x)) = (f(x) \cdot g(x)) + (f(x) \cdot h(x)),$$

$$(f(x) + g(x)) \cdot h(x) = (f(x) \cdot h(x)) + (g(x) \cdot h(x)).$$

(iv) Identity elements. The additive and multiplicative identities of F, considered as constant polynomials, serve as identity elements.

(v) Additive inverses. For each polynomial $f(x)$ over F, the polynomial $-f(x)$ serves as an additive inverse.

In the formula above, two polynomials are multiplied by multiplying each term of the first by each term of the second, and then collecting similar terms. The next example shows an efficient way to do this, by arranging the similar terms in columns.

$$
\begin{array}{r}
 \quad 2x^4 \qquad\qquad -3x^2 \quad +5x \quad +1 \\
\times \qquad\qquad\qquad\qquad\qquad x^2 \quad -4x \quad -2 \\
\hline
-4x^4 \qquad\qquad +6x^2 \quad -10x \quad -2 \\
-8x^5 \qquad +12x^3 \quad -20x^2 \quad -4x \\
2x^6 \qquad -3x^4 \quad +5x^3 \quad +x^2 \\
\hline
2x^6 \quad -8x^5 \quad -7x^4 \quad +17x^3 \quad -13x^2 \quad -14x \quad -2
\end{array}
$$

4.1.5 Proposition.   If $f(x)$ and $g(x)$ are nonzero polynomials in $F[x]$, then $f(x)g(x)$ is nonzero and

$$\deg(f(x)g(x)) = \deg(f(x)) + \deg(g(x)).$$

Proof. Suppose that

$$f(x) = a_m x^m + \ldots + a_1 x + a_0$$

and

$$g(x) = b_n x^n + \ldots + b_1 x + b_0,$$

with $\deg(f(x)) = m$ and $\deg(g(x)) = n$. Thus $a_m \neq 0$ and $b_n \neq 0$. It follows from the general formula for multiplication of polynomials that the leading coefficient of $f(x)g(x)$ must be $a_m b_n$, which must be nonzero by Proposition 4.1.3(b). Thus the degree of $f(x)g(x)$ is $m + n$ since $a_m b_n$ is the coefficient of x^{m+n}. $\square$

Note that we could extend the statement of Proposition 4.1.5 to include the zero polynomial, provided we would use the convention that assigns to the zero polynomial the degree $-\infty$.

4.1.6 Corollary.   If $f(x)$, $g(x)$, $h(x) \in F[x]$ and $f(x)$ is nonzero, then $f(x)g(x) = f(x)h(x)$ implies $g(x) = h(x)$.

Proof. If $f(x)g(x) = f(x)h(x)$, then we can use the distributive law to rewrite the equation as $f(x)(g(x) - h(x)) = 0$, and since $f(x) \neq 0$, the previous proposition implies that $g(x) - h(x) = 0$, or simply $g(x) = h(x)$. $\square$

Having proved Proposition 4.1.5, we can make some further observations about $F[x]$. If $f(x)g(x) = 1$, then both $f(x)$ and $g(x)$ must be constant polynomials, since the sum of their degrees must be 0. This shows that the only polynomials that have multiplicative inverses are those of degree 0, which correspond to the nonzero elements of F. In this sense $F[x]$ is very far away from being a field itself, although all of the other properties of a field are satisfied.

4.1.7 Definition. Let $f(x)$, $g(x) \in F[x]$. If $f(x) = q(x)g(x)$ for some $q(x) \in F[x]$, then we say that $g(x)$ is a *factor* or *divisor* of $f(x)$, and we write $g(x) \mid f(x)$.

4.1.8 Lemma. For any element $c \in F$, and any positive integer k,

$$(x - c) \mid (x^k - c^k).$$

Proof. A direct computation shows that we have the following factorization:

$$x^k - c^k = (x - c)(x^{k-1} + cx^{k-2} + \ldots + c^{k-2}x + c^{k-1}).$$

Note that the quotient

$$x^{k-1} + cx^{k-2} + \ldots + c^{k-2}x + c^{k-1}$$

has coefficients in F. $\square$

The next theorem shows that for any polynomial $f(x)$, the remainder when $f(x)$ is divided by $x - c$ is $f(c)$. That is,

$$\frac{f(x)}{x - c} = q(x) + \frac{f(c)}{x - c},$$

or simply,

$$f(x) = q(x)(x - c) + f(c).$$

The *remainder* $f(c)$ and *quotient* $q(x)$ are unique.

4.1.9 Theorem (Remainder Theorem). Let $f(x) \in F[x]$ be a nonzero polynomial, and let $c \in F$. Then there exists a polynomial $q(x) \in F[x]$ such that

$$f(x) = q(x)(x - c) + f(c).$$

Moreover, if $f(x) = q_1(x)(x - c) + k$, where $q_1(x) \in F[x]$ and $k \in F$, then $q_1(x) = q(x)$ and $k = f(c)$.

Proof. If $f(x) = a_m x^m + \ldots + a_0$, then

$$f(x) - f(c) = a_m(x^m - c^m) + \ldots + a_1(x - c).$$

By Lemma 4.1.8, $x - c$ is a divisor of each term on the right-hand side of the equation, and so it must be a divisor of $f(x) - f(c)$. Thus

$$f(x) - f(c) = q(x)(x - c)$$

for some polynomial $q(x) \in F[x]$, or equivalently,

$$f(x) = q(x)(x - c) + f(c).$$

If $f(x) = q_1(x)(x - c) + k$, then

$$(q(x) - q_1(x))(x - c) = k - f(c).$$

If $q(x) - q_1(x) \neq 0$, then by Proposition 4.1.5 the left-hand side of the equation has degree ≥ 1, which contradicts the fact that the right-hand side of the equation is a constant. Thus $q(x) - q_1(x) = 0$, which also implies that $k - f(c) = 0$, and so the quotient and remainder are unique. $\square$

Example 4.1.4

Let us follow through the previous proof in the case $F = \mathbf{Q}, f(x) = x^2 + 5x - 2$ and $c = 5$:

$$\begin{aligned}
f(x) - f(5) &= (x^2 + 5x - 2) - (5^2 + 5 \cdot 5 - 2) \\
&= (x^2 - 5^2) + (5x - 25) \\
&= (x + 5)(x - 5) + 5(x - 5) \\
&= (x + 10)(x - 5).
\end{aligned}$$

Thus we obtain $x^2 + 5x - 2 = (x + 10)(x - 5) + 48$. $\square$

4.1.10 Definition.   Let $f(x) = a_m x^m + \ldots + a_0 \in F[x]$. An element $c \in F$ is called a *root* of the polynomial $f(x)$ if $f(c) = 0$, that is, if c is a solution of the polynomial equation $f(x) = 0$.

4.1.11 Corollary.   Let $f(x) \in F[x]$ be a nonzero polynomial, and let $c \in F$. Then c is a root of $f(x)$ if and only if $x - c$ is a factor of $f(x)$. That is, $f(c) = 0$ if and only if $(x - c) \mid f(x)$.

4.1.12 Corollary.   A polynomial of degree n with coefficients in the field F has at most n distinct roots in F.

Proof. The proof will use induction on the degree of the polynomial $f(x)$. The result is certainly true if $f(x)$ has degree 0, that is, if $f(x)$ is a nonzero constant. Now suppose that the result is true for all polynomials of degree $n - 1$. If c is a root of $f(x)$, then by Corollary 4.1.11 we can write $f(x) = q(x)(x - c)$, for some polynomial $q(x)$. If a is any root of $f(x)$, then substituting shows that $q(a)(a - c) = 0$, which implies that either $q(a) = 0$ or $a = c$. By assumption, $q(x)$ has at most

$n - 1$ distinct roots and so this shows that $f(x)$ can have at most $n - 1$ distinct roots which are different from c.   □

EXERCISES: SECTION 4.1

1. Let a be a nonzero element of a field F. Show that $(a^{-1})^{-1} = a$ and $(-a)^{-1} = -a^{-1}$.

2. Let a, b, c be elements of a field F. Prove that if $a \neq 0$, then the equation $ax + b = c$ has a unique solution.

3. Show that the set $\mathbf{Q}(\sqrt{3}) = \{a + b\sqrt{3} \mid a, b \in \mathbf{Q}\}$ is closed under addition, subtraction, multiplication, and division.

4. Show that the set of matrices of the form $\begin{bmatrix} a & b \\ -b & a \end{bmatrix}$ where a, $b \in \mathbf{R}$, is a field.

5. Complete the proof that the set of matrices in Example 4.1.3 forms a field.

6. Show that the following set of matrices over $\mathbf{Z}_2$ forms a field:

$$\begin{bmatrix} 0 & 0 & 0 \\ 0 & 0 & 0 \\ 0 & 0 & 0 \end{bmatrix}, \begin{bmatrix} 1 & 0 & 0 \\ 0 & 1 & 0 \\ 0 & 0 & 1 \end{bmatrix}, \begin{bmatrix} 0 & 1 & 0 \\ 0 & 0 & 1 \\ 1 & 1 & 0 \end{bmatrix}, \begin{bmatrix} 1 & 1 & 0 \\ 0 & 1 & 1 \\ 1 & 1 & 1 \end{bmatrix},$$

$$\begin{bmatrix} 0 & 0 & 1 \\ 1 & 1 & 0 \\ 0 & 1 & 1 \end{bmatrix}, \begin{bmatrix} 1 & 0 & 1 \\ 1 & 0 & 0 \\ 0 & 1 & 0 \end{bmatrix}, \begin{bmatrix} 0 & 1 & 1 \\ 1 & 1 & 1 \\ 1 & 0 & 1 \end{bmatrix}, \begin{bmatrix} 1 & 1 & 1 \\ 1 & 0 & 1 \\ 1 & 0 & 0 \end{bmatrix}.$$

7. Let $f(x)$, $g(x)$, $h(x) \in F[x]$. Show that the following properties hold:
   (a) If $g(x) \mid f(x)$ and $h(x) \mid g(x)$, then $h(x) \mid f(x)$.
   (b) If $h(x) \mid f(x)$ and $h(x) \mid g(x)$, then $h(x) \mid (f(x) \pm g(x))$.
   (c) If $g(x) \mid f(x)$, then $g(x) \cdot h(x) \mid f(x) \cdot h(x)$.
   (d) If $g(x) \mid f(x)$ and $f(x) \mid g(x)$, then $f(x) = kg(x)$ for some $k \in F$.

8. Let p be a prime number. How many polynomials are there of degree n over $\mathbf{Z}_p$?

9. For $f(x) = 2x^3 + x^2 - 2x + 1$, use the method of Theorem 4.1.9 to write $f(x) = q(x)(x - 1) + f(1)$.

10. For $f(x) = x^3 + 3x^2 - 10x + 5$, use the method of Theorem 4.1.9 to write $f(x) = q(x)(x - 2) + f(2)$.

11. Write $f(x) = q(x)(x - c) + f(c)$ for
    (a) $f(x) = 2x^3 + x^2 - 4x + 3$; $c = 1$.
    (b) $f(x) = x^3 - 5x^2 + 6x + 5$; $c = 2$.

12. Show that rational numbers correspond to decimals which are either repeating or terminating.

    Hint: If $q = m/n$, then when dividing m by n to put q into decimal form there are at most n different remainders. Conversely, if d is a repeating decimal, then find s, t such that $10^s d - 10^t d$ is an integer.

13. Let (x_0, y_0), (x_1, y_1), (x_2, y_2) be points in the Euclidean plane $\mathbf{R}^2$ such that x_0, x_1, x_2 are distinct. Show that the formula

$$f(x) = \frac{y_0(x - x_1)(x - x_2)}{(x_0 - x_1)(x_0 - x_2)} + \frac{y_1(x - x_0)(x - x_2)}{(x_1 - x_0)(x_1 - x_2)} + \frac{y_2(x - x_0)(x - x_1)}{(x_2 - x_0)(x_2 - x_1)}$$

defines a function $f(x)$ such that $f(x_0) = y_0, f(x_1) = y_1$, and $f(x_2) = y_2$. (This is a special case of Lagrange's interpolation formula.)

14. Use Lagrange's interpolation formula to find a polynomial $f(x)$ such that $f(1) = 0$, $f(2) = 1$, and $f(3) = 4$.

15. Find a polynomial $f(x)$ such that $f(1) = -15, f(0) = 3, f(2) = -3$, and $f(4) = 15$.

16. Show that

$$f(x) = b_1 + \frac{1}{2}(b_2 - b_0)\left(\frac{x - a}{h}\right) + \frac{1}{2}(b_2 - 2b_1 + b_0)\left(\frac{x - a}{h}\right)^2$$

has the property that $f(a - h) = b_0, f(a) = b_1, f(a + h) = b_2$.

17. For the polynomial $f(x)$ in the previous question, find $\int_{a-h}^{a+h} f(x)dx$.

18. Can a quadratic polynomial be defined whose graph contains the four points $(-1,-2)$, $(0,-2)$, $(1,0)$, and $(2,2)$?

19. (a) Extend the formula of Exercise 13 to the case of four points in the plane.
    (b) Extend the formula of Exercise 13 to the case of k points in the plane.

20. From Exercises 2 and 3 of Appendix D, it appears to be a reasonable conjecture that $\sum_{i=1}^{n} i^k$ is a polynomial of degree $k + 1$ in n, e.g., $\sum_{i=1}^{n} i = (n^2 + n)/2$ and $\sum_{i=1}^{n} i^2 = (2n^3 + 3n^2 + n)/6$. We will assume, for the moment, that $\sum_{i=1}^{n} i^k = P_{k+1}(n)$ is a polynomial in n of degree $k + 1$, and then attempt to find $P_{k+1}(n)$ by using the formula from the previous exercise. For this purpose we need $k + 2$ points that the polynomial passes through. We get these by evaluating the sums $\sum_{i=1}^{n} i^k = P_{k+1}(n)$ for $n = 1, 2, \ldots, k + 2$. (Any $k + 2$ values of n will do, but these are easy to compute.) For example, to find $\sum_{i=1}^{n} i = P_2(n)$ we have $P_2(1) = \sum_{i=1}^{1} i = 1$, $P_2(2) = \sum_{i=1}^{2} i = 3$, $P_2(3) = \sum_{i=1}^{3} i = 6$. Hence $P_2(n)$ passes through $(1,1)$, $(2,3)$, and $(3,6)$. Thus

$$P_2(n) = \frac{1(n - 2)(n - 3)}{(1 - 2)(1 - 3)} + \frac{3(n - 1)(n - 3)}{(2 - 1)(2 - 3)} + \frac{1(n - 1)(n - 2)}{(3 - 1)(3 - 2)} = \frac{1}{2}n^2 + \frac{1}{2}n.$$

Use the above method to find
(a) $\sum_{i=1}^{n} i^2$ (b) $\sum_{i=1}^{n} i^3$.
Note that the derivation was based on the unproved assumption that $P_{k+1}(n)$ is a polynomial. Once we have $P_{k+1}(n)$, we can prove that $P_{k+1}(n) = \sum_{i=1}^{n} i^k$ by induction.

4.2 FACTORS

Our aim in this section is to obtain, for polynomials, results analogous to some of the theorems we proved in Chapter 1 for integers. In particular, we are looking for a division algorithm, an analog of the Euclidean algorithm, and a prime factorization theorem. Many of the arguments in Chapter 1 involved finding the smallest number in some set of integers, and so the size, in terms of absolute value, was important. Very similar arguments can be given for polynomials, with the notion of degree replacing that of absolute value.

Our first goal is to formulate and prove a division algorithm. The following example is included just to remind you of the procedure that you probably learned in high-school algebra.

Example 4.2.1

In dividing the polynomial

$$6x^4 - 2x^3 + x^2 + 5x - 18$$

by $2x^2 - 3$, the first step is to divide $6x^4$ by $2x^2$, to get $3x^2$. The next step is to multiply $2x^2 - 3$ by $3x^2$ and subtract the result from $6x^4 - 2x^3 + x^2 + 5x - 18$. The algorithm for division of polynomials then proceeds much like the algorithm for division of integers.

$$
\begin{array}{r|rrrrr}
 & 3x^2 & -x & +5 & & \\
\hline
2x^2 \ \ -3 & 6x^4 & -2x^3 & +x^2 & +5x & -18 \\
 & 6x^4 & & -9x^2 & & \\
\hline
 & & -2x^3 & +10x^2 & +5x & \\
 & & -2x^3 & & +3x & \\
\hline
 & & & 10x^2 & +2x & -18 \\
 & & & 10x^2 & & -15 \\
\hline
 & & & & 2x & -3
\end{array}
$$

We finally obtain

$$6x^4 - 2x^3 + x^2 + 5x - 18 = (3x^2 - x + 5)(2x^2 - 3) + (2x - 3),$$

where the last term is the remainder.   □

The proof of Theorem 4.2.1 is merely a formal verification, using induction, that the procedure followed in Example 4.2.1 will always work. Notice that if we divide polynomials with coefficients in a given field, then the quotient and remainder must have coefficients from the same field.

The division algorithm for integers (Theorem 1.1.3) was stated for integers a and b, with $b > 0$. It is easily extended to a statement that is parallel to the next theorem for polynomials: For any $a, b \in \mathbf{Z}$ with $b \neq 0$, there exist unique integers q and r such that $a = bq + r$, with $0 \leq r < |b|$. The role played by the absolute value of an integer is now played by the degree of a polynomial. Note that assigning the degree $-\infty$ to the zero polynomial would simplify the statement of the division algorithm, requiring simply that the degree of the remainder be less than the degree of the divisor. For the sake of clarity we have not used this convention.

4.2.1 Theorem (Division Algorithm).   For any polynomials $f(x)$, $g(x) \in$ $F[x]$, with $g(x) \neq 0$, there exist unique polynomials $q(x)$, $r(x) \in F$ such that

$$f(x) = q(x)g(x) + r(x),$$

where either $\deg(r(x)) < \deg(g(x))$ or $r(x) = 0$.

Proof. Let

$$f(x) = a_m x^m + \ldots + a_1 x + a_0$$

and

$$g(x) = b_n x^n + \ldots + b_0,$$

where $a_m \neq 0$ and $b_n \neq 0$. In case $f(x)$ has lower degree than $g(x)$, we can simply take $q(x) = 0$ and $r(x) = f(x)$. The proof of the other case will use induction on the degree of $f(x)$.

If $f(x)$ has degree zero, it is easy to see that the theorem holds. Now assume that the theorem is true for all polynomials $f(x)$ of degree less than m. (We are assuming that $m \geq n$.) The reduction to a polynomial of lower degree is achieved by using the procedure outlined in Example 4.2.1. We divide $a_m x^m$ by $b_n x^n$ to get $a_m b_n^{-1} x^{m-n}$, then multiply by $g(x)$ and subtract from $f(x)$. This gives

$$f_1(x) = f(x) - a_m b_n^{-1} x^{m-n} g(x),$$

where $f_1(x)$ has degree less than m since the leading term of $f(x)$ has been cancelled by $a_m b_n^{-1} x^{m-n} b_n x^n$. Now by the induction hypothesis we can write

$$f_1(x) = q_1(x)g(x) + r(x),$$

where the degree of $r(x)$ is less than n, unless $r(x) = 0$. Since

$$f(x) = f_1(x) + a_m b_{n-1} x^{m-n} g(x),$$

substitution gives the desired result:

$$f(x) = (q_1(x) + a_m b_n^{-1} x^{m-n}) g(x) + r(x).$$

The quotient $q(x) = q_1(x) + a_m b_n^{-1} x^{m-n}$ has coefficients in F, since a_m, $b_n \in F$, and $q_1(x)$ has coefficients in F by the induction hypothesis. Finally, the remainder $r(x)$ has coefficients in F by the induction hypothesis.

To show that the quotient $q(x)$ and remainder $r(x)$ are unique, suppose that

$$f(x) = q_1(x)g(x) + r_1(x)$$

and

$$f(x) = q_2(x)g(x) + r_2(x).$$

Thus

$$(q_1(x) - q_2(x))g(x) = r_2(x) - r_1(x),$$

and if $q_2(x) - q_1(x) \neq 0$, then the degree of $(q_2(x) - q_1(x))g(x)$ is greater than or equal to the degree of $g(x)$ (by Proposition 4.1.5), whereas the degree of $r_2(x) - r_1(x)$ is less than the degree of $g(x)$. This is a contradiction, so we can conclude that $q_2(x) = q_1(x)$, and this forces

$$r_2(x) - r_1(x) = (q_1(x) - q_2(x))g(x) = 0,$$

completing the proof. □

Theorem 4.1.9 is a particular case of the general division algorithm. If $g(x)$ is the linear polynomial $x - c$, then the remainder must be a constant when $f(x)$ is divided by $x - c$. Substituting c into the equation $f(x) = q(x)(x - c) + r(x)$ shows that $r(c) = f(c)$, so the remainder on division by $x - c$ is the same element of F as $f(x)$ evaluated at $x = c$.

Example 4.2.4

In this example we illustrate the division algorithm for the polynomials in Example 4.2.1, over the finite field $\mathbf{Z}_7$. Our first step is to reduce coefficients modulo 7, to obtain

$$f(x) = 6x^4 + 5x^3 + x^2 + 5x + 3$$

and

$$g(x) = 2x^2 + 4.$$

Since it is much easier to divide by a monic polynomial, we multiply $g(x)$ by 4 (the inverse of 2 modulo 7) and work with $x^2 + 2$. Proceeding as in Example 4.2.1, we obtain

$$6x^4 + 5x^3 + x^2 + 5x + 3 = (6x^2 + 5x + 3)(x^2 + 2) + (2x - 3),$$

where the last term is the remainder.

$$
\begin{array}{r}
6x^2 \quad +5x \quad +3 \\
x^2 + 2 \,\overline{)\, 6x^4 \quad +5x^3 \quad +x^2 \quad +5x \quad +3} \\
6x^4 \qquad\quad\; +5x^2 \\
\hline
5x^3 \quad +3x^2 \quad +5x \\
5x^3 \qquad\quad +3x \\
\hline
3x^2 \quad +2x \quad +3 \\
3x^2 \qquad\quad +6 \\
\hline
2x \quad +4
\end{array}
$$

To take care of the fact that we divided by $g(x)/2$, we need to multiply the divisor $x^2 + 2$ by 2 and divide the quotient $6x^2 + 5x + 3$ by 2. This finally give us

$$6x^4 + 5x^3 + x^2 + 5x + 3 = (3x^2 + 6x + 5)(2x^2 + 4) + (2x - 3),$$

with the remainder being unchanged.   □

The next result is parallel to Theorem 1.1.4, which shows that every sub-group of $\mathbf{Z}$ is cyclic. It will play an important role in Chapters 5 and 6.

4.2.2 Theorem.   Let I be a subset of $F[x]$ that satisfies the following conditions:
  (i) I contains a nonzero polynomial;
  (ii) if $f(x)$, $g(x) \in I$, then $f(x) + g(x) \in I$;
  (iii) if $f(x) \in I$ and $q(x) \in F[x]$, then $q(x)f(x) \in I$.
If $d(x)$ is any nonzero polynomial in I of minimal degree, then

$$I = \{f(x) \in F[x] \mid f(x) = q(x)d(x) \text{ for some } q(x) \in F[x]\}.$$

Proof. If I contains a nonzero polynomial, then the set of all natural numbers n such that I contains a polynomial of degree n is a nonempty set, so by the well-ordering principle it must contain a smallest element, say m. Thus we can find a nonzero polynomial of minimal degree m in I, say $d(x)$.

Every multiple $q(x)d(x)$ of $d(x)$ must be in I by condition (iii). Next we need to show that $d(x)$ is a divisor of any other polynomial $h(x) \in I$. One way to proceed is to simply divide $h(x)$ by $d(x)$, using the division algorithm, and then show that the remainder must be zero.

We can write

$$h(x) = q(x)d(x) + r(x),$$

where $r(x)$ is either zero or has lower degree than $d(x)$. Solving for $r(x)$, we have

$$r(x) = h(x) + (-q(x))d(x).$$

This shows that $r(x) \in I$, since $h(x) \in I$ and I is closed under addition and under multiplication by any polynomial. But then $r(x)$ must be zero, since I cannot contain a nonzero polynomial of lower degree than the degree of $d(x)$. This shows that $h(x)$ is a multiple of $d(x)$.   □

We should note that in any set of polynomials of the form

$$\{f(x) \in F[x] \mid f(x) = q(x)d(x) \text{ for some } q(x) \in F[x]\},$$

the degree of $d(x)$ must be minimal (among degrees of nonzero elements). Multi-plying by the inverse of the leading coefficient of $d(x)$ gives a monic polynomial of the same degree that is still in the set.

4.2.3 Definition.   A monic polynomial $d(x) \in F[x]$ is called the *greatest common divisor* of $f(x)$, $g(x) \in F[x]$ if

 (i) $d(x) \mid f(x)$ and $d(x) \mid g(x)$, and

 (ii) if $h(x) \mid f(x)$ and $h(x) \mid g(x)$ for some $h(x) \in F[x]$, then $h(x) \mid d(x)$.

The greatest common divisor of $f(x)$ and $g(x)$ is denoted by $\gcd(f(x),g(x))$.

   If $\gcd(f(x),g(x)) = 1$, then the polynomials $f(x)$ and $g(x)$ are said to be *relatively prime*.

   We can show the uniqueness of the greatest common divisor as follows. Suppose that $c(x)$ and $d(x)$ are both greatest common divisors of $f(x)$ and $g(x)$. Then $c(x) \mid d(x)$ and $d(x) \mid c(x)$, say $d(x) = a(x)c(x)$ and $c(x) = b(x)d(x)$, and so we have $d(x) = a(x)b(x)d(x)$. Therefore $a(x)b(x) = 1$, and so Proposition 4.1.5 shows that $a(x)$ and $b(x)$ are both of degree zero. Thus $c(x)$ is a constant multiple of $d(x)$, and since both are monic, the constant must be 1, which shows that $c(x) = d(x)$.

4.2.4 Theorem.   For any nonzero polynomials $f(x)$, $g(x) \in F[x]$, the greatest common divisor $\gcd(f(x),g(x))$ exists and can be expressed as a linear combination of $f(x)$ and $g(x)$, in the form

$$\gcd((f(x),g(x)) = a(x)f(x) + b(x)g(x)$$

for some $a(x)$, $b(x) \in F[x]$.

   Proof. It is easy to check that

$$I = \{a(x)f(x) + b(x)g(x) \mid a(x), b(x) \in F[x]\}$$

satisfies the conditions of Theorem 4.2.2, and so I consists of all polynomial multiples of any monic polynomial in I of minimal degree, say $d(x)$.

   Since $f(x)$, $g(x) \in I$, we have $d(x) \mid f(x)$ and $d(x) \mid g(x)$. On the other hand, since $d(x)$ is a linear combination of $f(x)$ and $g(x)$, it follows that if $h(x) \mid f(x)$ and $h(x) \mid g(x)$, then $h(x) \mid d(x)$. Thus $d(x) = \gcd(f(x),g(x))$. $\square$

Example 4.2.3 (Euclidean Algorithm for Polynomials)

   Let $f(x),g(x) \in F[x]$ be nonzero polynomials. We can use the division algorithm to write $f(x) = q(x)g(x) + r(x)$, with $\deg(r(x)) < \deg(g(x))$ or $r(x) = 0$. If $r(x) = 0$, then $g(x)$ is a divisor of $f(x)$, and so $\gcd(f(x),g(x)) = g(x)$. If $r(x) \neq 0$, then it is easy to check that $\gcd(f(x),g(x)) = \gcd(g(x),r(x))$. This step reduces the degrees of the polynomials involved, and so repeating the procedure leads to the greatest common divisor of the two polynomials in a finite number of steps.

   The Euclidean algorithm for polynomials is similar to the Euclidean algorithm for finding the greatest common divisor of nonzero integers. The polynomials $a(x)$ and $b(x)$ for which $\gcd((f(x),g(x)) = a(x)f(x) + b(x)g(x)$ can be found just as for integers (see Example 1.1.2). $\square$

Example 4.2.4

To find

$$\gcd(2x^4 + x^3 - 6x^2 + 7x - 2, \quad 2x^3 - 7x^2 + 8x - 4)$$

over **Q,** divide the polynomial of higher degree by the one of lower degree, to get the quotient $x + 4$ and remainder $14x^2 - 21x + 14$. The answer will be unchanged by dividing through by a nonzero constant, so we can use the polynomial $2x^2 - 3x + 2$.

As for integers, we now have

$$\gcd(2x^4 + x^3 - 6x^2 + 7x - 2, \quad 2x^3 - 7x^2 + 8x - 4)$$
$$= \gcd(2x^3 - 7x^2 + 8x - 4, \quad 2x^2 - 3x + 2).$$

Dividing as before gives the quotient $x - 2$, with remainder zero. This shows that the greatest common divisor that we are looking for is $x^2 - (3/2)x + 1$ (we divided through by 2 to obtain a monic polynomial). □

4.2.5 Proposition.   Let $p(x)$, $f(x)$, $g(x) \in F[x]$. If $p(x) \mid f(x)g(x)$ and $\gcd(p(x), f(x)) = 1$, then $p(x) \mid g(x)$.

Proof. If $\gcd(p(x), f(x)) = 1$, then

$$1 = a(x)p(x) + b(x)f(x)$$

for some $a(x)$, $b(x) \in F[x]$. Thus

$$g(x) = a(x)g(x)p(x) + b(x)f(x)g(x),$$

which shows that if $p(x) \mid f(x)g(x)$, then $p(x) \mid g(x)$. □

4.2.6 Definition.   A nonconstant polynomial is said to be *irreducible over the field F* if it cannot be factored in $F[x]$ into a product of polynomials of lower degree. It is said to be *reducible* over F if such a factorization exists.

All polynomials of degree 1 are irreducible. On the other hand, any polynomial of greater degree that has a root in F is reducible over F, since by the remainder theorem it can be factored into polynomials of lower degree. To check that a polynomial is irreducible over a field F, in general it is not sufficient to merely check that it has no roots in F. For example, $x^4 + 4x^2 + 4 = (x^2 + 2)^2$ is reducible over **Q,** but it certainly has no rational roots. However, a polynomial of degree 2 or 3 can be factored into a product of polynomials of lower degree if and only if one of the factors is linear, which then gives a root. This remark proves the next proposition.

4.2.7 Proposition.   A polynomial of degree 2 or 3 is irreducible over the field F if and only if it has no roots in F.

The field F is crucial in determining irreducibility. The polynomial $x^2 + 1$ is irreducible over **R,** since it has no real roots, but considered as a polynomial over **C,** it factors as $x^2 + 1 = (x + i)(x - i)$. Over the field $\mathbf{Z}_2$, the polynomial $x^2 + x + 1$ is irreducible, since it has no roots in $\mathbf{Z}_2$. But on the other hand, it is reducible over the field $\mathbf{Z}_3$, since $x^2 + x + 1 = (x + 2)^2$ when the coefficients are viewed as representing congruence classes in $\mathbf{Z}_3$.

4.2.8 Lemma. The nonconstant polynomial $p(x) \in F[x]$ is irreducible over F if and only if for all $f(x)$, $g(x) \in F[x]$, $p(x) \mid (f(x)g(x))$ implies $p(x) \mid f(x)$ or $p(x) \mid g(x)$.

Proof. First assume that $p(x) \mid f(x)g(x)$. If $p(x)$ is irreducible and $p(x) \nmid f(x)$, then $\gcd(p(x), f(x)) = 1$, and so $p(x) \mid g(x)$ by Proposition 4.2.5.

Conversely, if the given condition holds, then $p(x) \neq f(x)g(x)$ for polynomials of lower degree, since $p(x) \nmid f(x)$ and $p(x) \nmid g(x)$.   □

Because of the similarity between Lemma 4.2.8 and Lemma 1.2.5, it is evident that irreducible polynomials should play a role analogous to that of prime numbers, and one of the results we should look for is a unique factorization theorem. The proof of Theorem 1.2.6 can be carried over to polynomials by using irreducible polynomials in place of prime numbers and by using the degree of a polynomial in place of the absolute value of a number. For this reason we have chosen to omit the proof of the next theorem, even though it is extremely important.

4.2.9 Theorem (Unique Factorization). Any nonconstant polynomial with coefficients in the field F can be expressed as an element of F times a product of monic polynomials, each of which is irreducible over the field F. This expression is unique except for the order in which the factors occur.

Example 4.2.5 (Irreducible Polynomials over R and C)

In Appendix E we presented the fundamental theorem of algebra, which states that any polynomial over **C** of positive degree has a root in **C.** (A proof is given in Theorem 8.3.10.) This implies that the irreducible polynomials in $\mathbf{C}[x]$ are precisely the linear polynomials.

In Appendix E, we showed as a consequence of the fundamental theorem of algebra that any polynomial over **R** can be factored into a product of linear and quadratic polynomials. Consider the case of a quadratic. Over **R,** a polynomial $ax^2 + bx + c$ with $a \neq 0$ has roots

$$x = \frac{-b \pm \sqrt{b^2 - 4ac}}{2a},$$

and these are real numbers if and only if $b^2 - 4ac \geq 0$. Since any factors of $ax^2 + bx + c$ must be linear and hence correspond to roots, we can see that the polynomial is reducible over **R** if and only if $b^2 - 4ac \geq 0$. For example, $x^2 + 1$ is irreducible over **R**. In summary, irreducible polynomials in **R**$[x]$ must have one of the forms $ax + b$, with $a \neq 0$ or $ax^2 + bx + c$, with $a \neq 0$ and $b^2 - 4ac < 0$. □

Polynomials cannot be factored as easily over the field of rational numbers as over the field of real numbers, so the theory of irreducible polynomials over the field of rational numbers is much richer than the corresponding theory over the real numbers. For example, $x^2 + 2$ and $x^4 + x^3 + x^2 + x + 1$ are irreducible over the field of rational numbers, the first since $\sqrt{2}$ is irrational and the second as a consequence of a criterion we will develop in the next section.

In studying roots and factors of polynomials, it is often of interest to know whether there are any repeated roots or factors. The derivative $p'(x)$ of the polynomial $p(x)$ can be used to check for repeated roots and factors. It is possible to formally define the derivative of a polynomial over any field, and we will do so in Section 8.2. For the moment we will restrict ourselves to the case of polynomials with real coefficients, so that we can feel free to use any formulas we might need from calculus.

4.2.10 Definition.   Let $f(x) \in F[x]$. An element $c \in F$ is said to be a *root of multiplicity* $n \geq 1$ of $f(x)$ if

$$(x - c)^n \mid f(x) \quad \text{but} \quad (x - c)^{n+1} \nmid f(x).$$

4.2.11 Proposition.   A nonconstant polynomial $f(x)$ over the field **R** of real numbers has no repeated factors if and only if $\gcd(f(x), f'(x)) = 1$.

Proof. Suppose that $\gcd(f(x), f'(x)) = d(x) \neq 1$ and that $p(x)$ is an irreducible factor of $d(x)$. Then $f(x) = a(x)p(x)$ and $f'(x) = b(x)p(x)$ for some $a(x), b(x) \in F[x]$. Using the product rule to differentiate gives

$$f'(x) = a'(x)p(x) + a(x)p'(x) = b(x)p(x).$$

This shows that $p(x) \mid a(x)p'(x)$, since

$$a(x)p'(x) = b(x)p(x) - a'(x)p(x),$$

and thus $p(x) \mid a(x)$ because $p(x)$ is irreducible and $p(x) \nmid p'(x)$. Therefore $f(x) = c(x)p(x)^2$ for some $c(x) \in F[x]$, and so $f(x)$ has a repeated factor, which is a contradiction.

Conversely, if $f(x)$ has a repeated factor, say $f(x) = g(x)^n q(x)$, with $n > 1$, then

$$f'(x) = ng(x)^{n-1}g'(x)q(x) + g(x)^n q'(x)$$

and $g(x)$ is a common divisor of $f(x)$ and $f'(x)$. □

EXERCISES: SECTION 4.2

1. Use the division algorithm to find the quotient and remainder when $f(x)$ is divided by $g(x)$ over the field of rational numbers $\mathbf{Q}$.
   (a) $f(x) = 2x^4 + 5x^3 - 7x^2 + 4x + 8$ $g(x) = 2x - 1$
   (b) $f(x) = 2x^7 - 5x^6 + 5x^5 - x^3 - x^2 + 4x - 5$ $g(x) = x^2 - x + 1$
   (c) $f(x) = 2x^4 + x^3 - 6x^2 - x + 2$ $g(x) = 2x^2 - 5$

2. Find the greatest common divisor of the given polynomials, over $\mathbf{Q}$.
   (a) $2x^3 + 2x^2 - x - 1$ and $2x^4 - x^2$
   (b) $4x^3 - 2x^2 - 3x + 1$ and $2x^2 - x - 2$
   (c) $x^{10} - x^7 - x^5 + x^3 + x^2 - 1$ and $x^8 - x^5 - x^3 + 1$
   (d) $x^5 + x^4 + 2x^2 - x - 1$ and $x^3 + x^2 - x$

3. In each part of the previous exercise, write the greatest common divisor as a linear combination of the given polynomials. That is, given $f(x)$ and $g(x)$, find $a(x)$ and $b(x)$ such that $d(x) = a(x)f(x) + b(x)g(x)$, where $d(x)$ is the greatest common divisor of $f(x)$ and $g(x)$.

4. Find the greatest common divisor of $f(x)$ and $f'(x)$, over $\mathbf{Q}$.
   (a) $f(x) = x^4 - x^3 - x + 1$
   (b) $f(x) = x^3 - 3x - 2$
   (c) $f(x) = x^3 + 2x^2 - x - 2$

5. Let $p(x) = a_n x^n + a_{n-1}x^{n-1} + \ldots + a_1 x + a_0$ be a polynomial with rational coefficients. Show that $p(x)$ is irreducible over the field of rational numbers if and only if $q(x) = a_0 x^n + a_1 x^{n-1} + \ldots + a_{n-1}x + a_n$ is irreducible over the field of rational numbers.

6. Show that for any real number $a \neq 0$, $x^n - a$ has no multiple roots in $\mathbf{R}$.

7. Use the division algorithm to find the quotient and remainder when $f(x)$ is divided by $g(x)$, over the indicated field.
   (a) $f(x) = x^4 + 1$ $g(x) = x + 1$ over $\mathbf{Z}_2$
   (b) $f(x) = x^5 + 4x^4 + 2x^3 + 3x^2$ $g(x) = x^2 + 3$ over $\mathbf{Z}_5$
   (c) $f(x) = x^5 + 2x^3 + 3x^2 + x - 1$ $g(x) = x^2 + 5$ over $\mathbf{Z}_7$

8. Find the greatest common divisor of the given polynomials, and write it as a linear combination of them.
   (a) $x^4 + x^3 + x + 1$ and $x^3 + x^2 + x + 1$ over $\mathbf{Z}_2$
   (b) $x^3 - 2x^2 + 3x + 1$ and $x^3 + 2x + 1$ over $\mathbf{Z}_5$

9. List all monic irreducible polynomials of degree ≤ 5 over $\mathbf{Z}_2$.
   Hint: First develop a criterion that allows you to tell at a glance whether or not a polynomial has no roots. Among the polynomials with no roots, use irreducible factors of degree $\leq n$ to find reducible polynomials of degree $n + 1$.

10. List all monic irreducible polynomials of degree ≤ 3 over $\mathbf{Z}_3$. Write each of the following polynomials as a product of irreducible polynomials:
    (a) $x^2 - 2x + 1$
    (b) $x^4 + 2x^2 + 2x + 2$
    (c) $2x^3 - 2x + 1$

11. Show that there are exactly $(p^2 - p)/2$ monic irreducible polynomials of degree 2 over $\mathbf{Z}_p$.

12. Let $f(x)/g(x)$ be a rational function, where $f(x)$, $g(x) \in \mathbf{R}[x]$.
   (a) Show that if $g(x) = h(x)k(x)$, with $\gcd(h(x), k(x)) = 1$, then there exist polynomials $s(x)$, $t(x)$ such that

$$\frac{f(x)}{g(x)} = \frac{s(x)}{h(x)} + \frac{t(x)}{k(x)}.$$

   (b) Show that if $h(x) = p(x)^m$, where $p(x)$ is irreducible, then there exist polynomials $q(x)$, $r_0(x)$, $r_1(x)$, $\ldots$, $r_{m-1}(x)$ such that for each i either $r_i(x) = 0$ or $\deg(r_i(x)) < \deg(p(x))$ and

$$\frac{s(x)}{h(x)} = q(x) + \frac{r_{m-1}(x)}{p(x)} + \frac{r_{m-2}(x)}{p(x)^2} + \cdots + \frac{r_0(x)}{p(x)^m}.$$

   (c) Show that $f(x)/g(x)$ can be expressed as a polynomial plus a sum of "partial fractions" of the form

$$\frac{c}{(x + a)^m} \quad \text{or} \quad \frac{c(x + d)}{(x^2 + ax + b)^m}.$$

Note: This result is used in calculus to prove that the indefinite integral of any rational function can be expressed in terms of algebraic, trigonometric, or exponential functions and their inverses.

Hint: If $p(x)$ and $q(x)$ are relatively prime, then there exist polynomials $a(x)$ and $b(x)$ with $a(x)p(x) + b(x)q(x) = 1$. Therefore

$$\frac{1}{p(x)q(x)} = \frac{a(x)}{q(x)} + \frac{b(x)}{p(x)}.$$

Definition. Let $p(x)$ be a nonzero polynomial in $F[x]$. For any $f(x)$, $g(x) \in F[x]$, we write $f(x) \equiv g(x) \pmod{p(x)}$ if $f(x)$ and $g(x)$ have the same remainder when divided by $p(x)$. That is, $f(x) \equiv g(x) \pmod{p(x)}$ if and only if $p(x) \mid (f(x) - g(x))$.

Use the above definition in the following problems.

13. Let $p(x)$ be a nonzero polynomial. Show that for each polynomial $f(x)$ there is a unique polynomial $r(x)$ with $r(x) = 0$ or $\deg(r(x)) < \deg(p(x))$ such that $r(x) \equiv f(x) \pmod{p(x)}$.

14. Let $p(x)$ be a nonzero polynomial. Show that if $f(x) \equiv c(x) \pmod{p(x)}$ and $g(x) \equiv d(x) \pmod{p(x)}$, then

$$f(x) + g(x) \equiv c(x) + d(x) \bmod p(x))$$

$$f(x)g(x) \equiv c(x)d(x) \pmod{(x)}.$$

15. Compute the following products. (Your answer should have degree 1.)
   (a) $(a + bx)(c + dx) \equiv$??? $\pmod{x^2 + 1}$ over $\mathbf{Q}$
   (b) $(a + bx)(c + dx) \equiv$??? $\pmod{x^2 - 2}$ over $\mathbf{Q}$
   (c) $(a + bx)(c + dx) \equiv$??? $\pmod{x^2 + x + 1}$ over $\mathbf{Z}_2$

16. Let $f(x)$ be a nonzero polynomial. Show that there exists a polynomial $g(x)$ with $f(x)g(x) \equiv 1 \pmod{p(x)}$ if and only if $\gcd(f(x), p(x)) = 1$.

17. Find a polynomial $q(x)$ such that
    (a) $(a + bx)q(x) \equiv 1 \pmod{x^2 + 1}$ over $\mathbf{Q}$
    (b) $(a + bx)q(x) \equiv 1 \pmod{x^2 - 2}$ over $\mathbf{Q}$
    (c) $(a + bx)q(x) \equiv 1 \pmod{x^2 + x + 1}$ over $\mathbf{Z}_2$
    (d) $(x^2 + 2x + 1)q(x) \equiv 1 \pmod{x^3 + x^2 + 1}$ over $\mathbf{Z}_3$

4.3 POLYNOMIALS WITH INTEGER COEFFICIENTS

In this section we will give several criteria for determining when polynomials with integer coefficients have rational roots or are irreducible over the field of rational numbers. We will use the notation $\mathbf{Z}[x]$ for the set of all polynomials with integer coefficients.

4.3.1 Proposition.   Let $f(x) = a_n x^n + a_{n-1} x^{n-1} + \ldots + a_1 x + a_0$ be a polynomial with integer coefficients. If r/s is a rational root of $f(x)$, with $(r, s) = 1$, then $r \mid a_0$ and $s \mid a_n$.

Proof. If $f(r/s) = 0$, then multiplying $f(r/s)$ by s^n gives the equation

$$a_n r^n + a_{n-1} r^{n-1} s + \ldots + a_1 r s^{n-1} + a_0 s^n = 0.$$

It follows that $r \mid a_0 s^n$ and $s \mid a_n r^n$, so $r \mid a_0$ and $s \mid a_n$ since $(r, s) = 1$.   □

Example 4.3.1

Suppose that we wish to find all integral roots of

$$f(x) = x^3 - 3x^2 + 2x - 6.$$

Using Proposition 4.3.1, all rational roots of $f(x)$ can be found by testing only a finite number of values. By considering the signs we can see that $f(x)$ cannot have any negative roots, so we only need to check the positive factors of 6. Substituting, we obtain $f(1) = -6$, $f(2) = -6$, and $f(3) = 0$. Thus 3 is a root of $f(x)$, and so we can use the division algorithm to show that

$$x^3 - 3x^2 + 2x - 6 = (x^2 + 2)(x - 3).$$

It is now clear that 6 is not a root, and we are done.   □

Example 4.3.2

Let $f(x) \in \mathbf{Z}[x]$. If c is a root of $f(x)$, then $f(x) = q(x)(x - c)$ for some polynomial $q(x)$. The proof of the remainder theorem (Theorem 4.1.9) shows that $q(x) \in \mathbf{Z}[x]$. For any integer n, we must have $f(n) = q(n)(c - n)$, and since $f(n)$, $q(n) \in \mathbf{Z}$, this shows that $(c - n) \mid f(n)$.

This observation can be combined with Proposition 4.3.1 to find the rational roots of equations such as

$$x^3 + 15x^2 - 3x - 6 = 0.$$

By Proposition 4.3.1, the possible rational roots are ± 1, ± 2, ± 3, and ± 6. Letting $f(x) = x^3 + 15x^2 - 3x - 6$, we find that $f(1) = 7$, so for any root c, $(c - 1)|7$. This eliminates all of the possible values except $c = 2$ and $c = -6$. We find that $f(2) = 56$, so 2 is not a root. This shows, in addition, that $(c - 2)|56$ for any root c, but -6 still passes this test. Finally, $f(-6) = 336$, and so this eliminates -6, and $f(x)$ has no rational roots. We have also shown, by Proposition 4.2.7, that the polynomial

$$f(x) = x^3 + 15x^2 - 3x - 6$$

is irreducible in $\mathbf{Q}[x]$. $\square$

4.3.2 Definition.   A polynomial with integer coefficients is called *primitive* if the greatest common divisor of all of its coefficients is 1.

Given any polynomial with integer coefficients, we can obtain a primitive polynomial by factoring out the greatest common divisor of its coefficients (called the *content* of the polynomial). For example, the polynomial $12x^2 - 18x + 30$ has content 6, and so we can write

$$12x^2 - 18x + 30 = 6(2x^2 - 3x + 5),$$

where $2x^2 - 3x + 5$ is a primitive polynomial.

4.3.3 Lemma (Gauss).   The product of two primitive polynomials is itself primitive.

Proof. Let

$$f(x) = a_m x^m + \ldots + a_1 x + a_0$$

and

$$g(x) = b_n x^n + \ldots + b_1 x + b_0$$

be primitive polynomials. If $f(x)g(x)$ were not primitive, then some prime number p would be a divisor of each coefficient of $h(x) = f(x)g(x)$. Let a_k and b_t be the coefficients of least index not divisible by p in $f(x)$ and $g(x)$, respectively. Then

$$c_{k+t} = a_0 b_{k+t} + a_1 b_{k+t-1} + \ldots + a_{k-1} b_{t+1} + a_k b_t + a_{k+1} b_{t-1} + \ldots + a_{k+t} b_0,$$

for the coefficient c_{k+t} of $h(x)$. By the way in which we chose a_k and b_t, each term on the right-hand side is divisible by p, with the exception of $a_k b_t$. This implies that c_{k+t} is not divisible by p, a contradiction. $\square$

4.3.4 Theorem.   A polynomial with integer coefficients that can be factored into polynomials with rational coefficients can also be factored into polynomials of the same degree with integer coefficients.

Proof. Let $h(x) \in \mathbf{Z}[x]$, and assume that $h(x) = f(x)g(x)$ in $\mathbf{Q}[x]$. By factoring out the appropriate least common multiples of denominators and greatest common divisors of numerators, we can assume that $h(x) = (m/n)f^*(x)g^*(x)$, where $(m,n) = 1$ and $f^*(x)$, $g^*(x)$ are primitive, with the same degrees as $f(x)$ and $g(x)$, respectively. If d_i is any coefficient of $f^*(x)g^*(x)$, then $n|md_i$ since $h(x)$ has integer coefficients, so $n|d_i$ since $(n,m) = 1$. By Gauss's lemma, $f^*(x)g^*(x)$ is primitive, so we have $n = 1$, and thus $h(x)$ has a factorization $h(x) = (mf^*(x))(g^*(x))$ into a product of polynomials in $\mathbf{Z}[x]$. The general result, for any number of factors, can be proved by using an induction argument. $\square$

4.3.5 Theorem (Eisenstein's Irreducibility Criterion). For a given prime p, let

$$f(x) = a_n x^n + a_{n-1} x^{n-1} + \ldots + a_0$$

be a polynomial with integer coefficients such that

$$a_{n-1} \equiv a_{n-2} \equiv \ldots \equiv a_0 \equiv 0 \ (\mathrm{mod}\ p)$$

but $a_n \not\equiv 0 \ (\mathrm{mod}\ p)$ and $a_0 \not\equiv 0 \ (\mathrm{mod}\ p^2)$. Then $f(x)$ is irreducible over the field of rational numbers.

Proof. Suppose that $f(x)$ can be factored as

$$f(x) = (b_s x^s + \ldots + b_0)(c_t x^t + \ldots + c_0).$$

By Theorem 4.3.4 we can assume that both factors have integer coefficients. Furthermore, we can assume that either b_0 or c_0 is not divisible by p, since $b_0 c_0 = a_0$ is not divisible by p^2. Let us assume that $p \nmid b_0$. Then $p|c_0$, but we must have $p \nmid c_i$ for some i, since by assumption $a_n = b_s c_t$ is not divisible by p. Let m be the smallest index such that c_m is not divisible by p. Then from the equation

$$a_m = b_0 c_m + b_1 c_{m-1} + \ldots + b_m c_0$$

we see that a_m is not divisible by p, since each term on the right-hand side is divisible by p, with the exception of $b_0 c_m$. Thus $m = n$, since a_i is divisible by p for $i < n$, so $f(x)$ is irreducible because it cannot be factored into a product of polynomials of lower degree. $\square$

In Theorem 4.3.5, the condition $a_{n-1} \equiv \ldots \equiv a_0 \equiv 0 \ (\mathrm{mod}\ p)$ can be summed up as saying that p is a divisor of the greatest common divisor of these coefficients. The theorem cannot always be applied. For example, if $f(x) = x^3 - 5x^2 - 3x + 6$, then $\gcd(5, 3, 6) = 1$ and no prime can be found for which the necessary conditions are satisfied. Yet $f(x)$ is irreducible, since Propositions 4.3.1 can be used to show that $f(x)$ has no rational roots.

To show that $p(x)$ is irreducible, it is sufficient to show that $p(x + c)$ is irreducible for some integer c, since if $p(x) = f(x)g(x)$, then $p(x + c) = f(x + c)g(x + c)$. For example, Eisenstein's criterion cannot be applied to $x^2 + 1$, but substituting $x + 1$ for x gives another proof that $x^2 + 1$ is irreducible over the field of rational numbers.

Often the easiest way to make the substitution of $x + c$ for x is to use Taylor's formula. Recall that for a polynomial $p(x)$ of degree n,

$$p(x) = \frac{p^{(n)}(c)}{n!}(x - c)^n + \frac{p^{(n-1)}(c)}{(n - 1)!}(x - c)^{n-1} + \ldots + p'(c)(x - c) + p(c),$$

where $p^{(k)}(x)$ denotes the kth derivative of $p(x)$. Thus

$$p(x + c) = \frac{p^{(n)}(c)}{n!}x^n + \ldots + p'(c)x + p(c),$$

and it may be possible to apply Eisenstein's criterion to the new set of coefficients.

4.3.6 Corollary.   If p is prime, then the polynomial

$$\phi(x) = x^{p-1} + x^{p-2} + \ldots + x + 1$$

is irreducible over the field of rational numbers.

Proof. Note that

$$\phi(x) = \frac{x^p - 1}{x - 1}$$

and consider

$$\phi(x + 1) = \frac{(x + 1)^p - 1}{x} = x^{p-1} + \binom{p}{1}x^{p-2} + \binom{p}{2}x^{p-3} + \ldots + p.$$

Eisenstein's criterion can now be applied to $\phi(x + 1)$, so $\phi(x)$ is irreducible.   □

If p is prime, Corollary 4.3.6 shows that

$$x^p - 1 = (x - 1)(x^{p-1} + \ldots + 1)$$

gives the factorization over $\mathbf{Q}$ of $x^p - 1$ into irreducible factors. This is not the case when the degree is composite. For example,

$$x^4 - 1 = (x - 1)(x + 1)(x^2 + 1)$$

and

$$x^{15} - 1 =$$

$$(x - 1)(x^2 + x + 1)(x^4 + x^3 + x^2 + x + 1)(x^8 - x^7 + x^5 - x^4 + x^3 - x + 1).$$

EXERCISES: SECTION 4.3

1. If $f(x)$ has integer coefficients and m is an integer root of $f(x)$, then for any n we can reduce both m and the coefficients of $f(x)$ modulo n, and the equation becomes a congruence that still holds. For the following equations, verify that the given roots modulo 3 and 5 are in fact all such roots. Use this information to eliminate some of the integer roots, and then find all integer roots.
   (a) $x^2 - 7x^2 + 4x - 28 = 0$ (roots are 1 (mod 3) and $\pm 1,2$ (mod 5)).
   (b) $x^3 - 9x^2 + 10x - 16 = 0$ (roots are 2 (mod 3) and 3 (mod 5)).

2. Find all integer roots of the following equations (use any method).
   (a) $x^3 + 8x^2 + 13x + 6 = 0$
   (b) $x^3 - 5x^2 - 2x + 24 = 0$
   (c) $x^3 - 10x^2 + 27x - 18 = 0$
   (d) $x^4 + 4x^3 + 8x + 32 = 0$
   (e) $x^7 + 2x^5 + 4x^4 - 8x^2 - 32 = 0$
   (f) $x^4 - 2x^3 - 21x^2 + 22x + 40 = 0$
   (g) $y^3 - 9y^2 - 24y + 216 = 0$
   (h) $x^5 + 47x^4 + 423x^3 + 140x^2 + 1213x - 420 = 0$
   (i) $x^5 - 34x^3 + 29x^2 + 212x - 300 = 0$
   (j) $x^4 - 23x^3 + 187x^2 - 653x + 936 = 0$

3. Use Eisenstein's criterion to show that each of these polynomials is irreducible over the field of rational numbers. (You may need to make a substitution.)
   (a) $x^4 - 12x^2 + 18x - 24$
   (b) $4x^3 - 15x^2 + 60x + 180$
   (c) $x^2 + 2x - 5$ (substitute $x - 1$ or $x + 1$)
   (d) $x^3 + 3x^2 + 5x + 5$ (substitute $x - 1$)
   (e) $x^3 - 3x^2 + 9x - 10$
   (f) $2x^{10} - 25x^3 + 10x^2 - 30$

4. Show that if the positive integer n is not a perfect square, then for some prime p and some integer k,

$$x^2 - \frac{n}{p^{2k}}$$

   satisfies Eisenstein's criterion. Conclude that $\sqrt{n}$ is not a rational number.

5. Let $f(x) = a_n x^n + a_{n-1} x^{n-1} + \ldots + a_1 x + a_0$ be a polynomial with rational coefficients. Show that if $b \neq 0$ and b is a root of $f(x)$, then $1/b$ is a root of $g(x) = a_0 x^n + a_1 x^{n-1} + \ldots + a_{n-1} x + a_n$.

6. Find the irreducible factors of $x^6 - 1$ over $\mathbf{Q}$.

4.4 EXISTENCE OF ROOTS

The polynomial $x^2 + 1$ has no roots in the field $\mathbf{R}$ of real numbers. However, we can obtain a root by constructing an element i for which $i^2 = -1$ and adding it (in some way) to the field $\mathbf{R}$. This leads to the field $\mathbf{C}$, which contains elements of the form $a + bi$, for $a, b \in \mathbf{R}$. The only problem is to find a way of constructing the root i.

In this section we will show that for any polynomial, over any field, it is possible to construct a larger field in which the polynomial has a root. To do this we will use congruence classes of polynomials. The construction is similar in many ways to the construction of the field $\mathbf{Z}_p$ as a set of congruence classes of $\mathbf{Z}$. By iterating the process, it is possible to find a field that contains all of the roots of the polynomial, so that over this field the polynomial factors into a product of linear polynomials.

4.4.1 Definition.   Let E and F be fields. If F is a subset of E and has the operations of addition and multiplication induced by E, then F is called a *subfield* of E, and E is called an *extension field* of F.

4.4.2 Definition.   Let F be a field, and let $p(x)$ be a fixed polynomial over F. If $a(x), b(x) \in F[x]$, then we say that $a(x)$ and $b(x)$ are *congruent modulo $p(x)$*, written $a(x) \equiv b(x) \pmod{p(x)}$, if $p(x) \mid (a(x) - b(x))$.

The set $\{b(x) \in F[x] \mid a(x) \equiv b(x) \pmod{p(x)}\}$ is called the *congruence class* of $a(x)$, and will be denoted by $[a(x)]$. The collection of all congruence classes modulo $p(x)$ will be denoted by $F[x]/\langle p(x) \rangle$.

The reason for the notation $F[x]/\langle p(x) \rangle$ will become clear in Chapter 5. We first note that congruence of polynomials defines an equivalence relation. Then since $a(x) \equiv b(x) \pmod{p(x)}$ if and only if $a(x) - b(x) = q(x)p(x)$ for some $q(x) \in F[x]$, the polynomials in the congruence class of $a(x)$ modulo $p(x)$ must be precisely the polynomials of the form $b(x) = a(x) + q(x)p(x)$, for some $q(x)$. We gave a similar description for the congruence classes of $\mathbf{Z}_n$. When working with congruence classes modulo n, we have often chosen to work with the smallest nonnegative number in the class. Similarly, when working with congruence classes of polynomials, the polynomial of lowest degree in the congruence class is a natural representative. The next proposition guarantees that this representative is unique.

4.4.3 Proposition.   Let F be a field, and let $p(x)$ be a nonzero polynomial in $F[x]$. For any $a(x) \in F[x]$, the congruence class $[a(x)]$ modulo $p(x)$ contains a unique representative $r(x)$ with $\deg(r(x)) < \deg(p(x))$ or $r(x) = 0$.

Proof. Given $a(x) \in F[x]$, we can use the division algorithm to write

$$a(x) = q(x)p(x) + r(x),$$

with $\deg(r(x)) < \deg(p(x))$ or $r(x) = 0$. Solving for $r(x)$ in the above equation shows it to be in the congruence class $[a(x)]$. The polynomial $r(x)$ is the only representative with this property, since if

$$b(x) \equiv a(x) \pmod{p(x)}$$

and $\deg(b(x)) < \deg(p(x))$, then

$$b(x) \equiv r(x) \ (\text{mod } p(x))$$

and so $p(x) \mid (b(x) - r(x))$. This is a contradiction unless $b(x) = r(x)$, since either $\deg(b(x) - r(x)) < \deg(p(x))$ or $b(x) - r(x) = 0$. $\quad\square$

4.4.4 Proposition. Let F be a field, and let $p(x)$ be a nonzero polynomial in $F[x]$. For any polynomials $a(x)$, $b(x)$, $c(x)$, and $d(x)$ in $F[x]$, the following conditions hold:

(a) If $a(x) \equiv c(x) \ (\text{mod } p(x))$ and $b(x) \equiv d(x) \ (\text{mod } p(x))$, then $a(x) + b(x) \equiv c(x) + d(x) \ (\text{mod } p(x))$ and $a(x)b(x) \equiv c(x)d(x) \ (\text{mod } p(x))$.

(b) If $a(x)b(x) \equiv a(x)c(x) \ (\text{mod } p(x))$ and $\gcd(a(x), p(x)) = 1$, then $b(x) \equiv c(x)$ $(\text{mod } p(x))$.

Proof. The proof is similar to that of Theorem 1.3.3 and will be omitted. $\quad\square$

Example 4.4.1

Let $F = \mathbf{R}$, the field of real numbers, and let $p(x) = x^2 + 1$. Then every congruence class in $\mathbf{R}[x]/\langle x^2 + 1\rangle$ can be represented by a linear polynomial of the form $a + bx$, by Proposition 4.4.3. Proposition 4.4.4 implies that if we add and multiply congruence classes by choosing representatives, just as we did in $\mathbf{Z}_n$, then addition and multiplication of congruence classes are well-defined. If we multiply the congruence classes represented by $a + bx$ and $c + dx$, we have

$$ac + (bc + ad)x + bdx^2.$$

Dividing by $x^2 + 1$ gives the remainder

$$(ac - bd) + (bc + ad)x,$$

which is a representative of the product of the two congruence classes. An easier way to make this computation is to note that

$$x^2 + 1 \equiv 0 \ (\text{mod } x^2 + 1),$$

and so

$$x^2 \equiv -1 \ (\text{mod } x^2 + 1),$$

which means that we can replace x^2 by -1 in the product

$$ac + (bc + ad)x + bdx^2.$$

This multiplication is the same as the multiplication of complex numbers, and gives another way to define $\mathbf{C}$. Note that the congruence class $[x]$ has the property that its square is the congruence class $[-1]$, and so if we identify the set of real numbers with the set of congruence classes of the form $[a]$, where $a \in \mathbf{R}$, then the class $[x]$ would be identified with i. We can formalize this identification after we define the concept of an isomorphism of fields.   □

4.4.5 Proposition.   Let F be a field, and let $p(x)$ be a nonzero polynomial in $F[x]$. For any $a(x) \in F[x]$, the congruence class $[a(x)]$ has a multiplicative inverse in $F[x]/\langle p(x) \rangle$ if and only if $\gcd(a(x), p(x)) = 1$.

Proof. To find a multiplicative inverse for $[a(x)]$ we must find a congruence class $[b(x)]$ with $[a(x)][b(x)] = [1]$. Since

$$a(x)b(x) \equiv 1 \;(\mathrm{mod}\; p(x))$$

if and only if there exists $t(x) \in F[x]$ with

$$a(x)b(x) = 1 + t(x)p(x),$$

this shows that $[a(x)]$ has a multiplicative inverse if and only if 1 can be written as a linear combination of $a(x)$ and $p(x)$, which occurs if and only if $\gcd(a(x), p(x)) = 1$. The inverse $[b(x)] = [a(x)]^{-1}$ can be found by using the Euclidean algorithm.   □

4.4.6 Theorem.   Let F be a field, and let $p(x)$ be a nonconstant polynomial over F. Then $F[x]/\langle p(x) \rangle$ is a field if and only if $p(x)$ is irreducible over F.

Proof. Proposition 4.4.4 shows that addition and multiplication of congruence classes are well-defined. The associative, commutative, and distributive laws follow easily from the corresponding laws for addition and multiplication of polynomials. For example,

$$[a(x)][b(x)] = [a(x)b(x)] = [b(x)a(x)] = [b(x)][a(x)]$$

for all $a(x)$, $b(x) \in F[x]$. The additive identity is $[0]$ and the multiplicative identity is $[1]$, while the additive inverse of $[a(x)]$ is $[-a(x)]$. All that remains to show that $F[x]/\langle p(x) \rangle$ is a field is to show that each nonzero congruence class has a multiplicative inverse. Since by Proposition 4.4.3 we can work with representatives of lower degree than $\deg(p(x))$, by Proposition 4.4.5 each nonzero congruence class $[a(x)]$ has a multiplicative inverse if and only if $\gcd(a(x), p(x)) = 1$ for all nonzero polynomials $a(x)$ with $\deg(a(x)) < \deg(p(x))$. This occurs if and only if $p(x)$ is irreducible, completing the proof.   □

We note that whenever $p(x)$ is irreducible, the congruence class $[a(x)]$ is invertible if $a(x) \neq 0$ and $\deg(a(x)) < \deg(p(x))$. Conversely, if $[a(x)]$ is invertible, then $[a(x)] = [r(x)]$ where $r(x) \neq 0$ and $\deg(r(x)) < \deg(p(x))$.

Since $F[x]$ is an abelian group under addition and $\langle p(x) \rangle$ is a subgroup, addition of congruence classes is just the operation defined in the corresponding factor group. Thus results from Section 3.8 also imply that $F[x]/\langle p(x) \rangle$ is an abelian group under addition. In Section 5.3 we will handle multiplication of cosets in the more general setting of commutative rings.

4.4.7 Definition.   Let F_1 and F_2 be fields. A function $\phi : F_1 \rightarrow F_2$ is called an *isomorphism of fields* if it is one-to-one and onto, $\phi(a + b) = \phi(a) + \phi(b)$, and $\phi(ab) = \phi(a)\phi(b)$ for all $a, b \in F_1$.

Example 4.4.2

We can now give the full story of Example 4.4.1. In Appendix E we defined **C** to be the set of all expressions of the form $a + bi$, where $a, b \in \mathbf{R}$ and $i^2 = -1$. Since $x^2 + 1$ is irreducible over **R**, we know that $\mathbf{R}[x]/\langle x^2 + 1 \rangle$ is a field. Its elements are in one-to-one correspondence with polynomials of the form $a + bx$. Furthermore, the mapping $\phi : \mathbf{R}[x]/\langle x^2 + 1 \rangle \rightarrow \mathbf{C}$ defined by $\phi([a + bx]) = a + bi$ can be shown to be an isomorphism. Since $x^2 \equiv -1$ (mod $x^2 + 1$), the congruence class $[x]$ of the polynomial x satisfies the condition $[x]^2 = -1$. Thus the construction of $\mathbf{R}[x]/\langle x^2 + 1 \rangle$ using congruence classes allows us to provide a concrete model for the construction we gave previously, in which we merely conjured up an element i for which $i^2 = -1$. □

4.4.8 Theorem (Kronecker).   Let F be a field, and let $f(x)$ be any nonconstant polynomial in $F[x]$. Then there exists an extension field E of F and an element $\alpha \in E$ such that $f(\alpha) = 0$.

Proof. The polynomial $f(x)$ can be written as a product of irreducible polynomials, and so we let $p(x)$ be one of the irreducible factors of $f(x)$. It is sufficient to find an extension field E containing an element α such that $p(\alpha) = 0$.

By Proposition 4.4.6, $F[x]/\langle p(x) \rangle$ is a field, which we will denote by E. The field F is easily seen to be isomorphic to the subfield of E consisting of all congruence classes of the form $[a]$, where $a \in F$. We make this identification of F with the corresponding subfield of E so that we can consider E to be an extension of F. Let α be the congruence class $[x]$. If $p(x) = a_n x^n + \ldots + a_0$, where $a_i \in F$, then we must compute $p(\alpha)$. We obtain

$$p(\alpha) = a_n([x])^n + \ldots + a_1([x]) + a_0 = [a_n x^n + \ldots + a_1 x + a_0] = [0]$$

since $p(x) \equiv 0$ (mod $p(x)$). Thus $p(\alpha) = 0$ and the proof is complete.   □

4.4.9 Corollary.   Let F be a field, and let $f(x)$ be any nonconstant polynomial in $F[x]$. Then there exists an extension field E over which $f(x)$ can be factored into a product of linear factors.

Proof. Factor out all linear factors of $f(x)$ and let $f_1(x)$ be the remaining factor. We can find an extension E_1 in which $f_1(x)$ has a root, say α_1. Then we can write $f_1(x) = (x - \alpha_1)f_2(x)$, and by considering $f_2(x)$ as an element of $E_1[x]$, we can continue the same procedure for $f_2(x)$. We will finally arrive at an extension E that contains all of the roots of $f(x)$, and over this extension, $f(x)$ can be factored into a product of linear factors.   □

Example 4.4.3

Consider the polynomial $x^4 - x^2 - 2$ of Example 4.1.2, with coefficients in $F = \mathbf{Q}$. It factors as $(x^2 - 2)(x^2 + 1)$, and as our first step we let $E_1 = \mathbf{Q}[x]/\langle x^2 - 2 \rangle$, which is isomorphic (see Exercise 6) to $\mathbf{Q}(\sqrt{2})$. Although E_1 contains the roots $\pm\sqrt{2}$ of the factor $x^2 - 2$, it does not contain the roots $\pm i$ of the factor $x^2 + 1$, and so we must obtain a further extension $E_2 = E_1[x]/\langle x^2 + 1 \rangle$. In Chapter 6 we will see that E_2 is isomorphic to the smallest subfield of $\mathbf{C}$ that contains $\sqrt{2}$ and i, which is denoted by $\mathbf{Q}(\sqrt{2}, i)$. $\square$

Example 4.4.4

Let $F = \mathbf{Z}_2$ and let $p(x) = x^2 + x + 1$. Then $p(x)$ is irreducible over $\mathbf{Z}_2$ since it has no roots in $\mathbf{Z}_2$, and so $\mathbf{Z}_2[x]/\langle x^2 + x + 1 \rangle$ is a field. The congruence classes modulo $x^2 + x + 1$ can be represented by $[0]$, $[1]$, $[x]$ and $[1 + x]$, since these are the only polynomials of degree less than 2 over $\mathbf{Z}_2$. Addition and multiplication are given in Tables 4.4.1 and 4.4.2. To simplify these tables, all brackets have been omitted in listing the congruence classes. $\square$

TABLE 4.4.1 Addition in $\mathbf{Z}_2[x]/\langle x^2 + x + 1 \rangle$

+	0	1	x	$1 + x$
0	0	1	x	$1 + x$
1	1	0	$1 + x$	x
x	x	$1 + x$	0	1
$1 + x$	$1 + x$	x	1	0

TABLE 4.4.2 Multiplication in $\mathbf{Z}_2[x]/\langle x^2 + x + 1 \rangle$

$\times$	0	1	x	$1 + x$
0	0	0	0	0
1	0	1	x	$1 + x$
x	0	x	$1 + x$	1
$1 + x$	0	$1 + x$	1	x

If $q(x)$ is irreducible over $\mathbf{Z}_p$, then $\mathbf{Z}_p[x]/\langle q(x) \rangle$ has p^n elements if $\deg(q(x)) = n$, since there are exactly $p^n - 1$ polynomials over $\mathbf{Z}_p$ of degree less than n (including 0 gives p^n elements). It is possible to show that there exist polynomials of degree n irreducible over $\mathbf{Z}_p$ for each integer $n > 0$. This guarantees the existence of a finite field having p^n elements, for each prime number p and each positive integer n. Finite fields will be investigated in greater detail in Section 6.5.

EXERCISES: SECTION 4.4

1. Prove that congruence of polynomials defines an equivalence relation on $F[x]$.

2. Let E be a field, and let F be a subfield of E. Prove that the multiplicative identity of F must be the same as that of E.

3. Let E be a field, and let F be a subset of E that contains a nonzero element. Prove that F is a subfield of E if and only if F is closed under the addition, subtraction, multiplication, and division of E.

4. Prove Proposition 4.4.4.

5. Prove that $\mathbf{R}[x]/\langle x^2 + 1 \rangle$ is isomorphic to $\mathbf{C}$.

6. Prove that $\mathbf{Q}[x]/\langle x^2 - 2 \rangle$ is isomorphic to $\mathbf{Q}(\sqrt{2})$, which is defined in Example 4.1.1.

7. Prove that the field given in Example 4.4.4 is isomorphic to the following field of the four matrices given in Example 4.1.3:

$$\left\{ \begin{bmatrix} 0 & 0 \\ 0 & 0 \end{bmatrix}, \begin{bmatrix} 1 & 0 \\ 0 & 1 \end{bmatrix}, \begin{bmatrix} 1 & 1 \\ 1 & 0 \end{bmatrix}, \begin{bmatrix} 0 & 1 \\ 1 & 1 \end{bmatrix} \right\}.$$

8. Find an irreducible polynomial $p(x)$ of degree 3 over $\mathbf{Z}_2$, and list all elements of $\mathbf{Z}_2[x]/\langle p(x) \rangle$. Give the identities necessary to multiply elements.

9. Give addition and multiplication tables for the field $\mathbf{Z}_3[x]/\langle x^2 + x + 2 \rangle$.

10. Find a polynomial of degree 3 irreducible over $\mathbf{Z}_3$, and use it to construct a field with 27 elements. List the elements of the field; give the identities necessary to multiply elements.

11. As in the previous exercise, construct a field having 125 elements.

12. Find multiplicative inverses of the given elements in the given fields:
    (a) $[a + bx]$ in $\mathbf{R}[x]/\langle x^2 + 1 \rangle$
    (b) $[a + bx]$ in $\mathbf{Q}[x]/\langle x^2 - 2 \rangle$
    (c) $[x^2 - 2x + 1]$ in $\mathbf{Q}[x]/\langle x^3 - 2 \rangle$
    (d) $[x^2 - 2x + 1]$ in $\mathbf{Z}_3[x]/\langle x^3 + x^2 + 2x + 1 \rangle$
    (e) $[x + 4]$ in $\mathbf{Z}_5[x]/\langle x^3 + x + 1 \rangle$

13. For which values of $a = 1, 2, 3, 4$ is $\mathbf{Z}_5[x]/\langle x^2 + a \rangle$ a field? Show your work.

14. For which values of $k = 2, 3, 5, 7, 11$ is $\mathbf{Z}_k[x]/\langle x^2 + 1 \rangle$ a field? Show your work.

15. Let F be a finite field. Show that $F[x]$ has irreducible polynomials of arbitrarily high degree.

    Hint: Imitate Euclid's proof that there exist infinitely many prime numbers.

APPENDIX F: SOLUTION BY RADICALS OF REAL CUBIC AND QUARTIC EQUATIONS

F.1 Solution of the General Quadratic Equation

To solve the equation

$$ax^2 + bx + c = 0$$

with a, b, $c \in \mathbf{R}$, we can divide through by a and make the substitution $x = y - b/2a$. (We have rearranged the usual approach of completing the square in order to parallel later work.) This gives

$$y^2 + p = 0,$$

where

$$-p = \frac{b^2 - 4ac}{4a^2},$$

and so

$$y = \frac{\pm\sqrt{b^2 - 4ac}}{2a}.$$

Thus the general solution is

$$x = \frac{-b \pm \sqrt{b^2 - 4ac}}{2a}.$$

F.2 Discriminant of a Quadratic Equation

For the real quadratic equation

$$ax^2 + bx + c = 0,$$

the discriminant

$$\Delta = b^2 - 4ac$$

determines whether the solutions are real numbers (when $\Delta \geq 0$) or imaginary numbers (when $\Delta < 0$). The equation has a multiple root if and only if $\Delta = 0$, and this occurs if and only if

$$ax^2 + bx + c = (mx + k)^2$$

for some $m, k \in \mathbf{R}$. If $a = 1$, then $\Delta = (x_1 - x_2)^2$, where x_1, x_2 are the solutions of the equation $x^2 + bx + c = 0$.

F.3 Discriminant of a Cubic Equation

In discussing the general cubic equation

$$ax^3 + bx^2 + cx + d = 0$$

with real coefficients, we will assume that $a = 1$. The discriminant Δ of the equation is then defined by

$$\Delta = (x_1 - x_2)^2(x_1 - x_3)^2(x_2 - x_3)^2$$

where x_1, x_2, x_3 are the roots of the equation, and it gives the following information: If the roots are all real, then $\Delta \geq 0$, and $\Delta = 0$ if and only if at least two of the
roots coincide. If not all of the roots are real, say x_2 is imaginary, then one of the
other roots must be its complex conjugate, say $x_3 = \bar{x}_2$, and the remaining root x_1 is
real. Since x_1 is real,

$$(x_1 - x_2)(x_1 - \bar{x}_2) = (x_1 - x_2)(\overline{x_1 - x_2})$$

is a real number, so

$$(x_1 - x_2)^2(x_1 - \bar{x}_2)^2 > 0.$$

But $x_2 - x_3 = x_2 - \bar{x}_2$ is purely imaginary, so $(x_2 - x_3)^2 < 0$, and therefore $\Delta < 0$.
Thus we have shown that $\Delta \geq 0$ when all roots are real, $\Delta < 0$ when two roots are
imaginary, and $\Delta = 0$ when there is a multiple root.

If we make the substitution $x = y - b/3$, which is the first step in solving the
general cubic equation in F.4, we obtain the reduced equation $y^3 + py + q = 0$.
This reduced equation has the same discriminant, which can now be computed as
$\Delta = -4p^3 - 27q^2$. This can be shown either by using the relations $y_1 + y_2 + y_3 = 0$,
$y_1y_2 + y_1y_3 + y_2y_3 = p$, and $y_1y_2y_3 = -q$ for the roots y_1, y_2, y_3 of the reduced
equation, or by a direct computation involving the roots (which will be determined
later). See the exercises at the end of this appendix for hints in making this
computation.

F.4 Solution of the General Cubic Equation

Summary: The roots of the reduced cubic equation

$$y^3 + py + q = 0$$

can be found from the roots of the resolvent equation

$$(z^3)^2 + qz^3 - \left(\frac{p}{3}\right)^3 = 0$$

where $y = z - p/3z$. The resolvent equation can be solved for z^3 by using the
quadratic formula.

The first step in solving the equation

$$x^3 + bx^2 + cx + d = 0$$

is to substitute $x = y - b/3$, which eliminates the quadratic term. This gives the
reduced cubic equation

$$y^3 + py + q = 0,$$

where

$$p = c - \frac{b^2}{3}$$

and

$$q = d - \frac{bc}{3} + \frac{2b^3}{27}.$$

The roots of the original equation can easily be found from those of the reduced equation. The method we will use to solve the reduced cubic equation is essentially the same as that used by Vieta in 1591. We let

$$\omega = -\frac{1}{2} + \frac{\sqrt{3}}{2} i$$

be a complex cube root of unity.

We next make the substitution $y = z - p/3z$, to obtain the equation

$$z^3 - \frac{p^3}{27z^3} + q = 0.$$

Multiplying through by z^3 gives

$$(z^3)^2 + q(z^3) - \left(\frac{p}{3}\right)^3 = 0;$$

which is a quadratic equation in z^3. Using the quadratic formula we obtain

$$z^3 = \frac{1}{2}\left(-q \pm \sqrt{q^2 + 4\left(\frac{p}{3}\right)^3}\right) = -\frac{q}{2} \pm \sqrt{\left(\frac{p}{3}\right)^3 + \left(\frac{q}{2}\right)^2}.$$

If z_1^3 and z_2^3 are the two solutions, then $z_1^3 z_2^3 = -(p/3)^3$. It is possible to choose the cube root in such a way that $z_1 z_2 = -p/3$, or $z_2 = -p/3z_1$, and then the other cube roots will be ωz_1, $\omega^2 z_1$, ωz_2, and $\omega^2 z_2$. Now substituting z_1 in the equation $y = z - p/3z$ to find the corresponding root y_1, we find that $y_1 = z_1 - p/3z_1 = z_1 + z_2$. Since $(\omega z_1)(\omega^2 z_2) = (\omega^2 z_1)(\omega z_2) = -p/3$, in the same way we have $y_2 = \omega z_1 + \omega^2 z_2$ and $y_3 = \omega^2 z_1 + \omega z_2$. Thus the roots of $z^6 + qz^3 - (p/3)^3 = 0$ give at most three distinct values when substituted into $y = z - p/3z$, and these are the roots of $y^3 + py + q = 0$.

Using the identities $\omega^3 = 1$ and $1 + \omega + \omega^2 = 0$ it is not difficult to check that

$$y_1 = z_1 + z_2,$$
$$y_2 = \omega z_1 + \omega^2 z_2,$$

and

$$y_3 = \omega^2 z_1 + \omega z_2$$

satisfy the relations

$$y_1 + y_2 + y_3 = 0,$$
$$y_1 y_2 + y_1 y_3 + y_2 y_3 = p,$$

and

$$y_1 y_2 y_3 = -q,$$

so they are the three roots of the reduced equation. We now give additional details for finding the roots in practice.

Finding the roots when $\Delta = 0$

If $-4p^3 - 27q^2 = 0$, then

$$\frac{p^3}{27} + \frac{q^2}{4} = 0,$$

and so $z_1 = z_2$. Thus

$$y_1 = z_1 + z_2 = 2\sqrt[3]{\frac{q}{2}} = \sqrt[3]{-4q}.$$

Since $\omega + \omega^2 = -1$, we have $y_2 = -z_1 = \sqrt[3]{q}/\sqrt[3]{2}$, and similarly $y_3 = \sqrt[3]{q}/\sqrt[3]{2}$. Thus if $\Delta = 0$ we have roots $y = \sqrt[3]{-4q}$ and $y = \sqrt[3]{q}/\sqrt[3]{2}$, the latter with multiplicity 2.

Finding the roots when $\Delta < 0$

If $-4p^3 - 27q^2 < 0$, then

$$\frac{p^3}{27} + \frac{q^2}{4} > 0$$

and

$$\sqrt{\left(\frac{p}{3}\right)^3 + \left(\frac{q}{2}\right)^2}$$

is a real number. Thus we can take

$$z_1 = \sqrt[3]{\frac{-q}{2} + \sqrt{\left(\frac{p}{3}\right)^3 + \left(\frac{q}{2}\right)^2}},$$

$$z_2 = \sqrt[3]{\frac{-q}{2} - \sqrt{\left(\frac{p}{3}\right)^3 + \left(\frac{q}{2}\right)^2}},$$

and $y_1 = z_1 + z_2$. Then

$$y_2 = -\frac{1}{2}(z_1 + z_2) + \frac{1}{2}(z_1 - z_2)\sqrt{3}i$$

and

$$y_3 = -\frac{1}{2}(z_1 + z_2) - \frac{1}{2}(z_1 - z_2)\sqrt{3}i.$$

The real root y_1 leads to the factorization

$$y^3 + py + q = (y - y_1)(y^2 + y_1y + (p + y_1^2)),$$

and using the quadratic formula gives the roots

$$\frac{-y_1}{2} \pm \frac{i}{2} \sqrt{3y_1^2 + 4p}.$$

Once it has been checked that y_1 is a root, this formula can be used to check y_2 and y_3.

Finding the roots when $\Delta > 0$

When $\Delta > 0$, all roots are real, but

$$\sqrt{\left(\frac{p}{3}\right)^3 + \left(\frac{q}{2}\right)^2}$$

is imaginary, so to compute the roots of the reduced equation we must first find the cube roots of two imaginary numbers. It can be proved that no formula involving only real radicals can be given.

In this case it is easier to use another method for finding roots, although we must admit that it does not yield a solution by radicals. Note that since $-4p^3 - 27q^2 > 0$, we have $-4p^3 > 27q^2$ and so p must be negative. The trigonometric formula for the cosine of the sum of two angles can be used to show that $\cos 3\theta = 4\cos^3 \theta - 3\cos \theta$, so that $z = \cos \theta$ is a root of the equation $4z^3 - 3z = k$ if $\cos 3\theta = k$, that is, if $\theta = (\arccos k)/3$. In this case, $\cos(\theta + 2\pi/3)$ and $\cos(\theta + 4\pi/3)$ are also solutions, since $\cos 3(\theta + 2n\pi/3) = k$. This gives a method for solving the equation $4z^3 - 3z = k$, if $|k| \leq 1$.

In the reduced equation $y^3 + py + q = 0$, we make the substitution

$$y = 2 \sqrt{\frac{-p}{3}} z.$$

(Remember that $-p > 0$ since $\Delta > 0$.) We obtain the equation

$$4\left(2 \sqrt{-\left(\frac{p}{3}\right)^3}\right) z^3 - 3\left(2 \left(-\frac{p}{3}\right) \sqrt{-\frac{p}{3}}\right) z + q = 0,$$

or

$$4z^3 - 3z = k,$$

where

$$k = \frac{-q}{2} \sqrt{-\left(\frac{3}{p}\right)^3} = \frac{-q}{2} \left(\frac{3}{-p}\right)^{3/2}.$$

Since $\Delta > 0$, we showed above that $27q^2 < -4p^3$, so $(27q^2)/(-4p^3) = k^2 < 1$, and thus $z_1 = \cos \theta$ is a root, where

$$\theta = \frac{1}{3} \arccos \left(\frac{-q}{2} \left(\frac{3}{-p}\right)^{3/2}\right).$$

Thus to solve $y^3 + py + q = 0$ when $\Delta > 0$, we find

$$\theta = \frac{1}{3} \arccos \left(\frac{-q}{2} \left(\frac{3}{-p} \right)^{3/2} \right).$$

Then the roots are

$$y_k = 2 \sqrt{\frac{-p}{3}} \cos \left(\theta + \frac{2k\pi}{3} \right),$$

for $k = 0, 1, 2$.

F.5 Solution of the General Quartic Equation (Ferrari)

The general quartic equation

$$x^4 + bx^3 + cx^2 + dx + e = 0$$

becomes

$$y^4 + py^2 + qy + r = 0$$

after substituting $x = y - (b/4)$. Then $y^4 = -py^2 - qy - r$, and adding $y^2 z + (z^2/4)$ to both sides of the equation gives

$$\left(y^2 + \frac{1}{2} z \right)^2 = (z - p)y^2 - qy + \left(\frac{1}{4} z^2 - r \right).$$

If the right-hand side can be put in the form $(my + k)^2$, then the roots of the reduced quartic are the roots of

$$y^2 + \frac{1}{2} z = my + k$$

and

$$y^2 + \frac{1}{2} z = -my - k.$$

Since the right-hand side is a quadratic in y, it can be put in the form $(my + k)^2$ if and only if its discriminant is zero. This occurs if and only if

$$q^2 - 4(z - p) \left(\frac{z^2}{4} - r \right) = 0,$$

or equivalently, if and only if

$$z^3 - pz^2 - 4rz + (4pr - q^2) = 0.$$

This latter equation in z is called the *resolvent cubic equation*, and any real root can be used to solve the reduced quartic.

For example, for the equation

$$y^4 + 3y^2 - 2y + 3 = 0,$$

the resolvent cubic is

$$z^3 - 3z^2 - 12z + 32 = 0,$$

which has the root $z = 4$. Thus

$$(y^2 + 2)^2 = y^2 + 2y + 1 = (y + 1)^2,$$

so either $y^2 + 2 = y + 1$ and then

$$y = \frac{1}{2} \pm \frac{\sqrt{3}}{2}\,i,$$

or else $y^2 + 2 = -y - 1$ and then

$$y = -\frac{1}{2} \pm \frac{\sqrt{11}}{2}\,i.$$

F.6 Solution of the General Quartic Equation (Descartes)

We will try to factor the reduced quartic

$$y^4 + py^2 + qy + r = 0$$

as

$$(y^2 + ky + m)(y^2 - ky + n) = y^4 + (m + n - k^2)y^2 + (kn - km)y + mn.$$

We must have $m + n - k^2 = p$, $k(n - m) = q$, and $mn = r$. Since $m + n = p + k^2$ and $n - m = q/k$, we have $2n = p + k^2 + (q/k)$ and $2m = p + k^2 - (q/k)$. Then

$$4r = 2n2m = \left(p + k^2 + \frac{q}{k}\right)\left(p + k^2 - \frac{q}{k}\right),$$

which leads to the equation

$$(k^2)^3 + 2p(k^2)^2 + (p^2 - 4r)k^2 - q^2 = 0.$$

Any root of this equation gives a factorization of the desired form, and then the roots can be found from the quadratic formula.

For example, the equation $y^4 - 3y^2 + 6y - 2 = 0$ leads to the equation

$$(k^2)^3 - 6(k^2)^2 + 17k^2 - 36 = 0,$$

which has the root $k^2 = 4$. Thus

$$y^4 - 3y^2 + 6y - 2 = (y^2 + 2y - 1)(y^2 - 2y + 2),$$

which gives the four roots $y_1 = -1 + \sqrt{2}$, $y_2 = -1 - \sqrt{2}$, $y_3 = 1 + i$, $y_4 = 1 - i$.

We have shown that any polynomial equation of degree less than 5 can be solved by radicals. It is possible to show that certain equations of degree 5 or higher cannot be solved by radicals. For example, the equation $2x^5 - 10x + 5 = 0$ cannot be solved by radicals. We can summarize this section with the following theorem.

F.7 Theorem.   Any polynomial equation of degree ≤ 4 with real coefficients is solvable by radicals.

EXERCISES: APPENDIX F

1. Show that $x^3 + ax + 2 = 0$ has three real roots if and only if $a \leq -3$.

2. To show that the discriminant of $y^3 + py + q = 0$ is $-4p^3 - 27q^2$, we can use the relations $y_1 + y_2 + y_3 = 0$, $y_1y_2 + y_1y_3 + y_2y_3 = p$, and $y_1y_2y_3 = -q$ for the roots y_1, y_2, y_3. Substituting $y_1 = -y_2 - y_3$ in the second equation gives $p = -y_2^2 - y_3^2 - y_2y_3$ and in the third equation gives $q = y_2^2y_3 + y_2y_3^2$. Compute $-27q^2 - 4p^3$ and check that it is equal to $(y_1 - y_2)^2(y_1 - y_3)^2(y_2 - y_3)^2$ when the same substitution $y_1 = -y_2 - y_3$ is made.

3. Give another proof that the discriminant of $y^3 + py + q = 0$ is $-4p^3 - 27q^2$ by using the following form of the roots: $y_1 = z_1 + z_2$; $y_2 = \omega z_1 + \omega^2 z_2$; $y_3 = \omega^2 z_1 + \omega z_2$; where $z_1z_2 = -p/3$.

   Hint: Since 1, ω, ω^2 are cube roots of unity, $\omega^3 = 1$, $\omega^2 + \omega + 1 = 0$, and $(x - 1)(x - \omega)(x - \omega^2) = x^3 - 1$, for any x. Substituting $x = z_1/z_2$ in the latter formula gives $(z_1 - z_2)(z_1 - \omega z_2)(z_1 - \omega^2 z_2) = z_1^3 - z_2^3$. Write out $(y_1 - y_2)(y_1 - y_3)(y_2 - y_3) = (1 - \omega)(z_1 - \omega^2 z_2) \ldots$, square both sides, and simplify.

4. In F.4 check that y_1, y_2, y_3 as given are solutions of $y^3 + py + q = 0$.

5. Verify that $\cos 3\theta = 4 \cos^3 \theta - 3 \cos \theta$.

5

Commutative Rings

Many of the algebraic properties of the set of integers are also valid for the set of all polynomials with coefficients in any field. By working with these properties in an abstract setting, it is sometimes possible to prove one theorem that can be applied to both situations, rather than proving two separate theorems, one for each case. As an additional bonus, the abstract setting for the theorem can then be applied to new situations.

For example, we have shown that any nonzero integer not equal to ± 1 can be written as a product of prime numbers. It is also true that any nonconstant polynomial with real coefficients can be written as a product of irreducible polynomials. Since the same basic principles are used in each proof, it should be possible to give one proof that would cover both cases. A mathematician should try to recognize such similarities and take advantage of them, and should also ask whether the techniques can be applied to other questions.

The interplay between specific examples and abstract theories is critical to the development of useful mathematics. As more and more is learned about various specific examples, it becomes necessary to synthesize the knowledge, so that it is easier to grasp the essential character of each example and to relate the examples to each other. By generalizing and abstracting from concrete examples, it may be possible to present a unified theory that can be more easily understood than seemingly unrelated pieces of information from a variety of situations.

In a general theory of algebraic objects of a particular type, the main problem is that of classifying and describing the objects. This often includes determining the simplest sort of building blocks and describing all of the ways in which they can be put together. In this chapter we give a very simple example of a theorem of this sort, when we show that, as a ring, $\mathbf{Z}_n$ is isomorphic to a direct sum of similar rings of prime power order. These rings have a particularly simple structure, and cannot themselves be decomposed.

The concept of an abstract commutative ring will be introduced in Section 5.1 to provide a common framework for studying a variety of questions. The set of integers modulo n will be shown to form a commutative ring. This is an example of an important procedure by which new rings can be constructed—the use of a congruence relation on a given ring. The notion of a factor group will be extended to that of a factor ring. In the construction of factor rings, the role of normal subgroups in forming factor groups will be played by subsets called ideals. Examples are $n\mathbf{Z}$ in $\mathbf{Z}$ and $\langle f(x) \rangle$ in $F[x]$. Then the notions of prime number and irreducible polynomial motivate the definition of a "prime ideal," which allows us to tie together a number of facts about integers and polynomials. In Section 5.4 we construct quotient fields for integral domains, and thus characterize all subrings (with identity) of fields.

5.1 COMMUTATIVE RINGS; INTEGRAL DOMAINS

In Chapter 1, we began our study of abstract algebra by concentrating on one of the most familiar algebraic structures, the set of integers. In both $\mathbf{Z}$ and $\mathbf{Z}_n$ we had two operations—addition and multiplication. Subtraction and division (when possible) were defined in terms of the two basic operations. After studying groups in Chapter 3, where we had only one operation to deal with, we returned to systems with two operations when we worked with fields and polynomials in Chapter 4.

We will now undertake a systematic study of systems in which there are two operations that generalize the familiar operations of addition and multiplication. The examples you should have in mind are these: the set of integers $\mathbf{Z}$; the set $\mathbf{Z}_n$ of integers modulo n; any field F (in particular the set $\mathbf{Q}$ of rational numbers and the set $\mathbf{R}$ of real numbers); the set $F[x]$ of all polynomials with coefficients in a field F. The axioms we will use are the same as those for a field, with two crucial exceptions. We have dropped the requirement that each nonzero element has a multiplicative inverse, in order to include integers and polynomials in the class of objects we want to study. We have also dropped the requirement that a multiplicative identity element exists, which allows us, for example, to include the set of all even integers in our definition.

5.1.1 Definition. Let R be a set on which two binary operations are defined, called addition and multiplication, and denoted by $+$ and $\cdot$. Then R is called a *commutative ring* with respect to these operations if the following properties hold:

(i) *Closure.* If $a, b \in R$, then the sum $a + b$ and the product $a \cdot b$ are uniquely defined and belong to R.

(ii) *Associative laws.* For all $a, b, c \in R$,

$$a + (b + c) = (a + b) + c \quad \text{and} \quad a \cdot (b \cdot c) = (a \cdot b) \cdot c.$$

(iii) *Commutative laws.* For all $a, b \in R$,

$$a + b = b + a \quad \text{and} \quad a \cdot b = b \cdot a.$$

(iv) *Distributive laws.* For all $a, b, c \in R$,

$$a \cdot (b + c) = a \cdot b + a \cdot c \quad \text{and} \quad (a + b) \cdot c = a \cdot c + b \cdot c.$$

(v) *Additive identity.* The set R contains an *additive identity element,* denoted by 0, such that for all $a \in R$,

$$a + 0 = a \quad \text{and} \quad 0 + a = a.$$

(vi) *Additive inverses.* For each $a \in R$, the equations

$$a + x = 0 \quad \text{and} \quad x + a = 0$$

have a solution x in R, called the *additive inverse* of a, and denoted by $-a$.

The commutative ring R is called a *commutative ring with identity* if it contains an element 1, assumed to be different from 0 such that for all $a \in R$,

$$a \cdot 1 = a \quad \text{and} \quad 1 \cdot a = a.$$

In this case, 1 is called a *multiplicative identity element,* or, more generally, simply an *identity element.*

Our first observation is that any commutative ring R determines an abelian group by just considering the set R together with the single operation of addition. We call this the underlying additive group of R. Although we require that multiplication in R is commutative, the set of nonzero elements certainly need not define an abelian group under multiplication.

A set with two binary operations that satisfy conditions (i)–(vi), with the exception of the commutative law for multiplication, is called a *ring.* Although we will not discuss them here, there are many interesting examples of noncommutative rings. From your work in linear algebra, you should already be familiar with one such example, the set of all 2×2 matrices over **R.** The standard rules for matrix arithmetic provide all of the axioms for a commutative ring, with the exception of the commutative law for multiplication. Although this is certainly an important example worthy of study, we have chosen to work only with commutative rings, with emphasis on integral domains, fields, and polynomial rings over them.

Before giving some further examples of commutative rings, it is helpful to have some additional information about them. Our observation that any commutative ring is an abelian group under addition implies that the cancellation law holds for addition. Just as for fields, various uniqueness statements follow from Proposition 3.1.2.

Let R be a commutative ring, with elements a, b, $c \in R$.
(a) If $a + c = b + c$, then $a = b$.
(b) If $a + b = 0$, then $b = -a$.
(c) If $a + b = a$ for some $a \in R$, then $b = 0$.
(d) If $a \cdot b = a$ for all $a \in R$, then R has an identity 1 and $b = 1$.

In Proposition 4.1.3 the following properties were shown to hold for any field. In fact, their proof did not require the existence of a multiplicative identity element, and so they remain valid for any commutative ring. Note that (a) and (c) involve connections between addition and multiplication. Their proofs make use of the distributive law, since it provides the only link between the two operations.

Let R be a commutative ring.
(a) For all $a \in R$, $a \cdot 0 = 0$.
(b) For all $a \in R$, $-(-a) = a$.
(c) For all $a, b \in R$, $(-a) \cdot (-b) = a \cdot b$.

We will follow the usual convention of performing multiplications before additions unless parentheses intervene.

Example 5.1.1

In Section 1.4 we listed the properties of addition and multiplication of congruence classes, which show that the set $\mathbf{Z}_n$ of integers modulo n is a commutative ring with identity. From our study of groups we know that $\mathbf{Z}_n$ is a factor group of $\mathbf{Z}$ (under addition), and so it is an abelian group under addition. To verify that the necessary properties hold for multiplication, it is necessary to use the corresponding properties for $\mathbf{Z}$. We checked the distributive law in Section 1.4. It is worth commenting on the proof of the associative law, to point out the crucial parts of the proof. For example, we can check that the associative law holds for all $[a]$, $[b]$, $[c] \in \mathbf{Z}_n$:

$$[a]([b][c]) = [a][bc] = [a(bc)] \quad \text{and} \quad ([a][b])[c] = [ab][c] = [(ab)c],$$

and so these two expressions are equal because the associative law holds for multiplication in $\mathbf{Z}$. In effect, as soon as we have established that multiplication is well-defined (Proposition 1.4.2), it is easy to show that the necessary properties are inherited by the set of congruence classes.

The rings $\mathbf{Z}_n$ form a class of commutative rings that is a good source of counterexamples. For instance, it provides an easy example showing that the cancellation law may fail for multiplication. In the commutative ring $\mathbf{Z}_6$ we have $[2][3] = [4][3]$, but $[2] \neq [4]$.   □

Example 5.1.2 (Polynomial Rings)

Let R be any commutative ring with 1. We let $R[x]$ denote the set of infinite tuples

$$(a_0, a_1, a_2, \ldots)$$

such that $a_i \in R$ for all i, and $a_i \neq 0$ for only finitely many terms a_i. Two sequences are equal if and only if all corresponding terms are equal. We introduce addition and multiplication as follows:

$$(a_0, a_1, a_2, \ldots) + (b_0, b_1, b_2, \ldots) = (a_0 + b_0, a_1 + b_1, a_2 + b_2, \ldots)$$

$$(a_0, a_1, a_2, \ldots) \cdot (b_0, b_1, b_2, \ldots) = (c_0, c_1, c_2, \ldots), \quad \text{for} \quad c_k = \sum_{i+j=k} a_i b_j.$$

With these operations it can be shown that $R[x]$ is a commutative ring.

We can identify $a \in R$ with $(a, 0, 0, \ldots) \in R[x]$, and so if R has an identity 1, then $(1, 0, 0, \ldots)$ is an identity for $R[x]$. If we let $x = (0, 1, 0, \ldots)$, then the elements of $R[x]$ can be expressed in the form

$$a_0 + a_1 x + \ldots + a_{m-1} x^{m-1} + a_m x^m,$$

allowing us to use our previous notation for the *ring of polynomials over R in the indeterminate x*. As before, if n is the largest nonnegative integer such that $a_n \neq 0$, then we say that the polynomial has *degree n*, and a_n is called the *leading coefficient* of the polynomial.

Once we know that $R[x]$ is a commutative ring, it is easy to work with polynomials in two indeterminates x and y. We can simply use the commutative ring $R[x]$, and consider all polynomials in the indeterminate y, with coefficients in $R[x]$. For example, by factoring out the appropriate terms we have

$$2x - 4xy + y^2 + xy^2 + x^2 y^2 - 3xy^3 + x^3 y^2 + 2x^2 y^3 =$$

$$2x + (-4x)y + (1 + x + x^2 + x^3)y^2 + (-3x + 2x^2)y^3.$$

The ring of polynomials in two indeterminates with coefficients in R is usually denoted by $R[x, y]$, rather than by $(R[x])[y]$.   □

The next proposition will make it easier for us to give examples, by giving a simple criterion for testing subsets of known commutative rings to determine whether they are also commutative rings. From now on it seems easiest to just use ab to denote the product $a \cdot b$. But you must remember that this can represent any operation that merely behaves in certain ways like ordinary multiplication.

5.1.2 Definition.   Let S be a commutative ring. A nonempty subset R of S is called a *subring* of S if it is a commutative ring under the addition and multiplication of S.

Let F and E be fields. If F is a subring of E, according to the above definition, then we usually say (more precisely) that F is a *subfield* of E. Of course,
there may be other subrings of fields that are not necessarily subfields. Any
subring is a subgroup of the underlying additive group of the larger ring, so the two
commutative rings must have the same zero element. If F is a subfield of E, then
more must be true: The set $F^{\times}$ of nonzero elements of F is a subgroup of $E^{\times}$, and
so they must share the same multiplicative identity element. In Chapter 6 we will
study subfields in much greater detail.

5.1.3 Proposition.   Let S be a commutative ring, and let R be a nonempty
subset of S. Then R is a subring of S if and only if
   (i) R is closed under addition and multiplication; and
   (ii) if $a \in R$, then $-a \in R$.

Proof. If R is a subring, then the closure axioms must certainly hold.
Suppose that z is the additive identity of R. Then $z + z = z = z + 0$, where 0 is the
additive identity of S, so $z = 0$, since the cancellation law for addition holds in S.
Finally, if $a \in R$ and b is the additive inverse of a in R, then $a + b = 0$, so $b = -a$
by Proposition 3.1.2, and this shows that $-a \in R$.

Conversely, suppose that the given conditions hold. Conditions (ii)–(iv) of
Definition 5.1.1 are inherited from S. Finally, since R is nonempty, it contains
some element, say $a \in R$. Then $-a \in R$, so $0 = a + (-a) \in R$ since R is closed
under addition. Thus conditions (v) and (vi) of Definition 5.1.1 are also satisfied.   □

Example 5.1.3

Let S be the commutative ring $\mathbf{Z}_6$ and let R be the subset $\{[0],[2],[4]\}$. Then
R is closed under addition and multiplication and contains the additive inverse of each element in R. Since $[4][0] = [0]$, $[4][2] = [2]$, and $[4][4] = [4]$, R
also has a multiplicative identity, namely $[4]$. This shows that a subring may
have a different multiplicative identity from that of the given commutative
ring.

If e is the multiplicative identity element of a subring T of S, then $e^2 =
e$. An element e such that $e^2 = e$ is said to be *idempotent*.

The idempotent elements of $\mathbf{Z}_6$ (other than $[0]$) are $[1]$, $[3]$, and $[4]$.
This can be used to show that the only subrings of $S = \mathbf{Z}_6$ that have an
identity element are S itself, R, and $T = \{[0],[3]\}$.   □

Example 5.1.4

Let $\mathbf{Z}[i]$ be the set of complex numbers of the form $m + ni$, where $m, n \in \mathbf{Z}$.
Since

$$(m + ni) + (r + si) = (m + r) + (n + s)i$$

$$\text{and} \quad (m + ni)(r + si) = (mr - ns) + (nr + ms)i,$$

for all $m, n, r, s \in \mathbf{Z}$, the usual sum and product of numbers in $\mathbf{Z}[i]$ have the correct form to belong to $\mathbf{Z}[i]$. This shows that $\mathbf{Z}[i]$ is closed under addition and multiplication of complex numbers. The negative of any element in $\mathbf{Z}[i]$ again has the correct form, as does $1 = 1 + 0i$, so $\mathbf{Z}[i]$ is a commutative ring with identity by Proposition 5.1.3.   □

Example 5.1.5

In Example 4.1.1 we verified that $\mathbf{Q}(\sqrt{2})$ is a field. It has an interesting subset

$$\mathbf{Z}[\sqrt{2}] = \{m + n\sqrt{2} \mid m, n \in \mathbf{Z}\}$$

which is obviously closed under addition. The product of two elements is given by

$$(m_1 + n_1\sqrt{2})(m_2 + n_2\sqrt{2}) = (m_1m_2 + 2n_1n_2) + (m_1n_2 + m_2n_1)\sqrt{2}$$

and so the set is also closed under multiplication. Proposition 5.1.3 can be applied to show that $\mathbf{Z}[\sqrt{2}]$ is a subring of $\mathbf{Q}(\sqrt{2})$.

Since $\mathbf{Q}(\sqrt{2})$ is a field, it contains $1/(m + n\sqrt{2})$ whenever $m + n\sqrt{2} \neq 0$, but $1/(m + n\sqrt{2}) \in \mathbf{Z}[\sqrt{2}]$ if and only if $m/(m^2 - 2n^2)$ and $n/(m^2 - 2n^2)$ are integers. It can be shown that this occurs if and only if $m^2 - 2n^2 = \pm 1$. (See Exercise 5.)   □

5.1.4 Definition. Let R be a commutative ring with identity element 1. An element $a \in R$ is said to be *invertible* if there exists an element $b \in R$ such that $ab = 1$. The element a is also called a *unit* of R, and its multiplicative inverse is usually denoted by a^{-1}.

Since $0 \cdot b = 0$ for all $b \in R$, it is impossible for 0 to be invertible. Furthermore, if $a \in R$ and $ab = 0$ for some nonzero $b \in R$, then a cannot be a unit since multiplying both sides of the equation by the inverse of a (if it existed) would show that $b = 0$. An element a such that $ab = 0$ for some $b \neq 0$ is called a *divisor of zero*.

Example 5.1.6

Let R be the set of all functions from the set of real numbers into the set of real numbers, with ordinary addition and multiplication of functions (not composition of functions). It is not hard to show that R is a commutative ring with identity, since addition and multiplication are defined pointwise, and the addition and multiplication of real numbers satisfy all of the field axioms. It is easy to find divisors of zero in this ring: Let $f(x) = 0$ for $x < 0$ and $f(x) = 1$ for $x \geq 0$, and let $g(x) = 0$ for $x \geq 0$ and $g(x) = 1$ for $x < 0$. Then $f(x)g(x) = 0$ for all x, which shows that $f(x)g(x)$ is the zero function.

The multiplicative identity of R is the function $f(x) = 1$ (for all x). Then a function $g(x)$ has a multiplicative inverse if and only if $g(x) \neq 0$ for all x.

Thus, for example, $g(x) = 2 + \sin(x)$ has a multiplicative inverse, but $h(x) = \sin(x)$ does not. $\square$

When thinking of the units of a commutative ring, here are some good examples to keep in mind. The only units of $\mathbf{Z}$ are 1 and -1. We showed in Proposition 1.4.3 that the set of units of $\mathbf{Z}_n$ consists of the congruence classes $[a]$ for which $(a,n) = 1$. We showed in Example 3.1.4 that $\mathbf{Z}_n^\times$ is a group under multiplication of congruence classes. We will use the notation $R^\times$ for the set of units of any commutative ring R with identity.

5.1.5 Proposition.   Let R be a commutative ring with identity. Then the set $R^\times$ of units of R is an abelian group under the multiplication of R.

Proof. If a, $b \in R^\times$, then a^{-1} and b^{-1} exist in R, and so $ab \in R^\times$ since $(ab)(b^{-1}a^{-1}) = 1$. We certainly have $1 \in R^\times$, and $a^{-1} \in R^\times$ since $(a^{-1})^{-1} = a$. Multiplication is assumed to be associative and commutative. $\square$

In the context of commutative rings we can give the following definition: A field is a commutative ring with identity in which every nonzero element is invertible. We can say, loosely, that a field is a set on which the operations of addition, subtraction, multiplication, and division can be defined. For example, we showed in Corollary 1.4.4 that $\mathbf{Z}_n$ is a field if and only if n is a prime number.

We have already observed that the cancellation law for addition follows from the existence of additive inverses. A similar result holds for multiplication. If $ab = ac$ and a is a unit, then multiplying by a^{-1} gives $a^{-1}(ab) = a^{-1}(ac)$, and then by using the associative law for multiplication, the fact that $a^{-1}a = 1$, and the fact that 1 is a multiplicative identity element, we see that $b = c$.

If the cancellation law for multiplication holds in a commutative ring R, then for any elements a, $b \in R$, $ab = 0$ implies that $a = 0$ or $b = 0$. Conversely, if this condition holds and $ab = ac$, then $a(b - c) = 0$, so if $a \neq 0$ then $b - c = 0$ and $b = c$. Thus the cancellation law for multiplication holds in R if and only if R has no nonzero divisors of zero.

It is precisely this property that is crucial in solving polynomial equations. If $(x - 2)(x - 3) = 0$ and x is an integer, then we can conclude that either $x - 2 = 0$ or $x - 3 = 0$, and so either $x = 2$ or $x = 3$. But if x represents an element of $\mathbf{Z}_6$, then we might have $x - [2] = [3]$ and $x - [3] = [2]$, since this still gives $(x - [2])(x - [3]) = [3][2] = [0]$ in $\mathbf{Z}_6$. In addition to the obvious roots $[2]$ and $[3]$, we can also substitute $[5]$ and $[0]$, respectively, so the polynomial $x^2 - [5]x + [6]$ has four distinct roots over $\mathbf{Z}_6$.

5.1.6 Definition.   A commutative ring R with identity is called an *integral domain* if for all a, $b \in R$, $ab = 0$ implies $a = 0$ or $b = 0$.

The ring of integers $\mathbf{Z}$ is the most fundamental example of an integral domain. The ring of all polynomials with real coefficients is also an integral domain, since the product of any two nonzero polynomials is again nonzero. As shown in Example 5.1.6, the larger ring of all real valued functions is not an integral domain.

Example 5.1.7

Let R be an integral domain. The ring $R[x]$ of all polynomials with coefficients in R is also an integral domain. To show this we note that if $f(x)$ and $g(x)$ are nonzero polynomials with leading coefficients a_m and b_n, respectively, then since R is an integral domain, the product $a_m b_n$ is nonzero. This shows that the leading coefficient of the product $f(x)g(x)$ is nonzero, and so $f(x)g(x) \neq 0$. Just as in Proposition 4.1.5, we have $f(x)g(x) \neq 0$ because the degree of $f(x)g(x)$ is equal to $\deg(f(x)) + \deg(g(x))$.   □

The next theorem gives a condition that is very useful in studying integral domains. It shows immediately, for example, that $\mathbf{Z}[i]$ and $\mathbf{Z}[\sqrt{2}]$ are integral domains. A converse to Theorem 5.1.7 will be given in Section 5.4, showing that all integral domains can essentially be viewed as being subrings of fields. Proving this converse involves constructing a field of fractions in much the same way that the field of rational numbers can be constructed from the domain of integers.

5.1.7 Theorem.   Let F be a field with identity 1. Any subring of F that contains 1 is an integral domain.

Proof. Suppose that R is a subring of the field F, with $a, b \in R$. If $ab = 0$ in R, then of course the same equation holds in F. Either $a = 0$ or $a \neq 0$, and in the latter case a has a multiplicative inverse a^{-1} in F, even though the inverse may not be in R. Multiplying both sides of the equation by a^{-1} gives $b = 0$.   □

Example 5.1.8

We have already seen, in Example 5.1.1, that $\mathbf{Z}_6$ is not an integral domain. On the other hand, we have seen that if p is a prime number, then $\mathbf{Z}_p$ is a field, so it is certainly an integral domain. Which of the rings $\mathbf{Z}_n$ are integral domains? If we use the condition that $ab \equiv 0 \pmod{n}$ implies that $a \equiv 0 \pmod{n}$ or $b \equiv 0 \pmod{n}$, or equivalently, the condition that $n|ab$ implies $n|a$ or $n|b$, then we can see that n must be prime. Why should the notions of field and integral domain be the same for the rings $\mathbf{Z}_n$? The next theorem gives an answer, at least from one point of view.   □

5.1.8 Theorem.   Any finite integral domain must be a field.

Proof. Let D be a finite integral domain, and let $D^\times$ be the set of nonzero elements of D. If $d \in D$ and $d \neq 0$, then multiplication by d defines a function from $D^\times$ into $D^\times$, since $ad \neq 0$ if $a \neq 0$. Let $f: D^\times \to D^\times$ be defined by $f(x) = xd$, for all $x \in D^\times$. Then f is a one-to-one function, since $f(x) = f(y)$ implies $xd = yd$, and so $x = y$ since the cancellation law holds in an integral domain. But then f must map $D^\times$ onto $D^\times$, since any one-to-one function from a finite set into itself must be onto, and so $1 = f(a)$ for some $a \in D^\times$. That is, $ad = 1$ for some $a \in D$, and so d is invertible. Since we have shown that each nonzero element of D is invertible, it follows that D is a field. $\square$

EXERCISES: SECTION 5.1

1. Which of the following sets are subrings of the field **Q** of rational numbers? Assume that m, n are integers with $n \neq 0$ and $(m,n) = 1$.
   (a) $\{m/n \mid n$ is odd$\}$
   (b) $\{m/n \mid n$ is even$\}$
   (c) $\{m/n \mid 4 \nmid n\}$
   (d) $\{m/n \mid (n,k) = 1\}$ where k is a fixed positive integer

2. Which of the following sets are subrings of the field **R** of real numbers?
   (a) $\{m + n\sqrt{2} \mid m, n \in \mathbf{Z}$ and n is even$\}$
   (b) $\{m + n\sqrt{2} \mid m, n \in \mathbf{Z}$ and m is odd$\}$
   (c) $\{a + b\sqrt[3]{2} \mid a, b \in \mathbf{Q}\}$
   (d) $\{a + b\sqrt[3]{3} + c\sqrt[3]{9} \mid a, b, c \in \mathbf{Q}\}$

3. Consider the following conditions on the set of all 2×2 matrices $\begin{bmatrix} a & b \\ c & d \end{bmatrix}$ with rational entries. Which conditions below define a commutative ring? If the set is a ring, find all units.
   (a) all matrices with $a = d$, $c = 0$
   (b) all matrices with $a = d$, $b = c$
   (c) all matrices with $a = d$, $b = -c$
   (d) all matrices with $a = d$, $b = -2c$

   Hint: Since the set of 2×2 matrices over **Q** satisfies all of the properties of Definition 5.1.1 except the commutative law for multiplication, it is sufficient to check the commutative law and the conditions of Proposition 5.1.3.

4. Find all subrings of $\mathbf{Z}_9$ and $\mathbf{Z}_{10}$ that contain a multiplicative identity element.

5. Let $R = \{m + n\sqrt{2} \mid m, n \in \mathbf{Z}\}$. Show that $m + n\sqrt{2}$ is a unit in R if and only if $m^2 - 2n^2 = \pm 1$.
   Hint: Show that if $(m + n\sqrt{2})(x + y\sqrt{2}) = 1$, then $(m - n\sqrt{2})(x - y\sqrt{2}) = 1$ and multiply the two equations.

6. Let R be a subring of an integral domain D. Prove that if R has a multiplicative identity element, then it must coincide with the multiplicative identity element of D.

7. Let R be a commutative ring such that $a^2 = a$ for all $a \in R$. Show that $a + a = 0$ for all $a \in R$.

8. An element a of a commutative ring R is called *nilpotent* if $a^n = 0$ for some positive integer n. Prove that if u is a unit in R and a is nilpotent, then $u - a$ is a unit in R.
   Hint: First try the case when $u = 1$.

9. Let R be the set of all continuous functions from the set of real numbers into itself.
   (a) Show that R is a commutative ring if the formulas

$$(f + g)(x) = f(x) + g(x) \quad \text{and} \quad (f \cdot g)(x) = f(x)g(x)$$

   for all x, are used to define addition and multiplication of functions.
   (b) Which properties in the definition of a commutative ring fail if the product of two functions is defined to be $(fg)(x) = f(g(x))$, for all x?

10. Let I be any set and let R be the collection of all subsets of I. Define addition and multiplication of subsets $A, B \subseteq I$ as follows:

$$A + B = A \cup B - A \cap B \quad \text{and} \quad A \cdot B = A \cap B$$

Show that R is a commutative ring under this addition and multiplication.

11. In the previous exercise, write out addition and multiplication tables for the case in which I has two elements and the case in which I has three elements.

12. A commutative ring is called a *Boolean ring* if $a^2 = a$ for all $a \in R$. Show that in a Boolean ring the commutative law follows from the other axioms and the above condition.

13. Let I be any set and let R be the collection of all subsets of I. Define addition and multiplication of subsets $A, B \subseteq I$ as follows:

$$A + B = A \cup B \quad \text{and} \quad A \cdot B = A \cap B$$

Is R a commutative ring under this addition and multiplication?

14. Define new operations on $\mathbf{Z}$ by letting $m \oplus n = m + n - 1$ and $m \odot n = m + n - mn$. Is $\mathbf{Z}$ a commutative ring under these operations?

15. Let R and S be commutative rings. Prove that the set of all ordered pairs (r,s) such that $r \in R$ and $s \in S$ can be given a ring structure by defining

$$(r_1,s_1) + (r_2,s_2) = (r_1 + r_2, s_1 + s_2) \quad \text{and} \quad (r_1,s_1) \cdot (r_2,s_2) = (r_1 r_2, s_1 s_2).$$

This is called the *direct sum* of R and S, denoted by $R \oplus S$.

16. Give addition and multiplication tables for $\mathbf{Z}_2 \oplus \mathbf{Z}_2$.

17. Find all units of $\mathbf{Z} \oplus \mathbf{Z}$ and $\mathbf{Z}_4 \oplus \mathbf{Z}_9$.

18. Generalizing to allow the direct sum of three commutative rings, give addition and multiplication tables for $\mathbf{Z}_2 \oplus \mathbf{Z}_2 \oplus \mathbf{Z}_2$.

19. Let R be a commutative ring. Set $R_1 = \{(r,n) \mid r \in R, n \in \mathbf{Z}\}$. Define $+$ on S by $(r,n) + (s,m) = (r + s, n + m)$ and define $\cdot$ on R_1 by $(r,n) \cdot (s,m) = (rs + ns + mr, nm)$. Show that R_1 is a commutative ring with identity $(0,1)$ and that $\{(r,0) \mid r \in R\}$ is a subring of R_1.

5.2 RING HOMOMORPHISMS

In studying groups in Chapter 3, we found that group homomorphisms played an important role. Now in studying commutative rings we have two operations to consider. As with groups, we will be interested in functions which preserve the

algebraic properties that we are studying. We begin the section with two examples, each of which involves an isomorphism.

Example 5.2.1

In Appendix E, in order to define the set of complex numbers we introduced
an element i such that $i^2 = -1$, and let

$$\mathbf{C} = \{a + bi \mid a, b \in \mathbf{R}\}.$$

In Section 4.4 we were able to give a more rigorous definition by using
$\mathbf{R}[x]/\langle x^2 + 1 \rangle$. To look for an alternate description of $\mathbf{C}$, we might try to find
such an element i with $i^2 = -1$ in some familiar setting. If we identify real
numbers with scalar 2×2 matrices over $\mathbf{R}$, then the matrix

$$\begin{bmatrix} 0 & 1 \\ -1 & 0 \end{bmatrix}$$

has the property that its square is equal to the matrix corresponding to -1.
This suggests that we should consider the set T of matrices of the form

$$a\begin{bmatrix} 1 & 0 \\ 0 & 1 \end{bmatrix} + b\begin{bmatrix} 0 & 1 \\ -1 & 0 \end{bmatrix} = \begin{bmatrix} a & b \\ -b & a \end{bmatrix}.$$

Verify that T is a commutative ring. To show the connection with $\mathbf{C}$, we
define $\phi : \mathbf{C} \to \mathbf{T}$ by

$$\phi(a + bi) = \begin{bmatrix} a & b \\ -b & a \end{bmatrix}.$$

To add complex numbers we just add the corresponding real and imaginary
parts, and since matrix addition is componentwise, it is easy to show that ϕ
preserves sums. To show that it preserves products we give the following
computations:

$$\phi((a + bi)(c + di)) = \phi((ac - bd) + (ad + bc)i) = \begin{bmatrix} ac - bd & ad + bc \\ -(ad + bc) & ac - bd \end{bmatrix}$$

$$\phi(a + bi)\phi(c + di) = \begin{bmatrix} a & b \\ -b & a \end{bmatrix}\begin{bmatrix} c & d \\ -d & c \end{bmatrix} = \begin{bmatrix} ac - bd & ad + bc \\ -(ad + bc) & ac - bd \end{bmatrix}.$$

Thus $\phi((a + bi)(c + di)) = \phi(a + bi)\phi(c + di)$, and since it is clear that ϕ is
one-to-one and onto, we could compute sums and products of complex
numbers by working with the corresponding matrices. This gives a concrete
model of the complex numbers. □

Example 5.2.2

In Section 4.3, in some cases in which Eisenstein's criterion for irreducibility
did not apply to a polynomial $f(x)$ directly, we were able to apply the criterion after making a substitution of the form $x + c$. It is useful to examine this

procedure more carefully. Let F be a field and let $c \in F$. We can define a function $\phi : F[x] \to F[x]$ by $\phi(f(x)) = f(x + c)$, for each polynomial $f(x) \in F[x]$. To show that ϕ is one-to-one and onto we only need to observe that it has an inverse function ϕ^{-1} defined by $\phi^{-1}(f(x)) = f(x - c)$, for each $f(x) \in F[x]$. Furthermore, it can be checked that ϕ preserves addition and multiplication, in the following sense: Adding or multiplying two polynomials first and then substituting $x + c$ is the same as first substituting $x + c$ into each polynomial and then adding or multiplying. In symbols,

$$\phi(f(x) + g(x)) = \phi(f(x)) + \phi(g(x)) \quad \text{and} \quad \phi(f(x)g(x)) = \phi(f(x))\phi(g(x))$$

for all $f(x), g(x) \in F[x]$.

Using the above correspondence, we see that if $f(x)$ has a nontrivial factorization $f(x) = g(x)h(x)$, then $\phi(f(x))$ has the nontrivial factorization $\phi(f(x)) = \phi(g(x))\phi(h(x))$. A similar condition holds with ϕ^{-1} in place of ϕ, and so we have shown that $f(x)$ is irreducible if and only if $\phi(f(x)) = f(x + c)$ is irreducible. $\square$

5.2.1 Definition.   Let R and S be commutative rings. A function $\phi : R \to S$ is called a *ring homomorphism* if

$$\phi(a + b) = \phi(a) + \phi(b) \quad \text{and} \quad \phi(ab) = \phi(a)\phi(b)$$

for all $a, b \in R$.

A ring homomorphism that is one-to-one and onto is called an *isomorphism*. If there is an isomorphism from R onto S, we say that R is *isomorphic* to S, and write $R \cong S$. An isomorphism from the commutative ring R onto itself is called an *automorphism* of R.

The condition that states that a ring homomorphism must preserve addition is equivalent to the statement that a ring homomorphism must be a group homomorphism of the underlying additive group of the ring. This means that we have at our disposal all of the results that we have obtained for group homomorphisms.

A word of warning similar to that given for group homomorphisms is probably in order. If $\phi : R \to S$ is a ring homomorphism, with $a, b \in R$, then we need to note that in the equation

$$\phi(a + b) = \phi(a) + \phi(b)$$

the sum $a + b$ occurs in R, using the addition of that ring, whereas the sum $\phi(a) + \phi(b)$ occurs in S, using the appropriate operation of S. A similar remark applies to the respective operations of multiplication in the equation

$$\phi(ab) = \phi(a)\phi(b)$$

where $ab \in R$ and $\phi(a)\phi(b) \in S$. We have chosen not to distinguish between the operations in the two rings, since there is generally not much chance for confusion.

Using the terminology of Definition 5.2.1, the function defined in Example 5.2.1 is an isomorphism providing a different way of looking at the set of complex numbers. The function defined in Example 5.2.2 is an automorphism of the ring $F[x]$ of polynomials over the field F, which preserves irreducibility of polynomials. We begin with some basic results on isomorphisms.

It follows from the next proposition that "is isomorphic to" is reflexive, symmetric, and transitive. Recall that a function is one-to-one and onto if and only if it has an inverse. Thus a ring isomorphism always has an inverse, but it is not evident that this inverse must preserve addition and multiplication.

5.2.2 Proposition.
(a) The inverse of a ring isomorphism is a ring isomorphism.

(b) The composition of two ring isomorphisms is a ring isomorphism.

Proof. (a) Let $\phi : R \to S$ be an isomorphism of commutative rings. We have shown in Proposition 3.4.2 that ϕ^{-1} is an isomorphism of the underlying additive groups. To show that ϕ^{-1} is a ring homomorphism, let $s_1, s_2 \in S$. Since ϕ is onto, there exist $r_1, r_2 \in R$ such that $\phi(r_1) = s_1$ and $\phi(r_2) = s_2$. Then $\phi^{-1}(s_1 s_2)$ must be the unique element $r \in R$ for which $\phi(r) = s_1 s_2$. Since ϕ preserves multiplication,

$$\phi^{-1}(s_1 s_2) = \phi^{-1}(\phi(r_1)\phi(r_2)) = \phi^{-1}(\phi(r_1 r_2)) = r_1 r_2 = \phi^{-1}(s_1)\phi^{-1}(s_2).$$

(b) If $\phi : R \to S$ and $\theta : S \to T$ are isomorphisms of commutative rings, then

$$\theta\phi(ab) = \theta(\phi(ab)) = \theta(\phi(a)\phi(b)) = \theta(\phi(a)) \cdot \theta(\phi(b)) = \theta\phi(a) \cdot \theta\phi(b).$$

The remainder of the proof follows immediately from the corresponding result for group homomorphisms. $\square$

To show that commutative rings R and S are isomorphic, we usually construct the isomorphism. To show that they are not isomorphic, it is necessary to show that no isomorphism can possibly be constructed. Sometimes this can be done by just considering the sets involved. For example, if n and m are different positive integers, then $\mathbf{Z}_n$ is not isomorphic to $\mathbf{Z}_m$ since no one-to-one correspondence can be defined between $\mathbf{Z}_n$ and $\mathbf{Z}_m$.

One way to show that two commutative rings are not isomorphic is to find an algebraic property that is preserved by all isomorphisms and that is satisfied by one commutative ring but not the other. We now look at several very elementary examples of such properties.

If $\phi : R \to S$ is an isomorphism, then from our results on group isomorphisms we know that ϕ must map 0 to 0 and must preserve additive inverses. If R has an identity element 1, then for any $s \in S$, there exists $r \in R$ such that $\phi(r) = s$, since ϕ is onto. Thus we have

$$\phi(1) \cdot s = \phi(1) \cdot \phi(r) = \phi(1 \cdot r) = \phi(r) = s,$$

showing that $\phi(1)$ must be an identity element for S. It is also easy to show that if

r is a unit, then so is s, and so ϕ preserves units. This implies that R is a field if and only if S is a field.

As an immediate consequence of the remarks in the previous paragraph, we can see that $\mathbf{Z}$ is not isomorphic to $\mathbf{Q}$, since $\mathbf{Q}$ is a field, while $\mathbf{Z}$ is not. Another interesting problem is to show that $\mathbf{R}$ and $\mathbf{C}$ are not isomorphic. Since both are fields, we cannot use the previous argument. Let us suppose that we could define an isomorphism $\phi : \mathbf{C} \to \mathbf{R}$. Then we would have

$$\phi(i)^2 = \phi(i^2) = \phi(-1) = -\phi(1) = -1,$$

and so $\mathbf{R}$ would have a square root of -1, which we know to be impossible. Thus $\mathbf{R}$ and $\mathbf{C}$ cannot be isomorphic.

In many important cases we will be interested in functions that preserve addition and multiplication of commutative rings but are not necessarily one-to-one and onto. Example 5.2.4 will be particularly important in later work.

Example 5.2.3

Consider the mapping $\phi : \mathbf{Q}[x] \to \mathbf{R}$ defined by $\phi(f(x)) = f(\sqrt{2})$, for all polynomials $f(x) \in \mathbf{Q}[x]$. That is, the mapping ϕ is defined on a polynomial with rational coefficients by substituting $x = \sqrt{2}$. It is easy to check that adding (or multiplying) two polynomials first and then substituting in $x = \sqrt{2}$ is the same as substituting first in each polynomial and then adding (or multiplying). Thus ϕ preserves sums and products and is a ring homomorphism. Note that ϕ is not one-to-one, since $\phi(x^2 - 2) = 0$. We will show in Chapter 6 that the image of ϕ is $\mathbf{Q}(\sqrt{2}) = \{a + b\sqrt{2} \mid a, b \in \mathbf{Q}\}$, and so ϕ is not onto. $\square$

Example 5.2.4

The previous example can be generalized, since there is nothing special about the particular fields we chose or the particular element we worked with. Let F and E be fields, with F a subfield of E. For any element $\alpha \in E$ we can define a function $\phi_\alpha : F[x] \to E$ by letting $\phi_\alpha(f(x)) = f(\alpha)$, for each $f(x) \in F[x]$. Then ϕ_α preserves sums and products since

$$\phi_\alpha(f(x) + g(x)) = f(\alpha) + g(\alpha) = \phi_\alpha(f(x)) + \phi_\alpha(g(x))$$

and

$$\phi_\alpha(f(x) \cdot g(x)) = f(\alpha) \cdot g(\alpha) = \phi_\alpha(f(x)) \cdot \phi_\alpha(g(x)),$$

for all $f(x), g(x) \in F[x]$. Thus ϕ_α is a ring homomorphism. $\square$

Example 5.2.5

The previous example may become even clearer in a more general context. We never used the fact that we were working with polynomials over a field, and so we can consider polynomials over any commutative ring, as in Example 5.1.2. Furthermore, one of the essential facts we used was that the inclusion mapping from a subfield into the field containing it is a homomorphism, and so we might just as well consider the situation in that generality.

Let R and S be commutative rings, let $\theta : R \rightarrow S$ be a ring homomorphism, and let s be any element of S. Then there exists a unique ring homomorphism $\hat{\theta} : R[x] \rightarrow S$ such that $\hat{\theta}(r) = \theta(r)$ for all $r \in R$ and $\hat{\theta}(x) = s$. We will first show the uniqueness. If $\phi : R[x] \rightarrow S$ is any homomorphism with the required properties, then for any polynomial

$$f(x) = a_0 + a_1 x + \ldots + a_m x^m$$

in $R[x]$ we must have

$$\begin{aligned}
\phi(a_0 + a_1 x + \ldots + a_m x^m) &= \phi(a_0) + \phi(a_1 x) + \ldots + \phi(a_m x^m) \\
&= \phi(a_0) + \phi(a_1)\phi(x) + \ldots + \phi(a_m)(\phi(x))^m \\
&= \phi(a_0) + \phi(a_1)s + \ldots + \phi(a_m)s^m.
\end{aligned}$$

This shows that the only possible way to define $\hat{\theta}$ is the following:

$$\hat{\theta}(a_0 + a_1 x + \ldots + a_m x^m) = \theta(a_0) + \theta(a_1)s + \ldots + \theta(a_m)s^m.$$

Given this definition, we must show that $\hat{\theta}$ is a homomorphism. Since addition of polynomials is defined componentwise, and θ preserves sums, it is easy to check that $\hat{\theta}$ preserves sums of polynomials. If

$$g(x) = b_0 + b_1 x + \ldots + b_n x^n,$$

then the coefficient c_k of the product $h(x) = f(x)g(x)$ is given by the formula

$$c_k = \sum_{i+j=k} a_i b_j.$$

Applying θ to both sides gives

$$\theta(c_k) = \theta\left(\sum_{i+j=k} a_i b_j \right) = \sum_{i+j=k} \theta(a_i)\theta(b_j)$$

since θ preserves both sums and products. This formula is precisely what we need to check that

$$\hat{\theta}(f(x)g(x)) = \hat{\theta}(h(x)) = \hat{\theta}(f(x))\hat{\theta}(g(x)).$$

This finishes the proof that $\hat{\theta}$ is a ring homomorphism. $\square$

Let $\phi : R \rightarrow S$ be a ring homomorphism. By elementary results on group theory, we know that ϕ must map the additive identity of R onto the additive identity of S, that ϕ must preserve additive inverses, and that $\phi(R)$ must be an additive subgroup of S. It is convenient to list these results formally in the next proposition. Part (c) of the proposition follows immediately from the observation that if 1 is an identity element for R, then we must have $\phi(1) \cdot \phi(1) = \phi(1 \cdot 1) = \phi(1)$. To complete the proof of part (d), the fact that $\phi(R)$ is closed under multiplication follows since ϕ preserves multiplication.

5.2.3 Proposition.   Let $\phi : R \to S$ be a ring homomorphism. Then
(a) $\phi(0) = 0$;
(b) $\phi(-a) = -\phi(a)$ for all $a \in R$;
(c) if 1 is an identity element for R, then $\phi(1)$ is an idempotent element of S;
(d) $\phi(R)$ is a subring of S.

Example 5.2.6

Any isomorphism $\phi : R \to S$ of commutative rings with identity induces a group isomorphism from $R^\times$ onto $S^\times$. To establish this result, we first need to show that such an isomorphism maps identity to identity. Since ϕ is onto, given $1 \in S$, we have $1 = \phi(r)$ for some $r \in R$. Then

$$\phi(1) = \phi(1) \cdot 1 = \phi(1)\phi(r) = \phi(1 \cdot r) = \phi(r) = 1.$$

Furthermore, for any $a \in R^\times$ we have

$$\phi(a)\phi(a^{-1}) = \phi(aa^{-1}) = \phi(1) = 1,$$

and so ϕ maps $R^\times$ into $S^\times$. Applying this argument to $(\phi)^{-1}$, which is also a ring homomorphism, shows that it maps $S^\times$ into $R^\times$. We can conclude that ϕ is one-to-one and onto when restricted to $R^\times$, and this shows that $R^\times \cong S^\times$. □

Before giving some additional examples, we need to give a definition analogous to one for groups. In fact, the kernel of a ring homomorphism is just the kernel of the mapping when viewed as a group homomorphism of the underlying additive groups of the rings.

5.2.4 Definition.   Let $\phi : R \to S$ be a ring homomorphism. The set $\{a \in R \mid \phi(a) = 0\}$ is called the *kernel* of ϕ, denoted by $\ker(\phi)$.

Example 5.2.7

We already know by results on factor groups in Section 3.8 that the mapping $\pi : \mathbf{Z} \to \mathbf{Z}_n$ given by $\pi(x) = [x]_n$, for all $x \in \mathbf{Z}$, is a group homomorphism. The formula $[x]_n[y]_n = [xy]_n$, which defines multiplication of congruence classes, shows immediately that π also preserves multiplication. Thus π is a ring homomorphism with $\ker(\pi) = n\mathbf{Z}$. □

Example 5.2.8

Let $\pi : \mathbf{Z} \to \mathbf{Z}_n$ be the natural projection considered in the previous example. In Example 5.1.2 we discussed the ring of polynomials with coefficients in a commutative ring, and so in this context we can consider the polynomial rings $\mathbf{Z}[x]$ and $\mathbf{Z}_n[x]$. Define $\hat{\pi} : \mathbf{Z}[x] \to \mathbf{Z}_n[x]$ as follows: For any polynomial

$$f(x) = a_0 + a_1 x + \ldots + a_m x^m$$

set

$$\hat{\pi}(f(x)) = \pi(a_0) + \pi(a_1)x + \ldots + \pi(a_m)x^m.$$

That is, $\hat{\pi}$ simply reduces all of the coefficients of $f(x)$ modulo n. This is actually a special case of the result obtained in Example 5.2.5. We can think of π as a homomorphism from $\mathbf{Z}$ into $\mathbf{Z}_n[x]$, and then we have extended π to $\mathbf{Z}[x]$ by mapping $x \in \mathbf{Z}[x]$ to $x \in \mathbf{Z}_n[x]$. It follows that $\hat{\pi}$ is a homomorphism. Furthermore, it is easy to see that the kernel of $\hat{\pi}$ is the set of all polynomials for which each coefficient is divisible by n.

To illustrate the power of homomorphisms, suppose that $f(x)$ has a nontrivial factorization $f(x) = g(x)h(x)$ in $\mathbf{Z}[x]$. Then

$$\hat{\pi}(f(x)) = \hat{\pi}(g(x)h(x)) = \hat{\pi}(g(x))\hat{\pi}(h(x)).$$

If $\deg(\hat{\pi}(f(x))) = \deg(f(x))$, then this gives a nontrivial factorization of $\hat{\pi}(f(x))$ in $\mathbf{Z}_n[x]$. This means that a polynomial $f(x)$ with integer coefficients can be shown to be irreducible over the field $\mathbf{Q}$ of rational numbers by finding a positive integer n such that when the coefficients of $f(x)$ are reduced modulo n, the new polynomial has the same degree and cannot be factored nontrivially in $\mathbf{Z}_n[x]$. Eisenstein's irreducibility criterion (Theorem 4.3.5) for the prime p can be interpreted as a condition that states that the polynomial cannot be factored when reduced modulo p^2. $\quad\square$

5.2.5 Proposition.   Let $\phi : R \to S$ be a ring homomorphism.
  (a) If $a, b \in \ker(\phi)$ and $r \in R$, then $a + b$, $a - b$, and ra belong to $\ker(\phi)$.
  (b) The homomorphism ϕ is an isomorphism if and only if $\ker(\phi) = \{0\}$ and $\phi(R) = S$.

Proof. (a) If $a, b \in \ker(\phi)$, then

$$\phi(a \pm b) = \phi(a) \pm \phi(b) = 0 \pm 0 = 0,$$

and so $a \pm b \in \ker(\phi)$. If $r \in R$, then

$$\phi(ra) = \phi(r) \cdot \phi(a) = \phi(r) \cdot 0 = 0,$$

showing that $ra \in \ker(\phi)$.

(b) This part follows from the fact that ϕ is a group homomorphism, since ϕ is one-to-one if and only if $\ker(\phi) = 0$ and onto if and only if $\phi(R) = S$. $\quad\square$

Example 5.2.9

Let $\phi : \mathbf{Z}_8 \to \mathbf{Z}_{12}$ be the ring homomorphism defined by $\phi([x]_8) = [9x]_{12}$, for all $[x]_8 \in \mathbf{Z}_8$. Then $\phi(\mathbf{Z}_8)$ consists of all multiples of $[9]_{12}$, and determines the subring $\{[0]_{12}, [3]_{12}, [6]_{12}, [9]_{12}\}$ of $\mathbf{Z}_{12}$. We also have $\ker(\phi) = \{[0]_8, [4]_8\}$. □

Let $\phi : R \to S$ be a ring homomorphism. The fundamental homomorphism theorem for groups implies that the abelian group $R/\ker(\phi)$ is isomorphic to the abelian group $\phi(R)$, which is a subgroup of S. In order to obtain a homomorphism theorem for commutative rings we need to consider the cosets of $\ker(\phi)$. Intuitively, the situation may be easiest to understand if we consider the cosets to be defined by the equivalence relation $\sim$ given by $a \sim b$ if $\phi(a) = \phi(b)$, for all $a, b \in R$. The sum of equivalence classes $[a]$ and $[b]$ in $R/\ker(\phi)$ is well-defined, using the formula $[a] + [b] = [a + b]$, for all $a, b \in R$. The product of equivalence classes $[a]$ and $[b]$ in $R/\ker(\phi)$ is defined by the expected formula $[a] \cdot [b] = [ab]$, for all $a, b \in R$. To show that this multiplication is well-defined, we note that if $a \sim c$ and $b \sim d$, then $ab \sim cd$ since

$$\phi(ab) = \phi(a)\phi(b) = \phi(c)\phi(d) = \phi(cd).$$

Since our earlier results on groups imply that $R/\ker(\phi)$ is an abelian group, to show that $R/\ker(\phi)$ is a commutative ring we only need to verify the distributive law and the associative and commutative laws for multiplication. We have

$$[a]([b] + [c]) = [a][b + c] = [a(b + c)] = [ab + ac] = [ab] + [ac] = [a][b] + [a][c],$$

showing that the distributive law follows directly from the definitions of addition and multiplication for equivalence classes, and the distributive law in R. The proofs that the associative and commutative laws hold are similar.

5.2.6 Theorem (Fundamental Homomorphism Theorem for Rings). Let $\phi : R \to S$ be a ring homomorphism. Then $R/\ker(\phi) \cong \phi(R)$.

Proof. Since we have already shown that $R/\ker(\phi)$ is a commutative ring, we can apply the fundamental homomorphism theorem for groups to show that the mapping $\overline{\phi}$ given by $\overline{\phi}([a]) = \phi(a)$, for all $a \in R$, defines an isomorphism of the abelian groups $R/\ker(\phi)$ and $\phi(R)$. We only need to show that $\overline{\phi}$ preserves multiplication, and this follows from the computation

$$\overline{\phi}([a][b]) = \overline{\phi}([ab]) = \phi(ab) = \phi(a)\phi(b) = \overline{\phi}([a])\overline{\phi}([b]). □$$

The coset notation $a + \ker(\phi)$ is usually used for the equivalence class $[a]$. With this notation, addition and multiplication of cosets are expressed by the formulas

$$(a + \ker(\phi)) + (b + \ker(\phi)) = (a + b) + \ker(\phi)$$

and

$$(a + \ker(\phi)) \cdot (b + \ker(\phi)) = (ab) + \ker(\phi).$$

It is important to remember that these are additive cosets, not multiplicative cosets.

5.2.7 Proposition. Let $R_1, R_2, \ldots, R_n$ be commutative rings. The set of n-tuples $(a_1, a_2, \ldots, a_n)$ such that $a_i \in R_i$ for each i is a commutative ring under componentwise addition and multiplication.

Proof. The proof that componentwise addition defines a group is an easy extension of the proof of Proposition 3.3.4. The remainder of the proof is left as an exercise.   □

5.2.8 Definition. Let $R_1, R_2, \ldots, R_n$ be commutative rings. The set of n-tuples $(a_1, a_2, \ldots, a_n)$ such that $a_i \in R_i$ for each i is called the *direct sum* of the commutative rings $R_1, R_2, \ldots, R_n$, and is denoted by

$$R_1 \oplus R_2 \oplus \cdots \oplus R_n.$$

Let $R_1, R_2, \ldots, R_n$ be commutative rings with identity. Then $(1, 1, \ldots, 1)$ is the identity of the direct sum $R = R_1 \oplus R_2 \oplus \cdots \oplus R_n$. Furthermore, an element $(a_1, a_2, \ldots, a_n)$ in the direct sum is a unit if and only each component a_i is a unit in R_i. This can be shown by observing that

$$(a_1, a_2, \ldots, a_n)(b_1, b_2, \ldots, b_n) = (1, 1, \ldots, 1)$$

if and only if $a_i b_i = 1$ for each i. It then follows easily that $R^{\times}$ and $R_1^{\times} \times R_2^{\times} \times \cdots \times R_n^{\times}$ are isomorphic as groups. (This is part of Exercise 7.)

Example 5.2.10

Let n be a positive integer with prime decomposition $n = p_1^{\alpha_1} p_2^{\alpha_2} \cdots p_m^{\alpha_m}$. Then the isomorphism

$$\mathbf{Z}_n \cong \mathbf{Z}_{p_1^{\alpha_1}} \oplus \mathbf{Z}_{p_2^{\alpha_2}} \oplus \cdots \oplus \mathbf{Z}_{p_m^{\alpha_m}}.$$

can be shown easily by referring to Theorem 3.5.4. The mapping ϕ defined by

$$\phi([x]_n) = ([x]_{p_1^{\alpha_1}}, [x]_{p_2^{\alpha_2}}, \ldots, [x]_{p_m^{\alpha_m}})$$

for all $[x]_n \in \mathbf{Z}_n$ is easily seen to preserve multiplication.

Recall that if n is a positive integer with prime decomposition

$$n = p_1^{\alpha_1} p_2^{\alpha_2} \ldots p_m^{\alpha_m},$$

then

$$\varphi(n) = n\left(1 - \frac{1}{p_1}\right)\left(1 - \frac{1}{p_2}\right) \cdots \left(1 - \frac{1}{p_m}\right),$$

where $\varphi(n)$ is the number of positive integers less than or equal to n and relatively prime to n. This was proved in Corollary 3.5.5 by counting the generators of the direct product of the component groups rather than counting the generators of $\mathbf{Z}_n$ directly. We are now in a position to give a proof based on results for commutative rings. From the general remarks preceding this example, we have

$$\mathbf{Z}_n^\times \cong \mathbf{Z}_{p_1^{\alpha_1}}^\times \times \mathbf{Z}_{p_2^{\alpha_2}}^\times \times \cdots \times \mathbf{Z}_{p_m^{\alpha_m}}^\times.$$

Then $\varphi(n)$, which is the order of $\mathbf{Z}_n^\times$, can be found by multiplying the orders of the factor groups $\mathbf{Z}_{p_i^{\alpha_i}}^\times$. From here this argument proceeds just as the one in Corollary 3.5.5. $\square$

As an application of the results in this section we study the characteristic of a commutative ring with identity.

5.2.9 Definition.   Let R be a commutative ring with identity. The smallest positive integer n such that $n \cdot 1 = 0$ is called the *characteristic* of R, denoted by char(R). If no such positive integer exists, then R is said to have *characteristic zero*.

The characteristic of a commutative ring R with identity is just the order of 1 in the underlying additive group. If char(R) = n, then it follows from the distributive law that $n \cdot a = (n \cdot 1) \cdot a = 0 \cdot a = 0$, and so the exponent of the underlying abelian group is n.

A more sophisticated way to view the characteristic is to define a ring homomorphism $\phi : \mathbf{Z} \to R$ by $\phi(n) = n \cdot 1$. The rules we developed in Chapter 3 for considering multiples of an element in an abelian group show that ϕ is a homomorphism. The characteristic of R is just the (nonnegative) generator of ker(ϕ).

5.2.10. Proposition.   An integral domain has characteristic 0 or p, for some prime number p.

Proof. Let D be an integral domain, and consider the mapping $\phi : \mathbf{Z} \to D$ defined by $\phi(n) = n \cdot 1$. The fundamental homomorphism theorem for rings shows that $\mathbf{Z}/\ker(\phi)$ is isomorphic to the subring $\phi(\mathbf{Z})$. Since any subring of an integral domain inherits the property that it has no nontrivial divisors of zero, this shows that $\mathbf{Z}/\ker(\phi)$ must be an integral domain. Thus either $\ker(\phi) = 0$, in which case $\mathrm{char}(D) = 0$, or $\ker(\phi) = p\mathbf{Z}$ for some prime number p, in which case $\mathrm{char}(D) = p$. $\square$

EXERCISES: SECTION 5.2

1. Let F be a field and let $\phi : F \to R$ be a ring homomorphism. Show that ϕ is either zero or one-to-one.

2. Let F, E be fields, with a homomorphism $\phi : F \to E$. Show that if ϕ is onto, then ϕ must be an isomorphism.

3. Show that taking complex conjugates defines an automorphism of $\mathbf{C}$. That is, for $z \in \mathbf{C}$, define $\phi(z) = \bar{z}$, and show that ϕ is an automorphism.

4. Show that the only ring automorphism of $\mathbf{Z}$ is the identity mapping.

5. Define $\phi : \mathbf{Q}(\sqrt{2}) \to \mathbf{Q}(\sqrt{2})$ by $\phi(a + b\sqrt{2}) = a - b\sqrt{2}$. Show that ϕ is an automorphism. Recall that $\mathbf{Q}(\sqrt{2}) = \{a + b\sqrt{2} \mid a, b \in \mathbf{Q}\}$.

6. Let R be a commutative ring with identity, and let D be an integral domain. Show that $\phi(1) = 1$ for any nonzero ring homomorphism $\phi : R \to D$.

7. Let $R_1, R_2, \ldots, R_n$ be commutative rings. Complete the proof of Proposition 5.2.7, to show that $R_1 \oplus R_2 \oplus \cdots \oplus R_n$ is a commutative ring. Show that if $R = R_1 \oplus R_2 \oplus \cdots \oplus R_n$ and each R_i has an identity, then

$$R^{\times} \cong R_1^{\times} \times R_2^{\times} \times \cdots \times R_n^{\times}.$$

8. Find all ring homomorphisms from $\mathbf{Z} \oplus \mathbf{Z}$ into $\mathbf{Z}$. That is, find all possible formulas, and show why no others are possible.

   Hint: Any ring homomorphism is in particular a group homomorphism. Show that it must be completely determined by the images of $(1, 0)$ and $(0, 1)$.

9. Find all ring homomorphisms from $\mathbf{Z} \oplus \mathbf{Z}$ into $\mathbf{Z} \oplus \mathbf{Z}$.

10. Show that the composition of two ring homomorphisms is a ring homomorphism.

11. Show that the direct sum of two nonzero rings is never an integral domain.

12. Find all ring homomorphisms from $\mathbf{Z}_8$ into $\mathbf{Z}_{12}$, and from $\mathbf{Z}_{12}$ into $\mathbf{Z}_8$.

13. Show that any ring homomorphism $\phi : \mathbf{Z}_n \to \mathbf{Z}_k$ has the following form: $\phi([x]_n) = [ex]_k$, where $k \mid (e^2 - e)$ and $n \mid x$ implies $k \mid ex$. Conversely, if $\phi : \mathbf{Z}_n \to \mathbf{Z}_k$ is defined as above, then show that ϕ is a ring homomorphism.

    Hint: See Example 3.7.5, since any ring homomorphism must be a homomorphism on the underlying abelian groups.

14. Define $\phi : R \oplus S \to R$ by $\phi(a, b) = a$ for $a \in R$, $b \in S$. Show that ϕ is a ring homomorphism.

15. Define $\phi : \mathbf{Z} \to \mathbf{Z}_m \oplus \mathbf{Z}_n$ by $\phi(x) = ([x]_m, [x]_n)$. Find the kernel and image of ϕ. Show that ϕ is onto if and only if $(m, n) = 1$.

16. Let R be the ring given by defining new operations on $\mathbf{Z}$ by letting $m \oplus n = m + n - 1$

and $m \odot n = m + n - mn$. Define $\phi: \mathbf{Z} \to R$ by $\phi(n) = 1 - n$. Show that ϕ is an isomorphism.

17. Let I be any set and let R be the collection of all subsets of I. Define addition and multiplication of subsets $A, B \subseteq I$ as follows:

$$A + B = A \cup B - A \cap B \quad \text{and} \quad A \cdot B = A \cap B$$

Show that if I has two elements, then R is isomorphic to $\mathbf{Z}_2 \oplus \mathbf{Z}_2$, and if I has three elements, then I is isomorphic to $\mathbf{Z}_2 \oplus \mathbf{Z}_2 \oplus \mathbf{Z}_2$.

18. Show that $\mathbf{Q}(\sqrt{2})$ is isomorphic to the set of all matrices over $\mathbf{Q}$ of the form

$$\begin{bmatrix} a & 2b \\ b & a \end{bmatrix}.$$

19. Let R be an integral domain. Show that R contains a subring isomorphic to $\mathbf{Z}_p$ for some prime number p if and only if $\operatorname{char}(R) = p$.

20. Show that if R is an integral domain with characteristic p, then for all $a, b \in R$ we must have $(a + b)^p = a^p + b^p$. Show by induction that we must also have $(a + b)^{p^n} = a^{p^n} + b^{p^n}$ for all positive integers n.

5.3 IDEALS AND FACTOR RINGS

We have shown that the kernel of any ring homomorphism is closed under sums, differences, and "scalar" multiples (see Proposition 5.2.5). In other words, the kernel of a ring homomorphism is an additive subgroup that is closed under multiplication by any element of the ring. For any integer n, the subset $n\mathbf{Z}$ of the ring of integers satisfies the same properties. The same is true for the subset $\langle f(x) \rangle$ of the ring $F[x]$ of polynomials over a field F, where $f(x)$ is any polynomial. In each of these examples we have been able to make the cosets of the given set into a commutative ring, and this provides the motivation for the next definition and several subsequent results.

5.3.1 Definition.   Let R be a commutative ring. A nonempty subset I of R is called an *ideal* of R if (i) $a \pm b \in I$ for all $a, b \in I$ and (ii) $ra \in I$ for all $a \in I$ and $r \in R$.

For any commutative ring R, it is clear that the set $\{0\}$ is an ideal, which we will refer to as the *trivial* ideal. The set R is also always an ideal. Among commutative rings with identity, fields are characterized by the property that these two ideals are the only ideals of the ring.

5.3.2 Proposition.   Let R be a commutative ring with identity. Then R is a field if and only if it has no proper nontrivial ideals.

Proof. First assume that R is a field, and let I be any ideal of R. Either $I = \{0\}$, or else there exists $a \in I$ such that $a \neq 0$. In the second case, since R is a field, there exists an inverse a^{-1} for a, and then for any $r \in R$ we have $r = r \cdot 1 = r(a^{-1}a) = (ra^{-1})a$, so by the definition of an ideal we have $r \in I$. We have shown that either $I = \{0\}$ or $I = R$.

Conversely, assume that R has no proper nontrivial ideals, and let a be a nonzero element of R. We will show that the set

$$I = \{x \in R \mid x = ra \text{ for some } r \in R\}$$

is an ideal. First, I is nonempty since $a = 1 \cdot a \in I$. If $r_1a, r_2a \in I$, then we have $r_1a \pm r_2a = (r_1 \pm r_2)a$, showing that I is an additive subgroup of R. Finally, if $x = ra \in I$, then for any $s \in R$ we have $sx = (sr)a \in I$, and so I is an ideal. By assumption we must have $I = R$, since $I \neq \{0\}$, and since $1 \in R$, we have $1 = ra$ for some $r \in R$. This implies that a is invertible, and so we have shown that R is a field. $\square$

Let R be a commutative ring with identity 1. For any $a \in R$, let Ra denote the set $I = \{x \in R \mid x = ra \text{ for some } r \in R\}$. The proof of the previous theorem shows that Ra is an ideal of R that contains a. It is obvious from the definition of an ideal that any ideal that contains a must also contain Ra, and so we are justified in saying that Ra is the smallest ideal that contains a.

Furthermore, $R \cdot 1$ consists of all of R, since every element $r \in R$ can be expressed in the form $r \cdot 1$. Thus R is the smallest ideal (in fact, the only ideal) that contains 1.

5.3.3 Definition. Let R be a commutative ring with identity, and let $a \in R$. The ideal

$$Ra = \{x \in R \mid x = ra \text{ for some } r \in R\}$$

is called the *principal ideal* generated by a. The notation $\langle a \rangle$ will also be used.

An integral domain in which every ideal is a principal ideal is called a *principal ideal domain*.

Example 5.3.1

In the terminology of the above definition, we see that Theorem 1.1.4 shows that the ring of integers $\mathbf{Z}$ is a principal ideal domain. Moreover, given any nonzero ideal I of $\mathbf{Z}$, the smallest positive integer in I is a generator for the ideal. $\square$

Example 5.3.2

We showed in Theorem 4.2.2 that if F is any field, then the ring $F[x]$ of polynomials over F is a principal ideal domain. If I is any nonzero ideal of $F[x]$, then $f(x)$ is a generator for I if and only if it has minimal degree among the nonzero elements of I. Since a generator of I is a divisor of every element of I, it is easy to see that there is only one monic generator for I.

We have also considered the ring of polynomials $R[x]$ over any commutative ring R. However, if the coefficients do not come from a field, then the proof of the division algorithm is no longer valid, and so we should not expect $R[x]$ to be a principal ideal domain. The ring $\mathbf{Z}[x]$ of polynomials with integer coefficients is a domain, but not every ideal is principal. (See Exercise 20.) $\square$

Let I be an ideal of the commutative ring R. Then I is a subgroup of the underlying additive group of R, and so by Theorem 3.8.4 the cosets of I in R determine a factor group R/I, which is again an abelian group. The cosets of I are usually denoted additively, in the form $a + I$, for all elements $a \in R$. We know from Proposition 3.8.1 that elements $a, b \in R$ determine the same coset of I if and only if $a - b \in I$. Since the cosets of I partition R, they determine an equivalence relation on R, by defining two elements of R to be equivalent if their difference is in I.

In Chapter 1 we used the notion of congruence modulo n to define the sets $\mathbf{Z}_n$, which we now know to be commutative rings. Since $a \equiv b \pmod{n}$ if and only if $a - b$ is a multiple of n, this is precisely the equivalence relation determined by the ideal $n\mathbf{Z}$. In Section 4.4, given an irreducible polynomial $p(x)$, we used the notion of congruence modulo $p(x)$ to construct a new field from the congruence classes, in which we could find a root of $p(x)$. Again in this case, the congruence classes are precisely those determined by the principal ideal $\langle p(x) \rangle$ of $F[x]$, consisting of all polynomial multiples of $p(x)$. We now extend this procedure to the equivalence classes defined by ideals in commutative rings. We have chosen to use the notation of congruence classes rather than additive cosets.

5.3.4 Definition.   Let I be an ideal of the commutative ring R. For $a, b \in R$ we define $a \equiv b \pmod{I}$ if $a - b \in I$. For any $a \in R$ the congruence class determined by a will be denoted by $[a]_I$, or simply $[a]$ when the context is clear.

The set of all congruence classes modulo I will be denoted by R/I.

5.3.5 Proposition.   Let I be an ideal of the commutative ring R, and let $a, b, c, d \in R$.

  (a) If $a \equiv c \pmod{I}$ and $b \equiv d \pmod{I}$, then $a + b \equiv c + d \pmod{I}$.

  (b) If $a \equiv c \pmod{I}$ and $b \equiv d \pmod{I}$, then $ab \equiv cd \pmod{I}$.

  (c) $[a]_I = \{r \in R \mid r = a + x \text{ for some } x \in I\}$.

Proof. Part (a) follows from Proposition 3.8.3, and part (c) follows from Proposition 3.8.1. We include a proof of part (b).

Assume that $a - c \in I$ and $b - d \in I$. Multiplying $a - c$ by b and $b - d$ by c gives us elements that still belong to I. Then using the distributive law and adding gives $(ab - cb) + (cb - cd) = ab - cd$, and this is an element of I, showing that $ab \equiv cd \pmod{I}$. $\square$

For an ideal I of a commutative ring R, the next theorem provides the justification for calling R/I the *factor ring* of R modulo I. For groups, we were only able to construct factor groups relative to normal subgroups, not all subgroups. For commutative rings, ideals play an analogous role. That is, among all subrings of a commutative ring, only those that are ideals can be used to construct factor rings.

5.3.6 Theorem.   If I is an ideal of the commutative ring R, then R/I is a commutative ring.

Proof. The previous proposition shows that addition and multiplication of congruence classes are well-defined. It follows from Theorem 3.8.4 that R/I is a group under the addition of congruence classes. To show that the distributive law holds, let a, b, $c \in R$. Then

$$[a]([b] + [c]) = [a][b + c] = [a(b + c)]$$

$$= [ab + ac] = [ab] + [ac]$$

$$= [a][b] + [a][c].$$

Note that we have used the definitions of addition and multiplication of equivalence classes, together with the fact that the distributive law holds in R. Similar computations show that the associative and commutative laws hold for both addition and multiplication, completing the proof that R/I is a commutative ring.   □

Let R be a commutative ring and let I be an ideal of R. For any element $a \in R$, the congruence class $[a]_I$ is just the additive coset of I determined by a. Using standard coset notation from group theory, we have $[a]_I = a + I$. With this notation the formulas for addition and multiplication in R/I take the following form, where a, $b \in R$:

$$(a + I) + (b + I) = (a + b) + I \qquad \text{and} \qquad (a + I) \cdot (b + I) = ab + I.$$

The most familiar factor ring is $\mathbf{Z}/n\mathbf{Z}$, for which we will continue to use the notation $\mathbf{Z}_n$. In this ring, multiplication can be viewed as repeated addition, and it is easy to show that any subgroup is an ideal. Thus we already know the lattice of ideals of $\mathbf{Z}_n$, in which ideals correspond to the divisors of n. Recall that in $\mathbf{Z}$ we have $m|k$ if and only if $m\mathbf{Z} \supseteq k\mathbf{Z}$, and so the lattice of ideals of $\mathbf{Z}_n$ corresponds to the lattice of all ideals of $\mathbf{Z}$ that contain $n\mathbf{Z}$. As shown by the next proposition, the analogous result holds in any factor ring.

5.3.7 Proposition.   Let I be an ideal of the commutative ring R.
  (a) The natural projection mapping $\pi : R \to R/I$ defined by $\pi(a) = [a]_I$ for all $a \in R$ is a ring homomorphism, and $\ker(\pi) = I$.

  (b) There is a one-to-one correspondence between the ideals of R/I and ideals of R that contain I.

Proof. The parts of the proposition that involve addition follow directly from Proposition 3.8.6. To prove part (a), the projection mapping must be shown to preserve multiplication, and this follows directly from the definition of multiplication of congruence classes.

To prove part (b), we can use the one-to-one correspondence between subgroups that is given by Proposition 3.8.6. We must show that this correspondence preserves ideals. If J is an ideal of R that contains I, then it corresponds to the additive subgroup

$$\pi(J) = \{[a] \mid a \in J\}.$$

For any element $[r] \in R/I$ and any element $[a] \in \pi(J)$, we have $[r][a] = [ra]$, and then $[ra] \in \pi(J)$ since $ra \in J$. On the other hand, if J is an ideal of R/I, then it corresponds to the subgroup

$$\pi^{-1}(J) = \{a \in R \mid [a] \in J\}.$$

For any $r \in R$ and $a \in \pi^{-1}(J)$, we have $ra \in \pi^{-1}(J)$ since $[ra] = [r][a] \in J$. $\square$

Example 5.3.3

Let $R = \mathbf{Q}[x]$ and let $I = \langle x^2 - 2x + 1 \rangle$. Using the division algorithm, it is possible to show that the congruence classes modulo I correspond to the possible remainders upon division by $x^2 - 2x + 1$, so that we only need to consider congruence classes of the form $[a + bx]$, for all $a, b \in \mathbf{Q}$. Since $x^2 - 2x + 1 \equiv 0 \pmod{I}$, we can use the formula $x^2 \equiv 2x - 1 \pmod{I}$ to simplify products. This gives us the following formulas:

$$[a + bx] + [c + dx] = [(a + c) + (b + d)x]$$

and

$$[a + bx] \cdot [c + dx] = [(ac - bd) + (bc + ad + 2bd)x].$$

By the previous proposition, the ideals of R/I correspond to the ideals of R that contain I. Since $\mathbf{Q}[x]$ is a principal ideal domain, these ideals are determined by the divisors of $x^2 - 2x + 1$, showing that there is only one proper nontrivial ideal in R/I, corresponding to the ideal generated by $x - 1$. $\square$

Example 5.3.4

Let $R = \mathbf{Q}[x, y]$, the ring of polynomials in two indeterminates with rational coefficients, and let $I = \langle y \rangle$. That is, I is the set of all polynomials that have y as a factor. In forming R/I we make the elements of I congruent to 0, and so in some sense we should be left with just polynomials in x. This is made precise in the following way: Define $\phi : \mathbf{Q}[x, y] \rightarrow \mathbf{Q}[x]$ by $\phi(f(x, y)) = f(x, 0)$. It is necessary to check that ϕ is a ring homomorphism. Then it is clear that $\ker(\phi) = \langle y \rangle$, and so we can conclude from the fundamental homomorphism theorem for rings that $R/I \cong \mathbf{Q}[x]$. $\square$

Example 5.3.5

Let $\phi : R \to S$ be an isomorphism of commutative rings, let I be any ideal of R, and let $J = \phi(I)$. We will show that R/I is isomorphic to S/J. To do this, let π be the natural projection from S onto S/J, and consider $\theta = \pi\phi$. Then θ is onto since both π and ϕ are onto, and

$$\ker(\theta) = \{r \in R \mid \phi(r) \in J\} = I,$$

so it follows from the fundamental homomorphism theorem for rings that R/I is isomorphic to S/J. $\square$

To motivate the next definition, consider the ring of integers $\mathbf{Z}$. We know that the proper nontrivial ideals of $\mathbf{Z}$ correspond to the positive integers, with $n\mathbf{Z} \subseteq m\mathbf{Z}$ if and only if $m|n$. Euclid's lemma (Lemma 1.2.5) states that an integer $p > 1$ is prime if and only if it satisfies the following property: If $p|ab$ for integers a, b, then either $p|a$ or $p|b$. In the language of ideals, this says that p is prime if and only if $ab \in p\mathbf{Z}$ implies $a \in p\mathbf{Z}$ or $b \in p\mathbf{Z}$, for all integers a, b.

Lemma 4.2.8 gives a similar characterization of irreducible polynomials, stating that a polynomial $p(x) \in F[x]$ is irreducible if and only if $p(x) \mid f(x)g(x)$ implies $p(x) \mid f(x)$ or $p(x) \mid g(x)$. When formulated in terms of the principal ideal $\langle p(x) \rangle$, this shows that $p(x)$ is irreducible over F if and only if $f(x)g(x) \in \langle p(x) \rangle$ implies $f(x) \in \langle p(x) \rangle$ or $g(x) \in \langle p(x) \rangle$. Thus irreducible polynomials play the same role in $F[x]$ as do prime numbers in $\mathbf{Z}$.

These two examples probably would not provide sufficient motivation to introduce a general definition of a "prime" ideal. In both $\mathbf{Z}$ and $F[x]$, where F is a field, we have shown that each element can be expressed as a product of primes or irreducibles, respectively. This is not true in general for all commutative rings. In fact, certain subrings of $\mathbf{C}$ which turn out to be important in number theory do not have this property. One of the original motivations for introducing the notion of an ideal (or "ideal number") was to be able to salvage at least the property that every ideal could be expressed as a product of prime ideals.

The prime numbers in $\mathbf{Z}$ can be characterized in another way. Since a prime p has no divisors except $\pm p$ and ± 1, there cannot be any ideals properly contained between $p\mathbf{Z}$ and $\mathbf{Z}$. We refer to ideals with this property as maximal ideals.

5.3.8 Definition.   Let I be a proper ideal of the commutative ring R. Then I is said to be a *prime ideal* of R if for all a, $b \in R$ it is true that $ab \in I$ implies $a \in I$ or $b \in I$.

The ideal I is said to be a *maximal ideal* of R if for all ideals J of R such that $I \subseteq J \subseteq R$, either $J = I$ or $J = R$.

Note that if R is a commutative ring with identity, then R is an integral domain if and only if the zero ideal of R is a prime ideal. This observation provides an example, namely, the zero ideal of $\mathbf{Z}$, of a prime ideal that is not maximal. On the other hand, the characterization of prime and maximal ideals given in the next proposition shows that in a commutative ring with identity any maximal ideal is also a prime ideal, since any field is an integral domain.

5.3.9 Proposition.   Let I be a proper ideal of the commutative ring R with identity.

(a) The factor ring R/I is a field if and only if I is a maximal ideal of R.

(b) The factor ring R/I is a integral domain if and only if I is a prime ideal of R.

(c) If I is maximal, then it is a prime ideal.

Proof. (a) By Proposition 5.3.2, R/I is a field if and only if it has no proper nontrivial ideals. Using the one-to-one correspondence between ideals given by Proposition 5.3.7 (b), this occurs if and only if there are no ideals properly between I and R, which is precisely the statement that I is a maximal ideal.

(b) Assume that R/I is an integral domain, and let $a, b \in R$ with $ab \in I$. Remember that the zero element of R/I is the congruence class (or coset) consisting of all elements of I. Thus in R/I we have a product $[a][b]$ of congruence classes that is equal to the zero congruence class, and so by assumption this implies that either $[a]$ or $[b]$ is the zero congruence class. This implies that either $a \in I$ or $b \in I$, and so I must be a prime ideal.

Conversely, assume that I is a prime ideal. If $a, b \in R$ with $[a][b] = [0]$ in R/I, then we have $ab \in I$, and so by assumption either $a \in I$ or $b \in I$. This shows that either $[a] = [0]$ or $[b] = [0]$, and so R/I is an integral domain.

(c) Every field is an integral domain. $\square$

Example 5.3.6

Let $\phi : R \to S$ be an isomorphism of commutative rings. The isomorphism gives a one-to-one correspondence between ideals of the respective rings, and it is not hard to show directly that prime (or maximal) ideals of R correspond to prime (or maximal) ideals of S. This can also be proved as an application of the previous proposition, since in Example 5.3.5 we observed that if I is any ideal of R, then $\phi(I)$ is an ideal of S with $R/I \cong S/\phi(I)$. It then follows immediately from Proposition 5.3.9 that I is prime (or maximal) in R if and only if $\phi(I)$ is prime (or maximal) in S. $\square$

We have already observed that in the ring **Z**, prime numbers determine maximal ideals. One reason behind this fact is that any finite integral domain is a field, which we proved in Theorem 5.1.8. It is also true that if F is a field, then irreducible polynomials in $F[x]$ determine maximal ideals. The next proposition gives a general reason applicable in both of these special cases.

5.3.10 Theorem.   Every nonzero prime ideal of a principal ideal domain is maximal.

Proof. Let P be a nonzero prime ideal of a principal ideal domain R, and let J be any ideal with $P \subseteq J \subseteq R$. Since R is a principal ideal domain, we can assume that $P = Ra$ and $J = Rb$ for some elements $a, b \in R$. Since $a \in P$, we have $a \in J$, and so there exists $r \in R$ such that $a = rb$. This implies that $rb \in P$, and so either $b \in P$ or $r \in P$. In the first case, $b \in P$ implies that $J = P$. In the second case, $r \in P$ implies that $r = sa$ for some $s \in R$, since P is generated by a. This gives $a = sab$, and using the assumption that R is an integral domain allows us to cancel a to get $1 = sb$. This shows that $1 \in J$, and so $J = R$. $\square$

Example 5.3.7 (Ideals of $F[x]$)

We can now summarize the information we have regarding polynomials over a field, using a ring theoretic point of view. Let F be any field. The nonzero ideals of $F[x]$ are all principal, of the form $\langle f(x) \rangle$, where $f(x)$ is the unique monic polynomial of minimal degree in the ideal. The ideal is prime (and hence maximal) if and only if $f(x)$ is irreducible. If $p(x)$ is irreducible, then the factor ring $F[x]/\langle p(x) \rangle$ is a field. $\square$

Example 5.3.8 (Evaluation Mapping)

Let F be a subfield of E, and for any element $\alpha \in E$, define the evaluation mapping $\phi_\alpha : F[x] \to E$ by $\phi_\alpha(f(x)) = f(\alpha)$, for all $f(x) \in F[x]$. We have already seen in Example 5.2.4 that ϕ_α defines a ring homomorphism. Since $\phi_\alpha(F[x])$ is a subring of E that contains 1, it follows from Theorem 5.1.7 that it must be an integral domain. By the fundamental homomorphism theorem for rings this image is isomorphic to $F[x]/\ker(\phi_\alpha)$, and so by Proposition 5.3.9, the kernel of ϕ_α must be a prime ideal. If $\ker(\phi_\alpha)$ is nonzero, then it follows from Theorem 5.3.10 that it is a maximal ideal. By Proposition 5.3.9, we know that $F[x]/\ker(\phi_\alpha)$ is a field, so it follows from the fundamental homomorphism theorem for rings that the image of ϕ_α is in fact a subfield of E. This fact will be very important in our study of fields in Chapter 6, where we denote $\phi_\alpha(F[x])$ by $F(\alpha)$. $\square$

EXERCISES: SECTION 5.3

1. Give a multiplication table for the ring $\mathbf{Z}_2[x]/\langle x^2 + 1 \rangle$.
2. Give a multiplication table for the ring $\mathbf{Z}_2[x]/\langle x^3 + x^2 + x + 1 \rangle$.
3. Give a multiplication table for the ring $\mathbf{Z}_3[x]/\langle x^2 - 1 \rangle$.
4. Let R be the ring $\mathbf{Q}[x]/\langle x^3 + 2x^2 - x - 3 \rangle$. Describe the elements of R and give the formulas necessary to describe the product of any two elements.
5. Let $R = F[x]$ and let I be any maximal ideal of R. Prove that $I = \langle p(x) \rangle$ for some irreducible polynomial $p(x)$.

6. Show that the intersection of two ideals of a commutative ring is again an ideal.

7. Let R be a commutative ring, with $a \in R$. The *annihilator* of a is defined by $\text{Ann}(a) = \{r \in R \mid ra = 0\}$. Prove that $\text{Ann}(a)$ is an ideal of R.

8. Let R and S be commutative rings with identity.
   (a) Show that if I is an ideal of R and J is an ideal of S, then $I \oplus J$ is an ideal of $R \oplus S$.
   (b) Show that any ideal of $R \oplus S$ must have the form of the ideals described in part (a).
   (c) Show that any prime ideal of $R \oplus S$ must have the form $P \oplus S$ for a prime ideal P of R or $R \oplus P$ for a prime ideal P of S.

9. An element of a commutative ring is said to be *nilpotent* if $a^n = 0$ for some positive integer n.
   (a) Show that the set N of all nilpotent elements of a commutative ring forms an ideal of the ring.
   (b) Show that R/N has no nonzero nilpotent elements.
   (c) Show that $N \subseteq P$ for each prime ideal P of R.

10. Let R be a commutative ring with ideals I, J. Let

$$I + J = \{x \in R \mid x = a + b \quad \text{for some} \quad a \in I, b \in J\}.$$

   (a) Show that $I + J$ is an ideal.
   (b) Determine $n\mathbf{Z} + m\mathbf{Z}$ in the ring of integers.

11. Let R be a commutative ring with ideals I, J. Define the product of the two ideals by

$$IJ = \left\{ \sum_{i=1}^{n} a_i b_i \mid a_i \in I, b_i \in J, n \in \mathbf{Z}^+ \right\}.$$

   (a) Show that IJ is an ideal that is contained in $I \cap J$.
   (b) Determine $(n\mathbf{Z})(m\mathbf{Z})$ in the ring of integers.

12. Let R be the set of all rational numbers m/n such that n is odd.
   (a) Show that R is a subring of $\mathbf{Q}$.
   (b) For any positive integer k, let $2^k R = \{m/n \in R \mid m$ is a multiple of $2^k\}$. Show that $2^k R$ is an ideal of R.
   (c) Show that $R/2^k R$ is isomorphic to $\mathbf{Z}_{2^k}$.

13. Let $R = \{m + n\sqrt{2} \mid m, n \in \mathbf{Z}\}$ and let

$$I = \{m + n\sqrt{2} \mid m, n \in \mathbf{Z} \text{ and } m \text{ is even }\}.$$

Show that I is an ideal of R. Find the well-known commutative ring to which R/I is isomorphic.

Hint: How many congruence classes does I determine?

14. Let R be the set of all matrices $\begin{bmatrix} a & b \\ c & d \end{bmatrix}$ over $\mathbf{Q}$ such that $a = d$ and $c = 0$. Let I be the set of all such matrices for which $a = d = 0$. Show that I is an ideal of R, and use the fundamental homomorphism theorem for rings to show that $R/I \cong \mathbf{Q}$.

15. Let R be a commutative ring with ideals I, J such that $I \subseteq J \subseteq R$.
   (a) Show that J/I is an ideal of R/I.
   (b) Show that the factor ring $(R/I)/(J/I)$ is isomorphic to R/J.

Hint: Define a homomorphism from R/I onto R/J and apply the fundamental homomorphism theorem for rings.

16. Use the previous problem together with Proposition 5.3.9 to determine all prime ideals and all maximal ideals of $\mathbf{Z}_n$.

17. Let R be the set of all continuous functions from the set of real numbers into itself. In Exercise 9 of Section 5.1, we have shown that R is a commutative ring if the following formulas

$$(f + g)(x) = f(x) + g(x) \quad \text{and} \quad (f \cdot g)(x) = f(x)g(x)$$

for all x, are used to define addition and multiplication of functions. Let I be the set of all functions $f(x) \in R$ such that $f(1) = 0$. Show that I is a maximal ideal of R.

18. Let P be a prime ideal of the commutative ring R. Prove that if I and J are ideals of R and $I \cap J \subseteq P$, then either $I \subseteq P$ or $J \subseteq P$.

19. Find a nonzero prime ideal of $\mathbf{Z} \oplus \mathbf{Z}$ that is not maximal.

20. Let I be the smallest ideal of $\mathbf{Z}[x]$ that contains both 2 and x. Show that I is not a principal ideal.

5.4 QUOTIENT FIELDS

In Section 5.1 we showed that any subring of a field is an integral domain (provided it contains the identity element of the field). In this section we will show that any integral domain is isomorphic to a subring of a field. This can be done in such a way that the subring and field are very closely connected.

The best example we can give is provided by considering the ring $\mathbf{Z}$ of integers as a subring of the field $\mathbf{Q}$ of rational numbers. Every rational number has the form m/n for integers m, n, with $n \neq 0$. Another way to say this is to observe that if q is any rational number, then there exists some nonzero integer n such that $nq \in \mathbf{Z}$, showing that $\mathbf{Z}$ and $\mathbf{Q}$ are closely related. A statement such as this is impossible to make when $\mathbf{Z}$ is viewed as a subring of $\mathbf{R}$, for example.

We start with any integral domain D. (You may want to keep in mind the ring $\mathbf{R}[x]$ of polynomials with real coefficients.) We have information only about D and the operations defined on it, so we must be very careful not to make implicit assumptions that are not warranted. The situation is something like that in constructing the complex numbers from the real numbers. At a naive level we can simply introduce a symbol i that does what we want it to do, namely, provide a root of the equation $x^2 + 1 = 0$. However, to give a completely rigorous development using only facts about the real numbers, we found it necessary to use the notion of a factor ring and work in $\mathbf{R}[x]/\langle x^2 + 1 \rangle$.

To formally construct fractions with numerator and denominator in D we will consider ordered pairs (a, b), where a is to represent the numerator and b is to represent the denominator. Of course, we need to require that $b \neq 0$. In $\mathbf{Q}$, two fractions m/n and r/s may be equal even though their corresponding numerators and denominators are not equal. We can express the fact that $m/n = r/s$ by writing $ms = nr$. This shows that we must make certain identifications within the set of ordered pairs (a, b), and the appropriate way to do this is to introduce an equivalence relation.

5.4.1 Lemma.   Let D be an integral domain, and let

$$W = \{(a, b) \mid a, b \in D \text{ and } b \neq 0\}.$$

The relation $\sim$ defined on W by $(a, b) \sim (c, d)$ if $ad = bc$ is an equivalence relation.

Proof. Given $(a, b) \in W$ we have $(a, b) \sim (a, b)$ since $ab = ba$. The symmetric law holds since if $(c, d) \in W$ with $(a, b) \sim (c, d)$, then $ad = bc$, and so $cb = da$, showing that $(c, d) \sim (a, b)$. To show the transitive law, suppose that $(a, b) \sim (c, d)$ and $(c, d) \sim (u, v)$ for ordered pairs in W. Then we have $ad = bc$ and $cv = du$, and so multiplying the first equation by v gives $adv = bcv$, while multiplying the second equation by b gives $bcv = bdu$. Thus we have $adv = bdu$, and since d is a nonzero element of an integral domain, we can cancel it to obtain $av = bu$, which shows that $(a, b) \sim (u, v)$.   □

5.4.2 Definition.   Let D be an integral domain. The equivalence classes of the set

$$\{(a, b) \mid a, b \in D \text{ and } b \neq 0\}$$

under the equivalence relation defined by $(a, b) \sim (c, d)$ if $ad = bc$ will be denoted by $[a, b]$.

The set of all such equivalence classes will be denoted by $Q(D)$.

Since our ultimate goal is to show that $Q(D)$ is a field that contains D as a subring, we must decide how to define addition and multiplication of equivalence classes. Returning to **Q**, we have

$$\frac{m}{n} + \frac{p}{q} = \frac{mq}{nq} + \frac{np}{nq} = \frac{mq + np}{nq}.$$

With this as motivation, we define

$$[a, b] + [c, d] = [ad + bc, bd],$$

since a and c represent the "numerators" of our equivalence classes, while b and d represent the "denominators," and by analogy with **Q** the "numerator" and "denominator" of the sum should be $ab + bc$ and bd, respectively. Similarly, since

$$\frac{m}{n} \cdot \frac{p}{q} = \frac{mp}{nq},$$

we define $[a, b] \cdot [c, d] = [ac, bd]$.

5.4.3 Lemma.   For any integral domain D, the following operations are well-defined on $Q(D)$. For $[a, b], [c, d] \in Q(D)$,

$$[a, b] + [c, d] = [ad + bc, bd] \quad \text{and} \quad [a, b] \cdot [c, d] = [ac, bd].$$

Proof. Let $[a, b] = [a', b']$ and $[c, d] = [c', d']$ be elements of $Q(D)$. To show that addition is well-defined, we must show that

$$[a, b] + [c, d] = [a', b'] + [c', d'],$$

or, equivalently, that

$$[ad + bc, bd] = [a'd' + b'c', b'd'].$$

In terms of the equivalence relation on $Q(D)$, this means that we must show that

$$(ad + bc)b'd' = bd(a'd' + b'c').$$

We are given that $ab' = ba'$ and $cd' = dc'$. Multiplying the first of these two equations by dd' and the second by bb' and then adding terms gives us the desired equation.

To show that multiplication is well-defined, we must show that $[ac, bd] = [a'c', b'd']$. This is left to the student, as Exercise 1. $\square$

5.4.4 Theorem. Let D be an integral domain. Then $Q(D)$ is a field that contains a subring isomorphic to D.

Proof. Showing that the commutative and associative laws hold for addition is left as Exercises 2 and 3, respectively. Let 1 be the identity element of D. For all $[a, b] \in Q(D)$, we have $[0, 1] + [a, b] = [a, b]$, and so $[0, 1]$ serves as an additive identity element. Furthermore,

$$[-a, b] + [a, b] = [-ab + ba, b^2] = [0, b^2],$$

which shows that $[-a, b]$ is the additive inverse of $[a, b]$ since $[0, b^2] = [0, 1]$.

Again, verifying the commutative and associative laws for multiplication is left as an exercise. The equivalence class $[1, 1]$ clearly acts as a multiplicative identity element. Note that we have $[1, 1] = [d, d]$ for any nonzero $d \in D$. If $[a, b]$ is a nonzero element of $Q(D)$, that is, if $[a, b] \neq [0, 1]$, then we must have $a \neq 0$. Therefore $[b, a]$ is an element of $Q(D)$, and since $[b, a] \cdot [a, b] = [ab, ab]$, we have $[b, a] = [a, b]^{-1}$. Thus every nonzero element of $Q(D)$ is invertible, and so to complete the proof that $Q(D)$ is a field we only need to check that the distributive law holds.

To show that the distributive law holds, let $[a, b]$, $[c, d]$, and $[u, v]$ be elements of $Q(D)$. We have

$$([a, b] + [c, d]) \cdot [u, v] = [(ad + bc)u, (bd)v]$$

and

$$[a, b] \cdot [u, v] + [c, d] \cdot [u, v] = [(au)(dv) + (bv)(cu), (bv)(dv)].$$

We can factor $[v, v]$ out of the second expression, showing equality.

Finally, consider the mapping $\phi : D \rightarrow Q(D)$ defined by $\phi(d) = [d, 1]$, for all $d \in D$. It is easy to show that ϕ preserves sums and products, and so it is a ring homomorphism. If $[d, 1] = [0, 1]$, then we must have $d = 0$, which shows that $\ker(\phi) = \{0\}$. By the fundamental homomorphism theorem for rings, $\phi(D)$ is a subring of $Q(D)$ that is isomorphic to D.   □

5.4.5 Definition.   Let D be an integral domain. The field $Q(D)$ defined in Definition 5.4.2 is called the *field of quotients* or *field of fractions of D*.

5.4.6 Theorem.   Let D be an integral domain, and let $\theta : D \to F$ be a one-to-one ring homomorphism from D into a field F. Then there exists a unique extension $\hat{\theta} : Q(D) \to F$ that is one-to-one and satisfies $\hat{\theta}(d) = \theta(d)$ for all $d \in D$.

$$D \quad \subseteq \quad Q(D)$$
$$\theta \searrow \quad \downarrow \quad \hat{\theta}$$
$$F$$

Proof. For $[a, b] \in Q(D)$, define $\hat{\theta}([a, b]) = \theta(a)\theta(b)^{-1}$. Since $b \neq 0$ and θ is one-to-one, $\theta(b)^{-1}$ exists in F, and the definition makes sense. We must show that $\hat{\theta}$ is well-defined. If $[a, b] = [a', b']$, then $ab' = ba'$, and applying θ to both sides of this equation gives $\theta(a)\theta(b') = \theta(b)\theta(a')$ since θ is a ring homomorphism. Both $\theta(b)^{-1}$ and $\theta(b')^{-1}$ exist, so we must have $\theta(a)\theta(b)^{-1} = \theta(a')\theta(b')^{-1}$.

The proof that $\hat{\theta}$ is a homomorphism is left as an exercise. To prove the uniqueness of $\hat{\theta}$, suppose that $\phi : Q(D) \to F$ with $\phi(d) = \theta(d)$ for all $d \in D$. Then for any element $[a, b] \in Q(D)$ we have

$$\phi([a, 1]) = \phi([a, b][b, 1]) = \phi([a, b])\phi([b, 1]),$$

and so

$$\theta(a) = \phi([a, b])\theta(b)$$

or

$$\phi([a, b]) = \theta(a)\theta(b)^{-1}. \quad □$$

We have continued to use the notation $[a, b]$ for elements of $Q(D)$ in order to emphasize that equivalence classes are involved. From this point on we will use the more familiar notation a/b in place of $[a, b]$. We identify an element $d \in D$ with the fraction $d/1$, and this allows us to assume that D is a subring of $Q(D)$. If $b \in D$ is nonzero, then $1/b \in Q(D)$, and $(1/b) \cdot (b/1) = b/b = 1$ shows that $1/b = b^{-1}$, (where we have identified b and $b/1$). Thus we can also write $a/b = (a/1) \cdot (1/b) = ab^{-1}$, for $a, b \in D$, with $b \neq 0$.

5.4.7 Corollary.   Let D be an integral domain that is a subring of a field F. If each element of F has the form ab^{-1} for some $a, b \in D$, then F is isomorphic to the field of quotients $Q(D)$ of D.

Proof. By Theorem 5.4.6, the inclusion mapping $\theta : D \to F$ can be extended to a one-to-one mapping $\hat{\theta} : Q(D) \to F$. The given condition is precisely the one necessary to guarantee that $\hat{\theta}$ is onto. $\square$

Example 5.4.1

Let D be the integral domain consisting of all fractions $m/n \in \mathbf{Q}$ such that n is odd. (See Exercise 12 in Section 5.3.) If a/b is any element of $\mathbf{Q}$ such that $\gcd(a, b) = 1$, then either b is odd, in which case $a/b \in D$, or b is even, in which case a is odd and $a/b = 1 \cdot (b/a)^{-1}$, with $b/a \in D$. Applying Corollary 5.4.7 shows that $\mathbf{Q} \cong Q(D)$. $\square$

Example 5.4.2

If F is any field, then we know that the ring of polynomials $F[x]$ is an integral domain. Applying Theorem 5.4.4 shows that we can construct a field that contains $F[x]$ by considering all fractions of the form $f(x)/g(x)$, where $f(x)$, $g(x)$ are polynomials with $g(x) \neq 0$. This field is called the *field of rational functions* in x. $\square$

5.4.8 Corollary. Any field contains a subfield isomorphic to $\mathbf{Q}$ or $\mathbf{Z}_p$, for some prime number p.

Proof. Let F be any field, and, as in Proposition 5.2.10, let ϕ be the homomorphism from $\mathbf{Z}$ into F defined by $\phi(n) = n \cdot 1$. If $\ker(\phi) \neq 0$, then as in Proposition 5.2.10, we have $\ker(\phi) = p\mathbf{Z}$ for some prime p, and so the image of ϕ is a subfield isomorphic to $\mathbf{Z}_p$. If ϕ is one-to-one, then by Theorem 5.4.6 it extends to an embedding of $\mathbf{Q}$ (the field of quotients of $\mathbf{Z}$) into F, and so the image of this extension is a subfield isomorphic to $\mathbf{Q}$. $\square$

EXERCISES: SECTION 5.4

Throughout these exercises, D will denote an integral domain.

1. Complete the proof in Lemma 5.4.3, to show that multiplication of equivalence classes in $Q(D)$ is well-defined.
2. Show that the associative law holds for addition in $Q(D)$.
3. Show that the commutative law holds for addition in $Q(D)$.
4. Show that the associative and commutative laws hold for multiplication in $Q(D)$.
5. In Theorem 5.4.6, verify that $\hat{\theta}$ is a ring homomorphism.
6. Determine $Q(D)$ for $D = \{m + n\sqrt{2} \mid m, n \in \mathbf{Z}\}$. (See Example 5.1.5.)
7. Let p be a prime number, and let $D = \{m/n \mid m, n \in \mathbf{Z} \text{ and } p \nmid n\}$. Verify that D is an integran domain and find $Q(D)$.
8. Determine $Q(D)$ for $D = \{m + ni \mid m, n \in \mathbf{Z}\} \subseteq \mathbf{C}$.

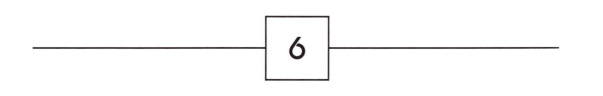

6

Fields

In this chapter we will show that for any polynomial over a field K, a larger field F can be constructed, which contains all of the roots of the given polynomial. The polynomial then "splits" into a product of linear polynomials with coefficients in F. Thus the roots of any polynomial can always be found in some extension of the base field. In Chapter 8 we will answer the question of when the roots can be obtained from the coefficients of the equation by allowing field operations and the extraction of nth roots. To do this we must study the interplay between the field of coefficients of the polynomial and the field of roots of the polynomial. This chapter therefore studies field extensions and splitting fields.

If F is an extension field of K (that is, K is a subfield of F), then F can be shown to have the structure of a vector space over K. We will exploit the concept of the dimension of a vector space to show that several geometric constructions are impossible. These were already studied by the Greeks in the fifth century B.C. It is impossible to find a general method to trisect an angle; given a circle, it is impossible to construct a square of the same area; and given a cube, it is impossible to construct a cube with double the given volume. The method of construction in each case is limited to using a straightedge and compass. Thus the points constructed at any stage of the procedure represent solutions of quadratic equations with coefficients from the smallest field containing the previously constructed numbers. We will show that any constructible number lies in an extension of $\mathbf{Q}$ whose dimension over $\mathbf{Q}$ is a power of 2, and thus to show that a particular number cannot be constructed, we only need to use a dimension argument.

A related problem is that of constructing (using only straightedge and compass) a regular polygon with n sides. The modern solution uses the methods of Galois theory and is beyond the scope of our book, but we can give a small part of the story at this point. The ancient Greeks already knew how to construct regular polygons of three, four, five, and six sides. To construct a regular pentagon, it is necessary to solve the equation $x^5 - 1 = 0$. It is certainly possible to give a trigonometric solution as a complex number of the form $\cos(2\pi/5) + i\sin(2\pi/5)$, but we need to know that this can be expressed in terms of rational numbers and their square roots. By results in Section 4.3 we can give the factorization

$$x^5 - 1 = (x - 1)(x^4 + x^3 + x^2 + x + 1)$$

of $x^5 - 1$ into factors irreducible over $\mathbf{Q}$. The primitive fifth root of unity that we need to construct is a root of the equation $x^4 + x^3 + x^2 + x + 1 = 0$, and since $x^5 = 1$, we can rewrite it in the form

$$x^{-1} + x^{-2} + x^2 + x + 1 = (x^2 + 1 + x^{-2}) + (x + x^{-1}) = 0.$$

Substituting $y = x + x^{-1}$ yields the equation

$$y^2 + y - 1 = 0$$

and from its solution we can find a solution (by radicals) of $x^5 - 1 = 0$.

When he was eighteen, Gauss discovered that the regular 17-gon is constructible. He published the solution in *Disquisitiones arithmeticae* as a special case of the general solution of the "cyclotomic equation" $x^n - 1 = 0$. He proved that a regular n-gon is constructible with straightedge and compass only if n has the form $n = 2^\alpha p_2 \cdots p_k$, where the numbers p_i are odd primes of the form $2^{2^n} + 1$. (The converse is also true, as was proved almost forty years later. We should note that not all numbers of the form $2^{2^n} + 1$ are prime, as conjectured by Fermat. Although $2^1 + 1 = 3, 2^2 + 1 = 5, 2^4 + 1 = 17, 2^8 + 1 = 257$, and $2^{16} + 1 = 65537$ are all prime, Euler showed that $2^{32} + 1 = 641 \cdot 6700417$.) The general solution of the cyclotomic equation, and his proof of the fundamental theorem of algebra, represent the most important contributions that Gauss made to the theory of algebraic equations.

As a further application we will be able to give a complete list of all finite fields. Such fields are used in algebraic coding theory, which provides one approach to the problem of encoding data for transmission. When data is transmitted over telephone lines or via satellite connections, there is a substantial chance that errors will occur. To make it possible to detect (and even correct) these errors, additional information can be sent along. In the most naive approach, the entire message could simply be repeated numerous times. The real task is to find efficient algorithms for encoding, and one successful algorithm involves the use of polynomials over finite fields, with the field of 256 elements often being used.

6.1 ALGEBRAIC ELEMENTS

We recall from Chapter 5 that a commutative ring with identity 1 (assumed to be different from the element 0) is called a field if every nonzero element is invertible. (See Definition 4.1.1. for a complete list of the properties of a field.) Thus the operations of addition, subtraction, multiplication, and division (by nonzero elements) are all possible within a field. We should also note that the elements of a field form an abelian group under addition, while the nonzero elements form an abelian group under multiplication. This observation allows us to make use of the results we have proved for groups in Chapter 3.

Our primary interest is to study roots of polynomials. This usually involves the interplay of two fields: one that contains the coefficients of the polynomial, and another determined by the roots of the polynomial. In many situations we will start with a known field K, and then construct a larger field F. We now give a restatement of Definition 4.4.1 in this context.

6.1.1 Definition.   The field F is said to be an *extension field* of the field K if K is a subset of F which is a field under the operations of F.

The above definition is equivalent to saying that K is a *subfield* of F. That is, the additive and multiplicative groups that determine K are subgroups of the corresponding additive and multiplicative groups of F.

If F is an extension field of K, then the elements of F that are roots of polynomials in $K[x]$ will be called algebraic elements. If $u \in F$ is a root of some polynomial $f(x) \in K[x]$, then let $p(x)$ have minimal degree among all polynomials of which u is a root. Using the division algorithm, we can write $f(x) = q(x)p(x) + r(x)$, where either $r(x) = 0$ or $\deg(r(x)) < \deg(p(x))$. Solving for $r(x)$ and substituting u shows that u is a root of $r(x)$, which violates the definition of $p(x)$ unless $r(x) = 0$, and so we have shown that $p(x)$ is a divisor of $f(x)$.

With the notation above, we next observe that $p(x)$ must be an irreducible polynomial. To show this, let $p(x) = g(x)h(x)$ for polynomials $g(x)$, $h(x) \in K[x]$. Then substituting u gives $g(u)h(u) = p(u) = 0$, and so either $g(u) = 0$ or $h(u) = 0$ since these are elements of F, which is a field. From the definition of $p(x)$ as a polynomial of minimal degree that has u as a root, we see that either $g(x)$ or $h(x)$ has the same degree as $p(x)$, and so we have shown that $p(x)$ is irreducible.

These facts about the polynomials that have a given element as a root can be proved in another way by using results from Chapter 5 on rings of polynomials. The approach using ring theory lends itself to more powerful applications, and so we will adopt that point of view. Recall that the nonzero ideals of $F[x]$ are all principal, of the form $\langle f(x) \rangle = \{q(x)f(x) \mid q(x) \in F[x]\}$, where $f(x)$ is any polynomial of minimal degree in the ideal. (See Example 5.3.7.) An ideal is prime (and hence maximal) if and only if its generator $f(x)$ is irreducible. Furthermore, if $p(x)$ is irreducible, then the factor ring $F[x]/\langle p(x) \rangle$ is a field.

6.1.2 Definition.   Let F be an extension field of K and let $u \in F$. If there exists a nonzero polynomial $f(x) \in K[x]$ such that $f(u) = 0$, then u is said to be *algebraic* over K. If there does not exist such a polynomial, then u is said to be *transcendental* over K.

The familiar constants e and π are transcendental. These are not easy results, and the proofs lie beyond the scope of this book. That e is transcendental was proved by Hermite in 1873, and the corresponding result for π was proved by Lindemann in 1882.

6.1.3 Proposition.   Let F be an extension field of K, and let $u \in F$ be algebraic over K. Then there exists a unique monic irreducible polynomial $p(x) \in K[x]$ such that $p(u) = 0$. It is characterized as the monic polynomial of minimal degree that has u as a root. Furthermore, if $f(x)$ is any polynomial in $K[x]$ with $f(u) = 0$, then $p(x) \mid f(x)$.

Proof. Let I be the set of all polynomials $f(x) \in K[x]$ such that $f(u) = 0$. It is easy to see that I is closed under sums and differences, and if $f(x) \in I$, then $g(x)f(x) \in I$ for all $g(x) \in K[x]$. Thus I is an ideal of $K[x]$, and so $I = \langle p(x) \rangle$ for any nonzero polynomial $p(x) \in I$ that has minimal degree. If $f(x)$, $g(x) \in K[x]$ with $f(x)g(x) \in I$, then we have $f(u)g(u) = 0$, which implies that either $f(u) = 0$ or $g(u) = 0$, and so we see that I is a prime ideal. This implies that the unique monic generator $p(x)$ of I must be irreducible. Finally, since $I = \langle p(x) \rangle$, we have $p(x) \mid f(x)$ for any $f(x) \in I$.   □

6.1.4 Definition.   Let F be an extension field of K, and let u be an algebraic element of F. The monic polynomial $p(x)$ of minimal degree in $K[x]$ such that $p(u) = 0$ is called the *minimal polynomial* of u over K. The degree of the minimal polynomial of u over K is called the *degree* of u over K.

Example 6.1.1

Considering the set of real numbers **R** as an extension field of the set of rational numbers **Q**, the number $\sqrt{2}$ has minimal polynomial $x^2 - 2$ over **Q**, and so it has degree 2 over **Q**.   □

Example 6.1.2

In this example we will compute the minimal polynomial of $\sqrt{2} + \sqrt{3}$ over **Q**. If we let $x = \sqrt{2} + \sqrt{3}$, then we must find a polynomial with rational coefficients that has x as a root. We begin by rewriting our equation as $x - \sqrt{2} = \sqrt{3}$. Squaring both sides gives $x^2 - 2\sqrt{2}x + 2 = 3$. Since we still need to eliminate the square root to obtain coefficients over **Q**, we can again rewrite the equation to obtain $x^2 - 1 = 2\sqrt{2}x$. Then squaring both sides and rewriting the equation gives $x^4 - 10x^2 + 1 = 0$.

To show that $x^4 - 10x^2 + 1$ is the minimal polynomial of $\sqrt{2} + \sqrt{3}$ over **Q**, we must show that it is irreducible. It is easy to check that there are no rational roots, so it could only be the product of two quadratic polynomials, which can be assumed to have integer coefficients. Unfortunately, Eisenstein's irreducibility criterion cannot be applied, and so we must try to verify directly that the polynomial is irreducible. A factorization over $\mathbf{Q}$ of the form

$$x^4 - 10x^2 + 1 = (x^2 + ax + b)(x^2 + cx + d)$$

leads to the equations $a + c = 0$, $b + ac + d = -10$, $ad + bc = 0$, and $bd = 1$. If $c = 0$, then $b + d = -10$ and $bd = 1$, so $b^2 + bd = -10b$, or $b^2 + 10b + 1 = 0$. The last equation has no rational roots, so we may assume that $c \neq 0$. Letting $c = -a$, we see that $b = d$, and so $b = \pm 1$. It follows that $a^2 = 8$ or $a^2 = 12$, a contradiction, since $a \in \mathbf{Q}$. We conclude that $x^4 - 10x^2 + 1$ is irreducible over $\mathbf{Q}$, and so it must be the minimal polynomial of $\sqrt{2} + \sqrt{3}$ over $\mathbf{Q}$.   □

In a field F, the intersection of any collection of subfields of F is again a subfield. In particular, if F is an extension field of K and $u \in F$, then the intersection of all subfields of F that contain both K and u is a subfield of F. This intersection is contained in any subfield that contains both K and u. This guarantees the existence of the field defined below.

6.1.5 Definition.   Let F be an extension field of K, and let u_1, u_2, $\ldots$, $u_n \in F$. The smallest subfield of F that contains K and u_1, u_2, $\ldots$, u_n will be denoted by $K(u_1, u_2, \ldots, u_n)$. It is called the *extension field of K generated by u_1, u_2, $\ldots$, u_n*. If $F = K(u)$ for a single element $u \in F$, then F is said to be a *simple extension of K*.

If F is an extension of K, and u_1, u_2, $\ldots$, $u_n \in F$, then it is possible to construct $K(u_1, u_2, \ldots, u_n)$ by adjoining one u_i at a time. That is, we first construct $K(u_1)$, and then consider the smallest subfield of F that contains $K(u_1)$ and u_2. This would be written as $K(u_1)(u_2)$, but it is clear from the definition of $K(u_1, u_2)$ that the two fields are equal. This procedure can be continued to construct $K(u_1, u_2, \ldots, u_n)$. It is thus sufficient to describe $K(u)$ for a single element, which we do in the next proposition.

6.1.6 Proposition.   Let F be an extension field of K, and let $u \in F$.
(a) If u is algebraic over K, then $K(u) \cong K[x]/\langle p(x)\rangle$, where $p(x)$ is the minimal polynomial of u over K.
(b) If u is transcendental over K, then $K(u) \cong K(x)$, where $K(x)$ is the quotient field of the integral domain $K[x]$.

Proof. Define $\phi : K[x] \to F$ by $\phi(f(x)) = f(u)$, for all polynomials $f(x) \in K[x]$. This defines a ring homomorphism, and $\ker(\phi)$ is the set of all polynomials $f(x)$ with $f(u) = 0$. The image of ϕ is a subring of F consisting of all elements of the form $a_0 + a_1 u + \ldots + a_n u^n$, and it must be contained in every subring of F that contains K and u. In particular, the image of ϕ must be contained in $K(u)$.

(a) If u is algebraic over K, then $\ker(\phi) = \langle p(x)\rangle$, for the minimal polynomial $p(x)$ of u over K. In this case the fundamental homomorphism theorem for rings implies that the image of ϕ is isomorphic to $K[x]/\langle p(x) \rangle$, which is a field since $p(x)$ is irreducible. But then the image of ϕ must in fact be equal to $K(u)$, since the image is a subfield containing K and u.

(b) If u is transcendental over K, then $\ker(\phi) = \{0\}$, and so the image of ϕ is isomorphic to $K[x]$. Since F is a field, by Theorem 5.4.6 there exists an isomorphism θ from the field of quotients of $K[x]$ into F. Since every element of the image of θ is a quotient of elements that belong to $K(u)$, it follows that this image must be contained in $K(u)$. Then since it is a field that contains u, it must be equal to $K(u)$. $\square$

To help understand the field $K(u)$, it may be useful to approach its construction from a more elementary point of view. Given an extension field F of K and an element $u \in F$, any subfield that contains K and u must be closed under sums and products, so it must contain all elements of the form $a_0 + a_1 u + \ldots + a_n u^n$. Furthermore, since it must be clo. ed under division, it must contain all elements of the form

$$\frac{a_0 + a_1 u + \ldots + a_n u^n}{b_0 + b_1 u + \ldots + b_m u^m}.$$

such that the denominator is nonzero. This set can be shown to be a subfield of F, and so it must be equal to $K(u)$.

If u is algebraic of degree n over K, let the minimal polynomial of u over K be $p(x) = c_0 + c_1 x + \ldots + c_n x^n$. Since $c_0 + c_1 u + \ldots + c_n u^n = 0$, we can solve for u^n and obtain a formula that allows us to reduce any expression of the form $a_0 + a_1 u + \ldots + a_m u^m$ to one involving only $u, u^2, \ldots, u^{n-1}$ and elements of K. Given any expression of the form $a_0 + a_1 u + \ldots + a_{n-1} u^{n-1}$, we let $f(x)$ be the corresponding polynomial $a_0 + a_1 x + \ldots + a_{n-1} x^{n-1}$. If $f(x)$ is nonzero, then it must be relatively prime to $p(x)$ since $p(x)$ is irreducible and $\deg(f(x)) < \deg(p(x))$. Thus there exist polynomials $g(x)$ and $q(x)$ such that $f(x)g(x) + p(x)q(x) = 1$. Substituting u gives $f(u)g(u) = 1$ since $p(u) = 0$, and so $g(u) = 1/f(u)$. Thus the denominators can be eliminated in our description of $K(u)$. We conclude that when u is algebraic over K, each element of $K(u)$ has the form $a_0 + a_1 u + \ldots + a_{n-1} u^{n-1}$ for elements $a_0, a_1, \ldots, a_{n-1} \in K$.

Example 6.1.3

Let u be a root of the polynomial $x^3 - 2$. The extension field $\mathbf{Q}(u)$ of $\mathbf{Q}$ is isomorphic to the factor ring $\mathbf{Q}[x]/\langle x^3 - 2 \rangle$, and so computations can be done in either field. For example, let us compute $(1 + u^2)^{-1}$. In $\mathbf{Q}(u)$ we can set up the equation $(1 + u^2)(a + bu + cu^2) = 1$. Using the identities $u^3 = 2$ and $u^4 = 2u$ to multiply out the left-hand side, we obtain the equations $a + 2b = 1$, $b + 2c = 0$, and $a + c = 0$. These lead to the solution $a = 1/5$, $b = 2/5$, and $c = -1/5$.

On the other hand, to find the multiplicative inverse of the element $[1 + x^2]$ in $\mathbf{Q}[x]/\langle x^3 - 2\rangle$, we can use the Euclidean algorithm to solve for

$$\gcd(x^2 + 1, x^3 - 2).$$

We obtain $x^3 - 2 = x(x^2 + 1) - (x + 2)$ and then $x^2 + 1 = (x - 2)(x + 2) + 5$. Solving for the linear combinations that give the greatest common divisor yields the equation

$$1 = \left(-\frac{1}{5}x^2 + \frac{2}{5}x + \frac{1}{5}\right)(x^2 + 1) + \left(\frac{1}{5}x - \frac{2}{5}\right)(x^3 - 2).$$

We can reduce the equation to a congruence modulo $x^3 - 2$ and use the isomorphism to obtain the same answer we got previously:

$$(1 + u^2)^{-1} = \frac{1}{5} + \frac{2}{5}u - \frac{1}{5}u^2. \quad \square$$

6.1.7 Proposition.   Let K be a field and let $p(x) \in K[x]$ be any irreducible polynomial. Then there exists an extension field F of K and an element $u \in F$ such that the minimal polynomial of u over K is $p(x)$.

Proof. This is simply a restatement of Kronecker's theorem (Theorem 4.4.8). Recall that the extension field F is constructed as $K[x]/\langle p(x)\rangle$, and K is viewed as isomorphic to the subfield consisting of all cosets determined by the constant polynomials. The element u is the coset determined by x, and it follows that $p(u) = 0$. Since $p(x)$ is irreducible, Proposition 6.1.3 shows that $p(x)$ is the minimal polynomial for u over K. $\square$

EXERCISES: SECTION 6.1

1. Show that the following complex numbers are algebraic over $\mathbf{Q}$:
   (a) $\sqrt{2}$
   (b) $\sqrt{n}$, for $n \in \mathbf{Z}^+$
   (c) $\sqrt{2} + \sqrt{3}$
   (d) $\sqrt{2 + \sqrt{3}}$
   (e) $\sqrt[3]{2} + \sqrt{2}$
   (f) $(-1 + \sqrt{3}i)/2$

2. (a) Show that $f(x) = x^3 + 3x + 3$ is irreducible over $\mathbf{Q}$.
   (b) Let θ be a zero of $f(x)$. Express θ^{-1} and $(\theta + 1)^{-1}$ in the form $a + b\theta + c\theta^2$, where $a, b, c \in \mathbf{Q}$.

3. Suppose that u is algebraic over K, and that $a \in K$. If $f(x)$ is the minimal polynomial for u over K, find the minimal polynomial for $u + a$ over K, and conclude that $u + a$ is algebraic over K.

4. Show that the intersection of any collection of subfields of a given field is again a subfield.

5. Let $F = K(u)$, where u is transcendental over K. If E is a field such that $K \subset E \subseteq F$, then show that u is algebraic over E.

6. Let F be an extension field of K.
   (a) Show that F is a vector space over K.
   (b) Let $u \in F$. Show that the subspace spanned by $\{1, u, u^2, \ldots\}$ is a field if and only if u is algebraic over K.

7. Let F be an extension field of K. If $u \in F$ is transcendental over K, then show that every element of $K(u)$ that is not in K is also transcendental over K.

8. A famous theorem of Gelfand and Schneider states that if u is algebraic over $\mathbf{Q}$, with $u \neq 0, 1$, and $r \notin \mathbf{Q}$ is algebraic over $\mathbf{Q}$, then u^r is transcendental over $\mathbf{Q}$. You may use this result to show that the following numbers are transcendental over $\mathbf{Q}$:
   (a) $\sqrt[3]{7}^{\sqrt{5}}$
   (b) $\sqrt[3]{7}^{\sqrt{5}} + 7$

6.2 FINITE AND ALGEBRAIC EXTENSIONS

If F is an extension field of K, then the multiplication of F defines a scalar multiplication, considering the elements of K as scalars and the elements of F as vectors. This gives F the structure of a vector space over K, and allows us to make use of the concept of the dimension of a vector space. (If you need to review results on dimension, see Appendix G at the end of this chapter.)

6.2.1 Proposition. Let F be an extension field of K and let $u \in F$ be an element algebraic over K. If the minimal polynomial of u over K has degree n, then $K(u)$ is an n-dimensional vector space over K.

Proof. Let $p(x) = c_0 + c_1 x + \ldots + c_n x^n$ be the minimal polynomial of u over K. We will show that the set $B = \{1, u, u^2, \ldots, u^{n-1}\}$ is a basis for $K(u)$ over K. By Proposition 6.1.6, $K(u) \cong K[x]/\langle p(x) \rangle$, and since each coset of $K[x]/\langle p(x) \rangle$ contains a unique representative of degree less than n, it follows from this isomorphism that each element of $K(u)$ can be represented uniquely in the form $a_0 1 + a_1 u + \ldots + a_{n-1} u^{n-1}$. Thus B spans $K(u)$, and the uniqueness of representations implies that B is also a linearly independent set of vectors. $\square$

6.2.2 Definition. Let F be an extension field of K. The dimension of F as a vector space over K is called the *degree* of F over K, denoted by $[F : K]$. If the dimension of F over K is finite, then F is said to be a *finite* extension of K.

In the next proposition, by using the notion of the degree of an extension, we are able to give a useful characterization of algebraic elements. The proposition implies, in particular, that every element of a finite extension must be algebraic.

6.2.3 Proposition.   Let F be an extension field of K and let $u \in F$. The following conditions are equivalent:

(1) u is algebraic over K;

(2) $K(u)$ is a finite extension of K;

(3) u belongs to a finite extension of K.

Proof. It is clear that (1) implies (2) and (2) implies (3). To prove (3) implies (1), suppose that $u \in E$ for an extension E with $K \subseteq E \subseteq F$ and $[E : K] = n$. The set $1, u, u^2, \ldots, u^n$ contains $n + 1$ elements, and these cannot be linearly independent in an n-dimensional vector space. Thus there exists a relation $a_0 + a_1 u + \ldots + a_n u^n = 0$ with scalars $a_i \in K$ that are not all zero. This shows that u is a root of a polynomial in $K[x]$. □

Counting arguments often provide very useful tools. In case we have extension fields $K \subseteq E \subseteq F$, we can consider the degree of E over K and the degree of F over E. The next theorem shows that there is a very simple relationship between these two degrees and the degree of F over K. Theorem 6.2.4 will play a very important role in our study of extension fields.

6.2.4 Theorem.   Let E be a finite extension of K and let F be a finite extension of E. Then F is a finite extension of K, and

$$[F : K] = [F : E][E : K].$$

Proof. Let $[F : E] = n$ and let $[E : K] = m$. Let $u_1, u_2, \ldots, u_n$ be a basis for F over E and let $v_1, v_2, \ldots, v_m$ be a basis for E over K. We claim that the set B of nm products $u_i v_j$ (where $1 \le i \le n$ and $1 \le j \le m$) is a basis for F over K.

We must first show that B spans F over K. If u is any element of F, then $u = \sum_{i=1}^n a_i u_i$ for elements $a_i \in E$. For each element a_i we have $a_i = \sum_{j=1}^m c_{ij} v_j$, where $c_{ij} \in K$. Substituting gives $u = \sum_{i=1}^n \sum_{j=1}^m c_{ij} v_j u_i$, and so B spans F over K.

To show that B is a linearly independent set, suppose that $\sum_{i,j} c_{ij} v_j u_i = 0$ for some linear combination of the elements of B, with coefficients in K. This expression can be written as $\sum_{i=1}^n (\sum_{j=1}^m c_{ij} v_j) u_i$. Since the elements $u_1, u_2, \ldots, u_n$ form a basis for F over E, each of the coefficients $\sum_{j=1}^m c_{ij} v_j$ (which belong to E) must be zero. Then since the elements $v_1, v_2, \ldots, v_m$ form a basis for E over K, for each i we must have $c_{ij} = 0$ for all j. This completes the proof. □

Example 6.2.1

In the previous section we showed that $\sqrt{2} + \sqrt{3}$ has degree 4 over $\mathbf{Q}$ by showing that it has the minimal polynomial $x^4 - 10x^2 + 1$. The previous theorem can be used to give an alternate proof, using the fact that $\mathbf{Q}(\sqrt{2} + \sqrt{3}) = \mathbf{Q}(\sqrt{2}, \sqrt{3})$. To show that the two field extensions are equal, we first observe that $\mathbf{Q}(\sqrt{2} + \sqrt{3}) \subseteq \mathbf{Q}(\sqrt{2}, \sqrt{3})$, since we have $\sqrt{2} + \sqrt{3} \in \mathbf{Q}(\sqrt{2}, \sqrt{3})$ and the field $\mathbf{Q}(\sqrt{2} + \sqrt{3})$ is defined by the property that it contains $\sqrt{2} + \sqrt{3}$ and is contained in any extension field that contains $\mathbf{Q}$ and $\sqrt{2} + \sqrt{3}$. On the other hand, $(\sqrt{3} - \sqrt{2})(\sqrt{3} + \sqrt{2}) = 1$, and so $\sqrt{3} - \sqrt{2} \in \mathbf{Q}(\sqrt{3} + \sqrt{2})$ since it is the multiplicative inverse of $\sqrt{3} + \sqrt{2}$. Because $\sqrt{3} = ((\sqrt{3} + \sqrt{2}) + (\sqrt{3} - \sqrt{2}))/2$, it follows that $\sqrt{3}$, $\sqrt{2} \in \mathbf{Q}(\sqrt{2} + \sqrt{3})$, and so we have the reverse inclusion $\mathbf{Q}(\sqrt{2}, \sqrt{3}) \subseteq \mathbf{Q}(\sqrt{2} + \sqrt{3})$.

The minimal polynomial of $\sqrt{2}$ over $\mathbf{Q}$ is $x^2 - 2$, which is irreducible, and so $\sqrt{2}$ has degree 2 over $\mathbf{Q}$. To compute $[\mathbf{Q}(\sqrt{2}, \sqrt{3}) : \mathbf{Q}(\sqrt{2})]$, we need to show that $x^2 - 3$ is irreducible over $\mathbf{Q}(\sqrt{2})$, in which case it will be the minimal polynomial of $\sqrt{3}$ over $\mathbf{Q}(\sqrt{2})$. The roots $\pm\sqrt{3}$ of the polynomial do not belong to $\mathbf{Q}(\sqrt{2})$, because they cannot be written in the form $a + b\sqrt{2}$, for $a, b \in \mathbf{Q}$. Thus

$$[\mathbf{Q}(\sqrt{2} + \sqrt{3}) : \mathbf{Q}] = [\mathbf{Q}(\sqrt{2}, \sqrt{3}) : \mathbf{Q}(\sqrt{2})][\mathbf{Q}(\sqrt{2}) : \mathbf{Q}] = 4.$$

This argument using degrees shows that any monic polynomial of degree 4 that has $\sqrt{2} + \sqrt{3}$ as a root must be its minimal polynomial. $\square$

6.2.5 Corollary.   Let F be a finite extension of K. Then the degree of any element of F is a divisor of $[F : K]$.

Proof. If $u \in F$, then $[F : K] = [F : K(u)][K(u) : K]$. $\square$

Example 6.2.2

The field $\mathbf{Q}(\sqrt[3]{2})$ does not contain $\sqrt{2}$, since the degree of $\sqrt{2}$ over $\mathbf{Q}$ is not a divisor of $[\mathbf{Q}(\sqrt[3]{2}) : \mathbf{Q}] = 3$. $\square$

6.2.6 Corollary.   Let F be an extension field of K, with algebraic elements $u_1, u_2, \ldots, u_n \in F$. Then the degree of $K(u_1, u_2, \ldots, u_n)$ over K is at most the product of the degrees of u_i over K, for $1 \leq i \leq n$.

Proof. We give a proof by induction on n. By Proposition 6.2.1, the result is true for $n = 1$. If the result is assumed to be true for the case $n - 1$, then let $E = K(u_1, u_2, \ldots, u_{n-1})$. Since

$$K(u_1, u_2, \ldots, u_{n-1}, u_n) = E(u_n),$$

the desired conclusion will follow from the equality

$$[E(u_n) : K] = [E(u_n) : E][E : K],$$

if we can show that $[E(u_n) : E]$ is less than the degree of u_n over K. Now the minimal polynomial of u_n over K is in particular a polynomial over E, and thus the minimal polynomial of u_n over E must be a divisor of it. Applying Proposition 6.2.1 completes the proof. $\square$

6.2.7 Corollary.   Let F be an extension field of K. The set of all elements of F that are algebraic over K forms a subfield of F.

Proof. If u, v are algebraic elements of F, then $K(u, v)$ is a finite extension of K by Theorem 6.2.4. Since $u + v$, $u - v$, and uv all belong to $K(u, v)$, these elements are algebraic by Proposition 6.2.3. The same argument applies to u/v, if $v \neq 0$. □

6.2.8 Definition.   An extension field F of K is said to be *algebraic* over K if each element of F is algebraic over K.

Example 6.2.3

Let $\overline{\mathbf{Q}}$ be the set of complex numbers $u \in \mathbf{C}$ such that u is algebraic over $\mathbf{Q}$. Then $\overline{\mathbf{Q}}$ is a subfield of $\mathbf{C}$ by Corollary 6.2.7, called the *field of algebraic numbers*. We note that $\overline{\mathbf{Q}}$ is an algebraic extension of $\mathbf{Q}$, but not a finite extension. We showed in Corollary 4.3.6 that for any prime p the polynomial $1 + x + \ldots + x^{p-2} + x^{p-1}$ is irreducible over $\mathbf{Q}$. The roots of this polynomial exist in $\mathbf{C}$ (they are the primitive pth roots of unity) and thus $\overline{\mathbf{Q}}$ contains algebraic numbers of arbitrarily large degree over $\mathbf{Q}$, which shows that it cannot have finite degree over $\mathbf{Q}$.   □

6.2.9 Proposition.   Let F be an algebraic extension of E and let E be an algebraic extension of K. Then F is an algebraic extension of K.

Proof. If F is algebraic over E, then any element $u \in F$ must satisfy some polynomial $f(x) = a_0 + a_1 x + \ldots a_n x^n$ over E. Since each element a_i, $0 \leq i \leq n$ is algebraic over K, it follows that $K(a_0, a_1, \ldots, a_n, u)$ is a finite extension of K, and so u is algebraic over K by Proposition 6.2.3.   □

EXERCISES: SECTION 6.2

1. Find the degree and a basis for each of the given field extensions:
   - (a) $\mathbf{Q}(\sqrt{3})$ over $\mathbf{Q}$
   - (b) $\mathbf{Q}(\sqrt{3}, \sqrt{7})$ over $\mathbf{Q}$
   - (c) $\mathbf{Q}(\sqrt{2}, \sqrt[3]{2})$ over $\mathbf{Q}$
   - (d) $\mathbf{Q}(\sqrt{3} + \sqrt{7})$ over $\mathbf{Q}$
   - (e) $\mathbf{Q}(\sqrt{2} + \sqrt[3]{2})$ over $\mathbf{Q}$
   - (f) $\mathbf{Q}(\sqrt{3}, \sqrt{21})$ over $\mathbf{Q}(\sqrt{7})$
   - (g) $\mathbf{Q}(\sqrt{3} + \sqrt{7})$ over $\mathbf{Q}(\sqrt{7})$
   - (h) $\mathbf{Q}(\sqrt{3}, \sqrt{7})$ over $\mathbf{Q}(\sqrt{3} + \sqrt{7})$
   - (i) $\mathbf{Q}(\omega)$ over $\mathbf{Q}$, where $\omega = (-1 + \sqrt{3}i)/2$

2. Let F be a finite extension of K such that $[F : K] = p$, a prime number. If $u \in F$ but $u \notin K$, show that $F = K(u)$.

3. Let $F \supseteq K$ be fields, and let R be a ring such that $F \supseteq R \supseteq K$. If F is an algebraic extension of K, show that R is a field. What happens if we do not assume that F is algebraic over K?

4. Determine $[\mathbf{Q}(\sqrt{n}) : \mathbf{Q}]$ for all $n \in \mathbf{Z}^+$.

5. For any positive integers a, b, show that $\mathbf{Q}(\sqrt{a} + \sqrt{b}) = \mathbf{Q}(\sqrt{a}, \sqrt{b})$.

6. Let F be an algebraic extension of K, and let S be a subset of F such that $S \supseteq K$, S is a vector space over K, and $s^n \in S$ for all $s \in S$ and all positive integers n. Prove that if char$(K) \neq 2$, then S is a subfield of F.

(The result is false in characteristic 2, and all of the tools necessary to construct a counterexample are at hand, but the counterexample is not an easy one.)

6.3 GEOMETRIC CONSTRUCTIONS

In this section we will exploit what we have learned about the degree of a finite extension defined in successive steps. We will show the impossibility of several geometric constructions which were first investigated by the ancient Greeks. One of the most elementary constructions taught in high-school geometry is how to bisect an angle. In this section, the word "construction" will be assumed to mean a geometric construction using only a compass and straightedge, not a ruler that can measure arbitrary lengths. But there is no general method for trisecting an angle. We will show this by proving that a 20° angle cannot be constructed, and so it is impossible for any general method to successfully trisect a 60° angle. Secondly, given a circle, it is impossible (in general) to construct a square with the same area as that of the circle. (This is known as "squaring the circle.") Finally, given a cube, it is not generally possible to construct a cube with double its original volume.

We will take as given a line segment that will be defined to be one unit in length. Using this line segment, lengths corresponding to all positive rational numbers can be constructed. Since the constructions we will allow must involve only a straightedge and compass, they will be limited to the following: (i) constructing a line through two points whose coordinates are known, (ii) constructing a circle with known radius and center at a point with known coordinates, and (iii) finding the points of intersection of given lines and circles.

6.3.1 Definition.   The real number a is said to be a *constructible number* if it is possible to construct a line segment of length $|a|$ by using only a straightedge and compass.

Note that the next proposition implies that all rational numbers are constructible.

6.3.2 Proposition.   The set of all constructible real numbers is a subfield of the field of all real numbers.

Proof. Let a, b be constructible real numbers, which we may assume to be positive. We must show that $a \pm b$ and ab are constructible, and that a/b is constructible, provided $b \neq 0$. Assuming $a > b$, it is obvious how to construct $a + b$ and $a - b$ on any given straight line. Furthermore, given positive constructible numbers y, z, w, by choosing any angle α we can easily construct a triangle with two sides of length z and w, as in Figure 6.3.1 below. Using the length y, we can construct a line parallel to the third side of the triangle, giving us two similar triangles with sides of length x, y, z, w that satisfy the relation $x/y = z/w$. (The diagram presumes that $y > w$, and can easily be modified if not.) To construct $x = ab$, choose $y = a$, $z = b$, and $w = 1$. To construct $x = a/b$, choose $y = 1$, $z = a$, and $w = b$.   □

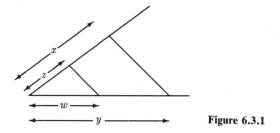

Figure 6.3.1

We now need further information about how a real number can actually be constructed. We will obtain it by considering intermediate extension fields between **Q** and the field of all constructible numbers.

6.3.3 Definition.   Let F be a subfield of **R**. The set of all points (x, y) in the Euclidean plane $\mathbf{R}^2$ such that $x, y \in F$ is called the *plane* of F. A straight line with an equation of the form $ax + by + c = 0$, for elements $a, b, c \in F$, is called a *line in F*. Any circle with an equation of the form $x^2 + y^2 + ax + by + c = 0$, for elements $a, b, c \in F$, is called a *circle in F*.

6.3.4 Lemma.   Let F be a subfield of **R**.
 (a) Any straight line joining two points in the plane of F is a line in F.
 (b) Any circle with its radius in F and its center in the plane of F is a circle in F.

Proof. (a) Let (a_1, b_1) and (a_2, b_2) be points in the plane of F. The case $a_1 = a_2$ is easily taken care of. Otherwise, the two-point form

$$y - b_1 = \frac{b_2 - b_1}{a_2 - a_1}(x - a_1)$$

of the equation of a line determines an equation of the form we need.
    (b) If $r \in F$ and (a, b) is in the plane of F, then the equation of the circle with radius r and center (a, b) is

$$(x - a)^2 + (y - b)^2 = r^2,$$

and expanding it gives the form we need, since F is a field.   □

6.3.5 Lemma. The points of intersection of lines in F and circles in F lie in the plane of $F(\sqrt{u})$, for some $u \in F$.

Proof. Given two lines in F, we can find their point of intersection by using elementary row operations on the associated matrix. Since we use only field operations, the point of intersection must still lie in the plane of F. Given two circles in F, subtracting one equation from the other reduces the question of their points of intersection to the question of the points of intersection of a line in F and a circle in F. To complete the proof, assume that we are given the equation of a line in F and the equation of a circle in F. In the equation of the line, one of x or y must have a nonzero coefficient, say y. Then we can solve for y and substitute into the equation of the given circle, yielding a quadratic equation in x. Unless the line and circle have no intersection, consider the solution produced by using the quadratic formula. Since F is closed under the operations of $\mathbf{R}$, each number involved in the solution, including the term u under the radical, is again an element of F. Substituting back into the equation of the line shows that the points of intersection lie in $F(\sqrt{u})$. $\square$

6.3.6 Theorem. The real number u is constructible if and only if there exists a finite set $u_1, u_2, \ldots, u_n$ of real numbers such that
  (i) $u_1^2 \in \mathbf{Q}$,
  (ii) $u_i^2 \in \mathbf{Q}(u_1, \ldots, u_{i-1})$, for $i = 2, \ldots, n$, and
  (iii) $u \in \mathbf{Q}(u_1, \ldots, u_n)$.

Proof. If u is constructible, then the construction can be done in a finite number of steps. Starting with $\mathbf{Q}$, the first step must consist of finding an intersection of lines or circles in $\mathbf{Q}$, so either this can be done in $\mathbf{Q}$, or else by Lemma 6.3.5 we obtain an extension of the form $\mathbf{Q}(\sqrt{v_1})$, for some $v_1 \in \mathbf{Q}$. We may assume that the next step in the construction involves lines and circles in $\mathbf{Q}(\sqrt{v_1})$, and so the points we obtain are in the plane of $\mathbf{Q}(\sqrt{v_1}, \sqrt{v_2})$ for some $v_2 \in \mathbf{Q}(\sqrt{v_1})$. Continuing in this manner allows us to obtain u as an element of a field of the required form $\mathbf{Q}(u_1, u_2, \ldots, u_n)$.

To show the converse, it suffices to show that if F is any subfield of the field of constructible numbers, then $\sqrt{u}$ is constructible for all $u \in F$, since this implies that every element of $F(\sqrt{u})$ is constructible. Given $u \in F$, we can construct a circle of diameter $1 + u$. Then, as in Figure 6.3.2, we can construct a perpendicular line at a distance of 1 from one end of the diameter. If x is the length on this line between the diameter and the intersection with the circle, then we have constructed similar triangles that yield the proportion $x/1 = u/x$. Thus $x = \sqrt{u}$, and the proof is complete. $\square$

6.3.7 Corollary. If u is a constructible real number, then u is algebraic over $\mathbf{Q}$, and the degree of its minimal polynomial over $\mathbf{Q}$ is a power of 2.

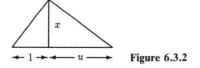

Figure 6.3.2

Proof. Assume that u is a constructible real number. Then by Theorem 6.3.6, u belongs to a field $F \subseteq \mathbf{R}$ of the form $F = \mathbf{Q}(u_1, u_2, \ldots u_n)$, where (i) $u_1^2 \in \mathbf{Q}$ and (ii) $u_i^2 \in \mathbf{Q}(u_1, \ldots, u_{i-1})$, for $i = 2, \ldots, n$. By Theorem 6.2.4, the degree of F over $\mathbf{Q}$ is a power of 2, since $[\mathbf{Q}(u_1) : \mathbf{Q}] \leq 2$ and

$$[\mathbf{Q}(u_1, \ldots, u_i) : \mathbf{Q}(u_1, \ldots, u_{i-1})] \leq 2,$$

for $1 < i \leq n$. The desired conclusion also follows from Theorem 6.2.4, since

$$[F : \mathbf{Q}] = [F : \mathbf{Q}(u)][\mathbf{Q}(u) : \mathbf{Q}],$$

and thus $[\mathbf{Q}(u) : \mathbf{Q}]$ is a divisor of a power of 2. $\square$

6.3.8 Theorem.   It is impossible to find a general construction for trisecting an angle, duplicating a cube, or squaring a circle.

Proof. If a 60° angle could be trisected, then it would be possible to construct a 20° angle, and so $u = \cos(20°)$ would be a constructible real number. We have the following trigonometric identity:

$$\begin{aligned}
\cos(3\theta) &= \cos(2\theta + \theta) \\
&= \cos(2\theta) \cos \theta - \sin(2\theta) \sin \theta \\
&= (2 \cos^2 \theta - 1) \cos \theta - (2 \sin \theta \cos \theta) \sin \theta \\
&= 2 \cos^3 \theta - \cos \theta - 2 \cos \theta + 2 \cos^3 \theta \\
&= 4 \cos^3 \theta - 3 \cos \theta
\end{aligned}$$

For $\theta = 20°$ we have $\cos 3\theta = 1/2$ and multiplying by 2 shows that u is a root of the polynomial $8x^3 - 6x - 1$. Using Proposition 4.2.1 it is easy to check that this polynomial has no rational roots, so it is irreducible over $\mathbf{Q}$ and therefore is the minimal polynomial of u over $\mathbf{Q}$. Since the degree of u over $\mathbf{Q}$ is not a power of 2, it cannot be constructible.

To construct a cube with double the volume of the unit cube requires the construction of a cube of volume 2. This requires constructing $\sqrt[3]{2}$, which is impossible since $\sqrt[3]{2}$ has degree 3 over $\mathbf{Q}$.

Finally, constructing a square with the same area as a circle of radius 1 requires that $\sqrt{\pi}$ be constructible. This is not true since π is not algebraic over $\mathbf{Q}$. For a proof of this fact we refer the student to a book by I. Niven called *Irrational Numbers*. Thus we have (almost) completed the proof. $\square$

EXERCISES: SECTION 6.3

1. In this exercise we outline how to construct a regular pentagon. Let $\zeta = \cos(2\pi/5) + i \sin(2\pi/5)$.
   (a) Show that ζ is a primitive fifth root of unity.
   (b) Show that $(\zeta + \zeta^{-1})^2 + (\zeta + \zeta^{-1}) - 1 = 0$.
   (c) Show that $\zeta + \zeta^{-1} = (-1 + \sqrt{5})/2$.
   (d) Show that $\cos(2\pi/5) = (-1 + \sqrt{5})/4$ and that $\sin(2\pi/5) = (\sqrt{10 + 2\sqrt{5}})/4$.
   (e) Conclude that a regular pentagon is constructible.
2. Prove that a regular heptagon is not constructible.
   Hint: Let $\zeta = \cos(2\pi/7) + i \sin(2\pi/7)$. Show that $[\mathbf{Q}(\zeta) : \mathbf{Q}]$ is not a power of 2.

6.4 SPLITTING FIELDS

We will be interested, ultimately, in the question of determining when a given equation is solvable by radicals. The answer to this question involves a comparison between the field generated by the coefficients of the polynomial and the field generated by the roots of the polynomial. This comparison must be done in some field that contains all roots of the polynomial. Over such a field the given polynomial can be factored (or "split") into a product of linear factors. Our task in this section is to study the existence and uniqueness of such fields. Recall that we have seen in Section 6.1 that given any field and any polynomial over that field, there exists an extension field in which the polynomial has a root. Now we simply need to iterate this procedure, to obtain all roots of the polynomial.

6.4.1 Definition. Let K be a field and let $f(x) = a_0 + a_1 x + \ldots + a_n x^n$ be a polynomial in $K[x]$ of degree $n > 0$. An extension field F of K is called a *splitting field for $f(x)$ over K* if there exist elements $r_1, r_2, \ldots, r_n \in F$ such that (i) $f(x) = a_n(x - r_1)(x - r_2) \cdots (x - r_n)$ and (ii) $F = K(r_1, r_2, \ldots, r_n)$.

In the above situation we usually say that $f(x)$ *splits* over tne field F. The elements $r_1, r_2, \ldots, r_n$ are roots of $f(x)$, and so F is obtained by adjoining to K a complete set of roots of $f(x)$.

6.4.2 Theorem. Let $f(x) \in K[x]$ be a polynomial of degree $n > 0$. Then there exists a splitting field F for $f(x)$ over K, with $[F : K] \leq n!$.

Proof. The proof is by induction on the degree of $f(x)$. If $\deg(f(x)) = 1$, then K itself is a splitting field. Assume that $\deg(f(x)) = n$ and that the theorem is true for any polynomial $g(x)$ with $\deg(g(x)) < n$ over any field K. Let $p(x)$ be an irreducible factor of $f(x)$. By Proposition 6.1.7 there exists an extension field E of K in which $p(x)$ has a root r. We now consider the field $K(r)$. Over this field $f(x)$ factors as

$$f(x) = p(x)q(x) = (x - r)g(x)$$

for some polynomial $g(x)$ of degree $n - 1$. Thus by the induction hypothesis there exists a splitting field F of $g(x)$ over $K(r)$, say $F = K(r)(r_1, r_2, \ldots, r_{n-1})$, with $[F : K(r)] \leq (n - 1)!$. Then $F = K(r, r_1, r_2, \ldots, r_{n-1})$ and it is clear that $f(x)$ splits over F. Finally,

$$[F : K] = [F : K(r)][K(r) : K] \leq (n - 1)!n = n!. \quad \square$$

Example 6.4.1

The splitting field of a polynomial in $K[x]$ depends on the field K. For example, considering $x^2 + 1$ as a polynomial with real coefficients means that we obtain $\mathbf{R}(i) = \mathbf{C}$ as a splitting field. On the other hand, if we view it as a polynomial with rational coefficients, then to obtain a splitting field we only need to adjoin i to $\mathbf{Q}$, and so a splitting field of $x^2 + 1$ over $\mathbf{Q}$ is $\mathbf{Q}(i)$. $\square$

Example 6.4.2

In this example we will construct a splitting field for the polynomial $x^3 - 2$ over $\mathbf{Q}$. We must adjoin to $\mathbf{Q}$ all solutions of the equation $x^3 = 2$. It is obvious that $x = \sqrt[3]{2}$ is a solution. But in addition, $\omega\sqrt[3]{2}$ is also a solution for any complex number ω such that $\omega^3 = 1$. Since we have

$$x^3 - 1 = (x - 1)(x^2 + x + 1)$$

we can let ω be a root of $x^2 + x + 1$; that is

$$\omega = \frac{-1 + \sqrt{-3}}{2} = -\frac{1}{2} + \frac{\sqrt{3}}{2} i.$$

Then $\sqrt[3]{2}$, $\omega\sqrt[3]{2}$, and $\omega^2\sqrt[3]{2}$ are the roots of $x^3 - 2$. Thus we obtain the splitting field as $\mathbf{Q}(\sqrt[3]{2}, \omega)$. The degree of the splitting field is 6 since $x^3 - 2$ is irreducible over $\mathbf{Q}$ and thus $[\mathbf{Q}(\sqrt[3]{2}) : \mathbf{Q}] = 3$. The polynomial $x^2 + x + 1$ is irreducible over $\mathbf{Q}$ and stays irreducible over $\mathbf{Q}(\sqrt[3]{2})$ since it has no root in that field. The degree of $x^2 + x + 1$ is not a divisor of $[\mathbf{Q}(\sqrt[3]{2}) : \mathbf{Q}]$. Since ω is a root of $x^2 + x + 1$, this implies that $[\mathbf{Q}(\sqrt[3]{2}, \omega) : \mathbf{Q}(\sqrt[3]{2})] = 2$. $\square$

Example 6.4.3

Our standard construction for a splitting field of the polynomial $x^2 + 1$ over $\mathbf{R}$ is to consider $\mathbf{R}[x]/\langle x^2 + 1 \rangle$. Then the field $\mathbf{R}$ is identified with the cosets determined by the constant polynomials. Another familiar construction is to simply use ordered pairs of the form $a + bi$, where $a, b \in \mathbf{R}$ and the "imaginary" number i is a square root of -1. This can be done rigorously by using the set of all ordered pairs (a,b) such that $a, b \in \mathbf{R}$, with componentwise addition and multiplication $(a,b) \cdot (c,d) = (ac - bd, ad + bc)$. In this case the field $\mathbf{R}$ is identified with ordered pairs of the form $(a,0)$.

Another construction of this splitting field uses the set of 2×2 matrices over **R**. We first identify **R** with the set of scalar matrices. Let

$$F = \left\{ \begin{bmatrix} a & b \\ -b & a \end{bmatrix} \mid a, b \in \mathbf{R} \right\}.$$

The function $\phi : \mathbf{C} \to F$ defined by

$$\phi(a + bi) = \begin{bmatrix} a & b \\ -b & a \end{bmatrix}$$

can be shown to be an isomorphism. We have adjoined to the set of scalar matrices the matrix $\begin{bmatrix} 0 & 1 \\ -1 & 0 \end{bmatrix}$, which satisfies the polynomial $x^2 + 1$.   □

Example 6.4.4

The technique of the previous example can be extended to give another proof of Proposition 6.1.7, which guarantees the existence of roots. If $p(x)$ is a monic irreducible polynomial of degree n over the field K, we first identify K with a field of $n \times n$ scalar matrices over K. There exists an $n \times n$ matrix C with $p(C) = 0$, called the *companion matrix* of the polynomial. (Refer to the exercises in this section for further details.) The set of all matrices of the form $a_0 I + a_1 C + \ldots + a_{n-1} C^{n-1}$, such that $a_0, a_1, \ldots, a_{n-1} \in K$, then defines an extension field F of K in which $p(x)$ has a root.   □

The previous examples show that splitting fields can be constructed in a variety of ways. One would hope that there would be some uniqueness involved. In fact, the next results show that splitting fields are unique up to isomorphism. This provides the necessary basis for our study of solvability by radicals. We will also use this fact in the next section to show that any two finite fields with the same number of elements must be isomorphic.

6.4.3 Lemma.   Let $\phi : K \to L$ be an isomorphism of fields. Let F be an extension field of K such that $F = K(u)$ for an algebraic element $u \in F$. Let $p(x)$ be the minimal polynomial of u over K. If v is any root of the image $q(x)$ of $p(x)$ under ϕ, and $E = L(v)$, then there is a unique way to extend ϕ to an isomorphism $\theta : F \to E$ such that $\theta(u) = v$ and $\theta(a) = \phi(a)$ for all $a \in K$.

$$
\begin{array}{ccc}
 & \theta & \\
K(u) & \to & L(v) \\
| & & | \\
K & \to & L \\
 & \phi &
\end{array}
$$

Proof. If $p(x)$ has degree n, then elements of $K(u)$ have the form $a_0 + a_1u + \ldots + a_{n-1}u^{n-1}$ for elements $a_0, a_1, \ldots, a_{n-1} \in K$. The required isomorphism $\theta : K(u) \to L(v)$ must have the form

$$\theta(a_0 + a_1u + \ldots + a_{n-1}u^{n-1}) = \phi(a_0) + \phi(a_1)v + \ldots + \phi(a_{n-1})v^{n-1}.$$

We could simply show by direct computation that this function is an isomorphism. However, it seems to be easier to show that θ is a composition of functions that we already know to be isomorphisms.

As in Example 5.2.5, the isomorphism ϕ can be extended uniquely to a ring isomorphism $\phi' : K[x] \to L[x]$. By assumption, $\phi'(p(x)) = q(x)$, and so ϕ' maps the ideal $\langle p(x) \rangle$ generated by $p(x)$ onto the ideal $\langle q(x) \rangle$ generated by $q(x)$. Example 5.3.5 shows that ϕ' induces an isomorphism $\theta' : K[x]/\langle p(x) \rangle \to L[x]/\langle q(x) \rangle$. Let $\eta : K[x]/\langle p(x) \rangle \to K(u)$ and $\varepsilon : L[x]/\langle q(x) \rangle \to L(v)$ be the isomorphisms defined by evaluation at u and v, respectively. Then $\theta = \varepsilon\theta'\eta^{-1}$ defines the required isomorphism from $K(u)$ onto $L(v)$. $\square$

6.4.4 Lemma.   Let F be a splitting field for the polynomial $f(x) \in K[x]$. If $\phi : K \to L$ is a field isomorphism that maps $f(x)$ to $g(x) \in L[x]$ and E is a splitting field for $g(x)$ over L, then there exists an isomorphism $\theta : F \to E$ such that $\theta(a) = \phi(a)$ for all $a \in K$.

Proof. The proof uses induction on the degree of $f(x)$. If $f(x)$ has degree zero or one, then $F = K$ and $E = L$, so there is nothing to prove. We now assume that the result holds for all polynomials of degree less than n and for all fields K. Let $p(x)$ be an irreducible factor of $f(x)$, which maps to the irreducible factor $q(x)$ of $g(x)$. All roots of $p(x)$ belong to F, so we may choose one, say u, which gives $K \subseteq K(u) \subseteq F$. Similarly, we may choose a root v of $q(x)$ in E, which gives $L \subseteq L(v) \subseteq E$. By Lemma 6.4.3 there exists an isomorphism $\phi' : K(u) \to L(v)$ such that $\phi'(u) = v$ and $\phi'(a) = \phi(a)$ for all $a \in K$. If we write $f(x) = (x - u)s(x)$ and $g(x) = (x - v)t(x)$, then the polynomial $s(x)$ has degree less than n, the extension F is a splitting field for $s(x)$ over $K(u)$, the polynomial $s(x)$ is mapped by ϕ to $t(x)$, and the extension E is a splitting field for $t(x)$ over $L(v)$. Thus the induction hypothesis can be applied, and so there exists an isomorphism $\theta : F \to E$ such that $\theta(w) = \phi'(w)$ for all $w \in K(u)$. In particular, $\theta(a) = \phi'(a) = \phi(a)$ for all $a \in K$, and the proof is complete. $\square$

The following theorem on the uniqueness of splitting fields is a special case of Lemma 6.4.4. In the induction argument in the proof of Lemma 6.4.4 we needed to change the base field. That made it necesary to use as an induction hypothesis the more general statement of that lemma.

6.4.5 Theorem.   Let $f(x)$ be a polynomial over the field K. The splitting field of $f(x)$ over K is unique up to isomorphism.

EXERCISES: SECTION 6.4

1. Find the splitting fields over $\mathbf{Q}$ for the following polynomials:
   (a) $x^2 - 2$
   (b) $x^2 + 3$
   (c) $x^4 + x^2 - 6$
   (d) $x^3 - 5$
   (e) $x^3 + 3x^2 + 3x - 4$
   (f) $x^3 - 1$
   (g) $x^4 - 1$

2. Let p be a prime number. Find the splitting field over $\mathbf{Q}$ for $x^p - 1$.

3. Find the splitting fields over $\mathbf{Z}_2$ for the following polynomials:
   (a) $x^2 + x + 1$
   (b) $x^2 + 1$
   (c) $x^3 + x + 1$
   (d) $x^3 + x^2 + 1$

4. Find the splitting field for $x^p - x$ over $\mathbf{Z}_p$.

5. Show that if F is an extension field of K of degree 2, then F is the splitting field over K for some polynomial.

6. For a monic polynomial $f(x) = a_0 + a_1x + \ldots + a_{n-1}x^{n-1} + x^n$, the following matrix C is called the companion matrix of $f(x)$:

$$\begin{bmatrix} 0 & 1 & 0 & \ldots & 0 \\ 0 & 0 & 1 & \ldots & 0 \\ 0 & 0 & 0 & \ldots & 0 \\ \vdots & \vdots & \vdots & & \vdots \\ 0 & 0 & 0 & \ldots & 1 \\ -a_0 & -a_1 & -a_2 & \ldots & -a_{n-1} \end{bmatrix}.$$

It is true in general that $f(C) = 0$. Do this computation for the case in which $\deg(f(x)) = 3$.

6.5 FINITE FIELDS

We first met finite fields in Chapter 1, in studying $\mathbf{Z}_p$, where p is a prime number. With the field theory we have developed it is now possible to give a complete description of the structure of all finite fields.

6.5.1 Proposition.   Let F be a finite field of characteristic p. Then F has p^n elements, for some positive integer n.

Proof. Recall that if F has characteristic p, then the ring homomorphism $\phi : \mathbf{Z} \to F$ defined by $\phi(n) = n \cdot 1$ for all $n \in \mathbf{Z}$ has kernel $p\mathbf{Z}$, and thus the image of ϕ is a subfield K of F isomorphic to $\mathbf{Z}_p$. Since F is finite, it must certainly have finite dimension as a vector space over K, say $[F : K] = n$. If $v_1, v_2, \ldots, v_n$ is a basis for F over K, then each element of F has the form $a_1v_1 + a_2v_2 + \ldots + a_nv_n$ for elements $a_1, a_2, \ldots, a_n \in K$. Thus to define an element of F there are n coefficients a_i, and for each coefficient there are p choices, since K has only p elements. Therefore the total number of ways in which an element in F can be defined is p^n. $\square$

If F is any field, then the smallest subfield of F that contains the identity element 1 is called the *prime subfield* of F. As noted above, if F is a finite field, then its prime subfield is isomorphic to $\mathbf{Z}_p$, where $p = \mathrm{char}(F)$ for some prime p.

6.5.2 Theorem.   Let F be a finite field with p^n elements. Then F is the splitting field of the polynomial $x^{p^n} - x$ over the prime subfield of F.

Proof. Let F be a finite field of characteristic p. Then as in Proposition 6.5.1, F is an extension of degree n of its prime subfield K, which is isomorphic to $\mathbf{Z}_p$. Since F has p^n elements, the order of the multiplicative group $F^\times$ of nonzero elements of F is $p^n - 1$. Therefore $x^{p^n-1} = 1$ for all $0 \neq x \in F$, and so $x^{p^n} = x$ for all $x \in F$. The polynomial $f(x) = x^{p^n} - x$ has at most p^n roots, and so its roots must be precisely the elements of F. Thus F is the splitting field of $f(x)$ over K. $\square$

Example 6.5.1

Let $p > 2$ be a prime number. The field $\mathbf{Z}_p$ is the splitting field of $x^p - x$, so we have

$$x^p - x = x(x - 1)(x - 2) \cdots (x - (p - 1)).$$

Thus

$$x^{p-1} - 1 = (x - 1)(x - 2) \cdots (x - (p - 1)),$$

and substituting $x = 0$ gives

$$-1 = (-1)(-2) \cdots (-(p - 1)).$$

There are an even number of factors on the right hand side, so the minus signs cancel out, leaving

$$(p - 1)! \equiv -1 \pmod{p},$$

which is Wilson's theorem. (See Exercise 24 of Section 1.4.) $\square$

6.5.3 Corollary.   Two finite fields are isomorphic if and only if they have the same number of elements.

Proof. Let F and E be finite field with p^n elements, containing prime subfields K and L, respectively. Then $K \cong \mathbf{Z}_p \cong L$ and so $F \cong E$ by Lemma 6.4.4, since by Theorem 6.5.2 both F and E are splitting fields of $x^{p^n} - x$ over K and L, respectively. $\square$

6.5.4 Lemma.   Let F be a field of prime characteristic p, and let $n \in \mathbf{Z}^+$. Then $\{a \in F \mid a^{p^n} = a\}$ is a subfield of F.

Proof. Let $E = \{a \in F \mid a^{p^n} = a\}$. With the exception of the coefficients of x^p and y^p, each binomial coefficient $(p!)/(k!(p - k)!)$ in the expansion of $(x \pm y)^p$ contains p in the numerator but not the denominator, because p is prime. Since char$(F) = p$, this implies that $(x \pm y)^p = x^p \pm y^p$, for all $x, y \in F$. Applying this formula inductively to $(a \pm b)^{p^k}$ for $k \le n$ shows that $(a \pm b)^{p^n} = a^{p^n} \pm b^{p^n}$, and it follows immediately that E is closed under addition and subtraction.

Since $(ab)^{p^n} = a^{p^n} b^{p^n}$, it is clear that E is closed under multiplication. To complete the proof, we only need to observe that if $a \in E$ is nonzero, then $(a^{-1})^{p^n} = (a^{p^n})^{-1} = a^{-1}$, and so $a^{-1} \in E$. $\square$

6.5.5 Proposition.   Let F be a field with p^n elements. Each subfield of F has p^m elements for some divisor m of n. Conversely, for each positive divisor m of n there exists a unique subfield of F with p^m elements.

Proof. Let K be the prime subfield of F. Any subfield E of F must have p^m elements, where $m = [E : K]$. Then $m \mid n$ since $n = [F : K] = [F : E][E : K]$.

Conversely, suppose that $m \mid n$ for some $m > 0$. Then $p^m - 1$ is a divisor of $p^n - 1$, and so $g(x) = x^{p^m-1} - 1$ is a divisor of $f(x) = x^{p^n-1} - 1$. Since F is the splitting field of $xf(x)$ over K, with distinct roots, it must contain all p^m distinct roots of $xg(x)$. By Lemma 6.5.4, these roots form a subfield of F. Furthermore, any other subfield with p^m elements must be a splitting field of $xg(x)$, and so it must consist of precisely the same elements. $\square$

6.5.6 Lemma.   Let F be a field of characteristic p. If n is a positive integer not divisible by p, then the polynomial $x^n - 1$ has no repeated roots in any extension field of F.

Proof. Let c be a root of $x^n - 1$ in an extension E of F. A direct computation shows that we must have the factorization

$$x^n - 1 = (x - c)(x^{n-1} + cx^{n-2} + c^2x^{n-3} + \ldots + c^{n-2}x + c^{n-1}).$$

With the notation $x^n - 1 = (x - c)f(x)$, we only need to show that $f(c) \ne 0$. Since $f(x)$ has n terms, we have $f(c) = nc^{n-1}$, and then $f(c) \ne 0$ since $p \nmid n$. $\square$

6.5.7 Theorem.   For each prime p and each positive integer n, there exists a field with p^n elements.

Proof. Let F be the splitting field of $f(x) = x^{p^n} - x$ over the field $\mathbf{Z}_p$. Since $x^{p^n} - x = x(x^{p^n-1} - 1)$ and $p^n - 1$ is not divisible by p, Lemma 6.5.6 implies that $f(x)$ has distinct roots. By Lemma 6.5.4, the set of all roots of $f(x)$ is a subfield of F, and so we conclude that F must consist of precisely the roots of $f(x)$, of which there are exactly p^n elements. $\square$

6.5.8 Definition.   Let p be a prime number and let $n \in \mathbf{Z}^+$. The field (unique up to isomorphism) with p^n elements is called the *Galois field of order p^n*, denoted by $GF(p^n)$.

We need the following lemma to show that the multiplicative group $F^\times$ of any finite field F is cyclic. For the sake of completeness we have included the proof, although we could have simply referred to Proposition 3.5.8.

6.5.9 Lemma.   Let G be a finite abelian group. If $a \in G$ is an element of maximal order in G, then the order of every element of G is a divisor of the order of a.

Proof. Let a be an element of maximal order in G, and let x be any element of G different from the identity. If $o(x) \nmid o(a)$, then in the prime factorizations of the respective orders there must exist a prime p that occurs to a higher power in $o(x)$ than in $o(a)$. Let $o(a) = p^\alpha n$ and $o(x) = p^\beta m$, where $\alpha < \beta$ and $p \nmid n$, $p \nmid m$. Now $o(a^{p^\alpha}) = n$ and $o(x^m) = p^\beta$, and so the orders are relatively prime since $p \nmid n$. It follows that the order of the product $a^{p^\alpha} x^m$ is equal to np^β, which is greater than $o(a)$, a contradiction. $\square$

6.5.10 Theorem.   The multiplicative group of nonzero elements of a finite field is cyclic.

Proof. Let F be a finite field with p^n elements, and let $k = p^n - 1$, so that $|F^\times| = k$. Let a be an element of $F^\times$ of maximal order, with $o(a) = m$. By Lemma 6.5.9, each element of $F^\times$ satisfies the polynomial $x^m - 1$. Since F is a field, there are at most m roots of this polynomial, and so $k \le m$. This implies that $m = k$ and $F^\times$ is cyclic. $\square$

An extension field F of K is said to be a *simple* extension of K if $F = K(u)$ for some $u \in F$.

6.5.11 Theorem.   Any finite field is a simple extension of its prime subfield.

Proof. Let F be a finite field with prime subfield K. By Theorem 6.5.10, the multiplicative group $F^\times$ is cyclic. It is clear that $F = K(u)$ for any generator u of $F^\times$. $\square$

6.5.12 Corollary.   For each positive integer n there exists an irreducible polynomial of degree n in $\mathbf{Z}_p[x]$.

Proof. Given $n \in \mathbf{Z}^+$, by Theorem 6.5.7 there exists an extension F of $\mathbf{Z}_p$ of degree n. By Theorem 6.5.11, F is a simple extension of $\mathbf{Z}_p$, and so if u is any generator of F, then the minimal polynomial $p(x)$ of u over $\mathbf{Z}_p$ must be an irreducible polynomial of degree n. $\square$

EXERCISES: SECTION 6.5

1. Give addition and multiplication tables for the finite fields $GF(2^2)$, $GF(2^3)$, and $GF(3^2)$. In each case find a generator for the cyclic group of nonzero elements under multiplication.

2. Find a generator for the cyclic group of nonzero elements of $GF(2^4)$.

3. Show that $x^3 - x - 1$ and $x^3 - x + 1$ are irreducible over $GF(3)$. Construct their splitting fields and explicitly exhibit the isomorphism between these splitting fields.

4. Show that if $g(x)$ is irreducible over $GF(p)$ and $g(x) \mid (x^{p^m} - x)$, then $\deg(g(x))$ is a divisor of m.

5. Let m, n be positive integers with $\gcd(m,n) = d$. Show that the greatest common divisor of $x^m - 1$ and $x^n - 1$ is $x^d - 1$.

6. If E and F are subfields of $GF(p^n)$ with p^e and p^f elements respectively, how many elements does $E \cap F$ contain? Prove your claim.

7. Let p be an odd prime.
   (a) Show that the set S of squares in $GF(p^n)$ contains $(p^n + 1)/2$ elements.
   (b) Given $a \in GF(p^n)$, let $T = \{a - x \mid x \in S\}$. Show that $T \cap S \neq \emptyset$.
   (c) Show that every element of $GF(p^n)$ is a sum of two squares.
   (d) What can be said about $GF(2^n)$?

8. Show that $x^p - x + a$ is irreducible over $GF(p)$ for all nonzero elements $a \in GF(p)$.

APPENDIX G: DIMENSION OF A VECTOR SPACE

We assume that the student has already taken an elementary course in linear algebra. It is quite possible that only real numbers were allowed as scalars, whereas we need to be able to use facts about vector spaces with scalars from any field. We have included the necessary definitions in this appendix, together with proofs of some facts about dimension, to make this book self-contained.

For our purposes, the main application of techniques from linear algebra occurs when we need to study the structure of an extension field F of a field K. We can consider F as a vector space with scalars from K, and this makes it possible to utilize the concept of the dimension of F. We can get useful information about other extensions of K by comparing their dimension with that of F.

G.1 Definition. A *vector space* over the field F is an additive abelian group V with a *scalar product* $a \cdot v \in V$ defined for all $a \in F$ and $v \in V$ such that the following conditions hold for all $a, b \in F$ and $u, v \in V$:
   (i) $a(b \cdot v) = (ab) \cdot v$;
   (ii) $(a + b) \cdot v = a \cdot v + b \cdot v$;
   (iii) $a \cdot (u + v) = a \cdot u + a \cdot v$; and
   (iv) $1 \cdot v = v$.

Note that scalar multiplication is not a binary operation on V. It must be defined by a function from $F \times V$ into V, rather than from $V \times V$ into V. As with groups, we will use the notation av rather than $a \cdot v$.

If we denote the additive identity of V by 0 and the additive inverse of $v \in V$ by $-v$, then we have the following results: $0 + v = v$, $a \cdot 0 = 0$, and $(-a)v = a(-v) = -(av)$ for all $a \in F$ and $v \in V$. The proofs are similar to those for the same results in a field, and involve the distributive laws, which provide the only connection between addition and scalar multiplication.

G.2 Definition. Let V be a vector space over the field F, and let $S = \{v_1, v_2, \ldots, v_n\}$ be a set of vectors in V. Any vector of the form $v = \sum_{i=1}^{n} a_i v_i$, for scalars $a_i \in F$, is called a *linear combination* of the vectors in S. The set S is said to *span* V if each element of V can be expressed as a linear combination of the vectors in S.

G.3 Definition. Let V be a vector space over the field F, and let $S = \{v_1, v_2, \ldots, v_n\}$ be a set of vectors in V. The vectors in S are said to be *linearly dependent* if one of the vectors can be expressed as a linear combination of the others. If not, then S is said to be a *linearly independent set*.

In the preceding definition, if S is linearly dependent and, for example, v_j can be written as a linear combination of the remaining vectors in S, then we can rewrite the resulting equation as

$$a_1 v_1 + \ldots + 1 \cdot v_j + \ldots + a_n v_n = 0$$

for some scalars $a_i \in F$. Thus there exists a nontrivial (i.e., at least one coefficient is nonzero) relation of the form

$$\sum_{i=1}^{n} a_i v_i = 0.$$

Conversely, if such a relation exists, then at least one coefficient, say a_j, must be nonzero. Since the coefficients are from a field, we can divide through by a_j and shift v_j to the other side of the equation to obtain v_j as a linear combination of the remaining vectors. If j is the largest subscript for which $a_j \neq 0$, then we can express v_j as a linear combination of $v_1, \ldots, v_{j-1}$.

From this point of view, S is linearly independent if and only if $\sum_{i=1}^{n} a_i v_i = 0$ implies $a_1 = a_2 = \ldots = a_n = 0$. This is the condition that is usually given as the definition of linear independence.

G.4 Theorem. Let V be a vector space, let $S = \{u_1, u_2, \ldots, u_m\}$ be a set that spans V, and let $T = \{v_1, v_2, \ldots, v_n\}$ be a linearly independent set. Then $n \leq m$, and V can be spanned by a set of m vectors that contains T.

Proof. Given $m > 0$, the proof will use induction on n. If $n = 1$, then of course $n \leq m$. Furthermore, v_1 can be written as a linear combination of vectors in S, so the set $S' = \{v_1, u_1, \ldots, u_m\}$ is linearly dependent. Using the remarks preceding the theorem, one of the vectors in S' can be written as a linear combination of the previous ones, so deleting it from S' leaves a set with m elements that still spans V and also contains v_1.

Now assume that the result holds for any set of n linearly independent vectors, and assume that T has $n + 1$ vectors. The first n vectors of T are still linearly independent, so by the induction assumption we have $n \leq m$ and V is spanned by some set $\{v_1, \ldots, v_n, w_{n+1}, \ldots, w_m\}$. Then v_{n+1} can be written as a linear combination of $\{v_1, \ldots, v_n, w_{n+1}, \ldots, w_m\}$. If $n = m$, then this contradicts the assumption that the set T is linearly independent, so we must have $n + 1 \leq m$. Furthermore, the set $S' = \{v_1, \ldots, v_n, v_{n+1}, w_{n+1}, \ldots, w_m\}$ is linearly dependent, so as before we can express one of the vectors v in S' as a linear combination of the previous ones. But v cannot be   of the vectors in T, and so we can omit one of the vectors $\{w_{n+1}, \ldots, w_m\}$, giving the desired set. $\square$

G.5 Corollary. Any two finite subsets that both span V and are linearly independent must have the same number of elements.

The above corollary justifies the following definition.

G.6 Definition. A subset of the vector space V is called a *basis* for V if it spans V and is linearly independent. If V has a finite basis, then it is said to be *finite dimensional,* and the number of vectors in the basis is called the *dimension* of V.

The proofs of the remaining results are left as exercises for the reader.

G.7 Corollary. Let V be an n-dimensional vector space. Then any set of more than n vectors is linearly dependent, and no set of fewer than n vectors can span V.

G.8 Theorem. Let V be a finite dimensional vector space. Then any linearly independent set of vectors is part of a basis for V, and any spanning set for V contains a subset that is a basis.

G.9 Corollary. Let V be an n-dimensional vector space, and let B be a set of n vectors in V. Then B is a basis for V if it is either linearly independent or spans V.

7

Structure of Groups

Our ultimate goal is to attach a group (called the Galois group) to any polynomial equation and show that the equation is solvable by radicals if and only if the group has a certain structure. (A group with this structure is called "solvable.") This requires a much more detailed knowledge of the structure of groups than we have already acquired. To show (in Chapter 8) that there exist polynomial equations over $\mathbf{C}$ of degree 5 that cannot be solved by radicals, our approach will be to find an equation whose Galois group is isomorphic to S_5, and then show that S_5 does not have the required property. This demands further study of permutation groups, in particular of A_n, which we do in Section 7.7.

Lagrange's theorem states that if H is a subgroup of a finite group G, then the order of H is a divisor of the order of G. In a cyclic group of order n, the converse is true by Proposition 3.5.3: For any divisor m of n, there exists a subgroup of order m. In Section 7.5 we will determine the structure of all finite abelian groups (originally done by Kronecker in 1870), and using this characterization we will be able to show that the converse of Lagrange's theorem holds for such groups. We will also be able to obtain some partial results in this direction, for arbitrary finite groups. The Sylow theorems (proved in Section 7.4) state that if the order of the group is divisible by a power of a prime, then there exists a subgroup whose order is the given prime power. These theorems were proved in 1872 by M.L. Sylow (1832–1918) in the context of permutation groups, and then in 1887 G. Frobenius (1849–1917) published a new proof based on the axiomatic definition of a group.

A finite group G is solvable if and only if it has a sequence of subgroups

$$G = N_0 \supseteq N_1 \supseteq \ldots \supseteq N_{k-1} \supseteq N_k = \{e\}$$

such that each subgroup is normal in the previous one and each of the factor groups N_{i-1}/N_i is cyclic of prime order. We study such chains of subgroups in Section 7.6. One way to study the structure of groups is to find suitable "building blocks" and then determine how they can be put together to construct groups. The appropriate building blocks are simple groups, which have no nontrivial normal subgroups. For any finite group it is possible to find a sequence of subgroups of the type given above, in which each factor group is simple (but not necessarily simple and abelian, as is the case for solvable groups).

The problem, then, is to determine the structure of all finite simple groups and to solve the "extension problem." That is, given a group G with normal subgroup N such that the structure of N and G/N are known, what is the structure of G? The extension problem is still open, but the classification of finite simple groups is generally accepted as complete.

A nontrivial abelian group is simple if and only if it is cyclic of prime order. We will show in Section 7.7 that the alternating group on n elements is simple, if $n \geq 5$. We can easily describe one other family of finite simple groups. Let F be any finite field, and let $G = GL_n(F)$, the group of invertible $n \times n$ matrices over F. Then G has a normal subgroup $N = SL_n(F)$, the subgroup of all matrices of determinant 1. (Note that N is normal because it is the kernel of the determinant mapping.) The center of N, which we denote by Z, may be nontrivial, in which case N is not simple. However, the factor group N/Z is simple except for the cases $n = 2$ and $F = \mathbf{Z}_2$ or $\mathbf{Z}_3$.

William Burnside (1852–1927) states in the second edition of his text *Theory of Groups of Finite Order* (published in 1911) that his research "suggests inevitably that simple groups of odd order do not exist." He had shown that the order of a simple finite group of odd order (nonabelian, of course) must have at least seven prime factors, and then he had checked all orders up to 40,000. This was finally shown to be true in 1963, by Walter Feit and John Thompson, in a 255-page paper that proved that all groups of odd order are solvable. This sparked a great deal of interest in the problem of classifying all finite simple groups, and the classification was finally completed in 1981. (The work required the efforts of many people, and is still being checked and simplified.) There are a number of infinite families of finite simple groups in addition to those mentioned above. In addition there are 26 "sporadic" ones, which do not fit into the other classes. The largest of these is known as the "monster," and has approximately 10^{54} elements.

7.1 ISOMORPHISM THEOREMS; AUTOMORPHISMS

We need to recall the fundamental homomorphism theorem for groups. If G_1 and G_2 are groups, and $\phi : G_1 \to G_2$ is a group homomorphism, then $\ker(\phi)$ is a normal subgroup of G_1, $\phi(G_1)$ is a subgroup of G_2, and the factor group $G_1/\ker(\phi)$ is isomorphic to the image $\phi(G_1)$. We will exploit this theorem in proving several

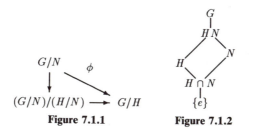

Figure 7.1.1   Figure 7.1.2

isomorphism theorems. Refer to Figure 7.1.1 for the first isomorphism theorem and Figure 7.1.2 for the second isomorphism theorem.

7.1.1 Theorem (First Isomorphism Theorem).   Let G be a group with normal subgroups N and H such that $N \subseteq H$. Then H/N is a normal subgroup of G/N, and

$$(G/N)/(H/N) \cong G/H.$$

Proof. By H/N we mean the set of all cosets of the form hN, where $h \in H$. Define $\phi : G/N \to G/H$ by $\phi(aN) = aH$ for all $a \in G$. Then ϕ is well-defined since if $aN = bN$ for $a, b \in G$, we have $b^{-1}a \in N$. Since $N \subseteq H$, this implies $b^{-1}a \in H$, and so $aH = bH$. It is clear that ϕ maps G/N onto G/H. To show that ϕ is a homomorphism we only need to note that

$$\phi(aNbN) = \phi(abN) = abH = aHbH = \phi(aN)\phi(bN).$$

Finally, $\ker(\phi) = \{aN \mid aH = H\} = H/N$, and so H/N is a normal subgroup of G/N. The fundamental homomorphism theorem for groups implies that $(G/N)/\ker(\phi) \cong G/H$, the desired result. $\square$

7.1.2 Theorem (Second Isomorphism Theorem).   Let G be a group, let N be a normal subgroup of G, and let H be any subgroup of G. Then HN is a subgroup of G, $H \cap N$ is a normal subgroup of H, and

$$(HN)/N \cong H/(H \cap N).$$

Proof. Define $\phi : H \to G/N$ by $\phi(a) = aN$, for all $a \in H$. Then ϕ is a homomorphism since $\phi(ab) = abN = aNbN = \phi(a)\phi(b)$. Therefore the image of ϕ is a subgroup of G/N, and its inverse image in G under the natural projection of G onto G/N is precisely HN, which must be a subgroup. Finally, $\ker(\phi) = H \cap N$, and so $H \cap N$ is a normal subgroup of H. By the fundamental homomorphism theorem we have $H/(H \cap N) \cong (HN)/N$. $\square$

The next theorem will be crucial in proving later theorems which describe the structure of all finite abelian groups. It is also useful in proving, for example, that $D_6 \cong S_3 \times \mathbf{Z}_2$, since we would only need to find normal subgroups of D_6 isomorphic to S_3 and $\mathbf{Z}_2$ which satisfy the conditions of the theorem.

7.1.3 Theorem.   Let G be a group with normal subgroups H, K such that $HK = G$ and $H \cap K = \{e\}$. Then $G \cong H \times K$.

Proof. We claim that $\phi : H \times K \to G$ defined by $\phi(h, k) = hk$, for all $(h, k) \in H \times K$ is a homomorphism. First, for all (h_1, k_1), $(h_2, k_2) \in H \times K$ we have

$$\phi((h_1, k_1)(h_2, k_2)) = \phi((h_1 h_2, k_1 k_2)) = h_1 h_2 k_1 k_2.$$

To show that this is equal to

$$\phi((h_1, k_1))\phi((h_2, k_2)) = h_1 k_1 h_2 k_2$$

it suffices to show that $h_2 k_1 = k_1 h_2$. For any elements $h \in H$ and $k \in K$, we have $hkh^{-1}k^{-1} \in H \cap K$ since hkh^{-1}, $k^{-1} \in K$ and h, $kh^{-1}k^{-1} \in H$. By assumption $H \cap K = \{e\}$, and so $hkh^{-1}k^{-1} = e$, or $hk = kh$. We have now verified our claim that ϕ is a homomorphism.

Since $HK = G$, it is clear that ϕ is onto. Finally, if $\phi((h, k)) = e$ for $(h, k) \in H \times K$, then $hk = e$ implies $h = k^{-1} \in H \cap K$, and so $h = e$ and $k = e$, which shows that $\ker(\phi)$ is trivial and hence ϕ is one-to-one. $\square$

We have been using the definition that a subgroup H of a group G is normal if $ghg^{-1} \in H$ for all $h \in H$ and $g \in G$. We now introduce a more sophisticated point of view using the notion of an inner automorphism. The more general notion of an automorphism of a group is also extremely important.

7.1.4 Proposition.   Let G be a group and let $a \in G$. The function $i_a : G \to G$ defined by $i_a(x) = axa^{-1}$ for all $x \in G$ is an isomorphism.

Proof. If x, $y \in G$, then

$$i_a(xy) = a(xy)a^{-1} = (axa^{-1})(aya^{-1}) = i_a(x)i_a(y),$$

and so i_a is a homomorphism. If $i_a(x) = e$, then $axa^{-1} = e$, so $x = e$ and i_a is one-to-one since its kernel is trivial. Given $y \in G$, we have $y = i_a(a^{-1}ya)$, and so i_a is also an onto mapping. $\square$

7.1.5 Definition.   Let G be a group. An isomorphism from G onto G is called an *automorphism* of G. An automorphism of G of the form i_a, for some $a \in G$, where $i_a(x) = axa^{-1}$ for all $x \in G$, is called an *inner automorphism* of G. The set of all automorphisms of G will be denoted by $\mathrm{Aut}(G)$ and the set of all inner automorphisms of G will be denoted by $\mathrm{Inn}(G)$.

The condition that a subgroup H of G is normal can be expressed by saying that $i_a(h) \in H$ for all $h \in H$ and all $a \in G$. Equivalently, H is normal if and only if $i_a(H) \subseteq H$ for all $a \in G$, and we can also express this by saying that H is invariant under all inner automorphisms of G.

7.1.6 Proposition.   Let G be a group. Then $\text{Aut}(G)$ is a group under composition of functions, and $\text{Inn}(G)$ is a normal subgroup of $\text{Aut}(G)$.

Proof. Composition of functions is always associative. We already know that the composition of two isomorphisms is again an isomorphism, and that the inverse of an isomorphism is an isomorphism, so it follows immediately that $\text{Aut}(G)$ is a group.

For any elements a, b, $x \in G$, we have

$$i_a i_b(x) = a(bxb^{-1})a^{-1} = (ab)x(ab)^{-1} = i_{ab}(x),$$

and so this yields the formula $i_a i_b = i_{ab}$. It follows easily that i_e is the identity mapping and $(i_a)^{-1} = i_{a^{-1}}$, so $\text{Inn}(G)$ is a subgroup of $\text{Aut}(G)$. To show that it is normal, let $\alpha \in \text{Aut}(G)$ and let $i_a \in \text{Inn}(G)$. For $x \in G$, we have

$$\alpha i_a \alpha^{-1}(x) = \alpha(a(\alpha^{-1}(x))a^{-1})$$

$$= (\alpha(a))(\alpha\alpha^{-1}(x))(\alpha(a^{-1}))$$

$$= (\alpha(a))(x)(\alpha(a))^{-1} = bxb^{-1}$$

$$= i_b(x)$$

for the element $b = \alpha(a)$. Thus $\alpha i_a \alpha^{-1} \in \text{Inn}(G)$, and so $\text{Inn}(G)$ is a normal subgroup of $\text{Aut}(G)$. $\square$

7.1.7 Definition.   For any group G, the subset

$$Z(G) = \{x \in G \mid xg = gx \text{ for all } g \in G\}$$

is called the *center* of G.

7.1.8 Proposition.   For any group G, we have $\text{Inn}(G) \cong G/Z(G)$.

Proof. Define $\phi : G \to \text{Inn}(G)$ by $\phi(a) = i_a$, for all $a \in G$. Then

$$\phi(ab) = i_{ab} = i_a i_b = \phi(a)\phi(b)$$

and we have defined a homomorphism. Since ϕ is onto by the definition of $\text{Inn}(G)$, we only need to compute $\ker(\phi)$. If i_a is the identity mapping, then for all $x \in G$ we have $axa^{-1} = x$, or $ax = xa$, so the kernel of ϕ is the center $Z(G)$. $\square$

Example 7.1.1

To compute $\text{Aut}(\mathbf{Z})$ and $\text{Inn}(\mathbf{Z})$ we first observe that all inner automorphisms of an abelian group are trivial (equal to the identity mapping.) Next, we observe that any isomorphism maps generators to generators, so if $\alpha \in \text{Aut}(\mathbf{Z})$, then $\alpha(1) = \pm 1$. Thus there are two possible automorphisms, with the formulas $\alpha(n) = n$ or $\alpha(n) = -n$, for all $n \in \mathbf{Z}$. $\square$

Example 7.1.2

The computation of $\mathrm{Aut}(\mathbf{Z}_n)$ is similar to that of $\mathrm{Aut}(\mathbf{Z})$. For any automorphism α of $\mathbf{Z}_n$, let $\alpha([1]) = [a]$. Since $[1]$ is a generator, $[a]$ must also be a generator, and thus $\gcd(a,n) = 1$. Then α must be given by the formula $\alpha([m]) = [am]$, for all $[m] \in \mathbf{Z}_n$. Since the composition of such functions corresponds to multiplying the coefficients, it follows that $\mathrm{Aut}(\mathbf{Z}_n) \cong \mathbf{Z}_n^{\times}$, where $\mathbf{Z}_n^{\times}$ is the multiplicative group of units of $\mathbf{Z}_n$. $\quad\square$

EXERCISES: SECTION 7.1

1. Prove that $D_6 \cong S_3 \times \mathbf{Z}_2$.
2. Let a, b be positive integers, and let $d = \gcd(a, b)$ and $m = \mathrm{lcm}[a, b]$. Prove that $\mathbf{Z}_a \times \mathbf{Z}_b \cong \mathbf{Z}_d \times \mathbf{Z}_m$.
3. Determine $\mathrm{Aut}(\mathbf{Z}_2 \times \mathbf{Z}_2)$.
4. Let G be a finite abelian group of order n, and let m be a positive integer with $(n, m) = 1$. Show that $\phi : G \to G$ defined by $\phi(g) = g^m$ for all $g \in G$ belongs to $\mathrm{Aut}(G)$.
5. Let $\phi : G \to G$ be the function defined by $\phi(g) = g^{-1}$ for all $g \in G$. Find conditions on G such that ϕ is an automorphism.
6. Show that for $G = S_3$, $\mathrm{Inn}(G) \cong G$.
7. Determine $\mathrm{Aut}(S_3)$.
8. For groups G_1 and G_2, determine the center of $G_1 \times G_2$.
9. Show that $G/Z(G)$ cannot be a nontrivial cyclic group. That is, if $G/Z(G)$ is cyclic, then G must be abelian, and hence $Z(G) = G$.
10. Describe the centers $Z(D_n)$ of the dihedral groups D_n, for all positive integers n.
11. In the group $GL_2(\mathbf{C})$ of all invertible 2×2 matrices with complex entries, let Q be the following set of matrices (the quaternion group, defined in Example 3.3.7):

$$\pm \begin{bmatrix} 1 & 0 \\ 0 & 1 \end{bmatrix}, \pm \begin{bmatrix} i & 0 \\ 0 & -i \end{bmatrix}, \pm \begin{bmatrix} 0 & 1 \\ -1 & 0 \end{bmatrix}, \pm \begin{bmatrix} 0 & i \\ i & 0 \end{bmatrix}.$$

   (a) Show that Q is not isomorphic to D_4.
   (b) Find $Z(Q)$.
12. Give another proof of Theorem 7.1.2 by constructing an isomorphism from $(HN)/N$ onto $H/(H \cap N)$.

   In our proof we constructed an isomorphism from $H/(H \cap N)$ onto $(HN)/N$. The point is that it may be much easier to define a function in one direction than in the other.

7.2 CONJUGACY

If G is a group with a subgroup H that is not normal in G, then there must exist at least one element $a \in G$ such that $aHa^{-1} \neq H$. The set aHa^{-1} is a subgroup of G, since it is the image of H under the inner automorphism i_a of G. It is important to study the subgroups related to H in this way. It is also important to study elements of the form aha^{-1}, for $h \in H$, since H is normal if and only if it contains all such elements.

7.2.1 Definition.   Let G be a group, and let $x, y \in G$. The element y is said to be a *conjugate* of the element x if there exists an element $a \in G$ such that $y = axa^{-1}$.

If H and K are subgroups of G, then K is said to be a *conjugate subgroup of* H if there exists $a \in G$ such that $K = aHa^{-1}$.

In an abelian group, elements or subgroups are only conjugate to themselves. More generally, an element x of a group G has no conjugates other than itself if and only if $axa^{-1} = x$ for all $a \in G$, and this holds if and only if x is a member of the center $Z(G)$. A subgroup H has no conjugate subgroups other than itself if and only if it is normal.

We can exploit the fact that a conjugate axa^{-1} of the element $x \in G$ is the image of x under the inner automorphism i_a to obtain valuable information. Since an automorphism preserves orders of elements, each conjugate of x must have the same order as x. Since an automorphism preserves inverses, if y is conjugate to x, then y^{-1} is conjugate to x^{-1}.

7.2.2 Proposition.   Conjugacy of elements defines an equivalence relation on any group G.

Conjugacy of subgroups defines an equivalence relation on the set of all subgroups of G.

Proof. For $x, y \in G$, write $x \sim y$ if y is a conjugate of x. Then for all $x \in G$, we have $x \sim x$ since $x = exe^{-1}$. If $x \sim y$, then $y = axa^{-1}$ for some $a \in G$, and it follows that $x = byb^{-1}$ for $b = a^{-1}$, which shows that $y \sim x$. Finally, if $z \in G$ and $x \sim y$, $y \sim z$, then there exist $a, b \in G$ with $z = aya^{-1}$ and $y = bxb^{-1}$. Thus $z = abxb^{-1}a^{-1}$, so $z = (ab)x(ab)^{-1}$ and we have $x \sim z$.

A similar proof shows that conjugacy of subgroups defines an equivalence relation on the set of all subgroups of G. $\square$

Example 7.2.1

If the group S_3 is given by generators a, b of order 3 and order 2, respectively, and $ba = a^2b$, then we can find the conjugacy classes as follows: Since a does not commute with b, we know that there is more than one element in the conjugacy class of a, and we can compute $bab^{-1} = a^2bb^{-1} = a^2$, so that a^2 is conjugate to a. We know that the subgroup $\{e, a, a^2\}$ is normal, so a cannot be conjugate to any element outside of the subgroup.

To find the conjugates of b, we have

$$aba^{-1} = aba^2 = aa^2ba = ba = a^2b,$$

which shows that a^2b is conjugate to b. Conjugating b by a^2b gives ab, and so the conjugacy classes of S_3 are the following:

$$\{e\}, \{a, a^2\}, \{b, ab, a^2b\}. \quad \square$$

Let x be an element of a group G. The computations in the preceding example show that we need to find an answer to the general question of when elements $a, b \in G$ determine the same conjugate of x. Since $axa^{-1} = bxb^{-1}$ if and only if $(b^{-1}a)x = x(b^{-1}a)$, it turns out that we need to find the elements that commute with x. Then the cosets of this subgroup correspond to distinct conjugates of x.

7.2.3 Definition.   Let G be a group. For any element $x \in G$, the set $\{a \in G \mid axa^{-1} = x\}$ is called the *centralizer* of x in G, denoted by $C(x)$.

For any subgroup H of G, the set $\{a \in G \mid aHa^{-1} = H\}$ is called the *normalizer* of H in G, denoted by $N(H)$.

Using the above definition, note that $C(x) = G$ if and only if $x \in Z(G)$, and $N(H) = G$ if and only if H is a normal subgroup of G.

7.2.4 Proposition.   Let G be a group and let $x \in G$. Then $C(x)$ is a subgroup of G.

Proof. If $a, b \in C(x)$, then

$$(ab)x(ab)^{-1} = a(bxb^{-1})a^{-1} = axa^{-1} = x,$$

and so $ab \in G$. Furthermore, $axa^{-1} = x$ implies that $a^{-1}xa = x$, and then $a^{-1} \in C(x)$ since $a^{-1}x(a^{-1})^{-1} = x$. Finally, it is clear that $e \in C(x)$. $\square$

Similarly, it can be shown that if H is a subgroup of the group G, then $N(H)$ is a subgroup of G. (This also follows from a general result we will prove in Section 7.3.) The normalizer of H is in fact the largest subgroup of G in which H is normal.

If the conjugacy classes of G are known, then it is possible to tell whether or not a subgroup H is normal by checking that each conjugacy class lies either entirely inside of H or entirely outside of H. Equivalently, H is normal if and only if it is a union of some of the conjugacy classes of G.

7.2.5 Proposition.   Let x be an element of the group G. Then the elements of the conjugacy class of x are in one-to-one correspondence with the left cosets of the centralizer $C(x)$ of x in G.

Proof. For $a, b \in G$, we have $axa^{-1} = bxb^{-1}$ if and only if $(b^{-1}a)x(b^{-1}a)^{-1} = x$, or equivalently, if and only if $b^{-1}a \in C(x)$. Thus a and b determine the same conjugate of x if and only if a and b belong to the same left coset of $C(x)$. $\square$

Example 7.2.2

We now compute the conjugacy class and centralizer of each element in the dihedral group D_4, described by the generators a and b of order 4 and order 2, respectively, with $ba = a^3b$. Since we have the relation $ba^i = a^{-i}b$, it is clear that $\{e, a^2\} \subseteq Z(D_4)$, and it can be shown that this is in fact the center. Thus the conjugacy classes with exactly one element are $\{e\}$ and $\{a^2\}$.

It is always true that $\langle x \rangle \subseteq C(x)$, since any power of x must commute with x. Thus $\{e, a, a^2, a^3\} \subseteq C(a)$, and we must have equality since $a \notin Z(D_4)$ implies $C(a) \neq D_4$. The conjugacy class of a is $\{a, a^3\}$. Similarly, $b \in C(b)$, and we also have $a^2 \in C(b)$ since a^2 commutes with b. Again, $C(b) \neq D_4$, and so we may conclude that $C(b) = \{e, b, a^2, a^2b\}$, and then we can easily show that the conjugacy class of b is $\{b, a^2b\}$ by conjugating b by any element not in $C(b)$. Remember that the conjugates of b correspond to the left cosets of $C(b)$. Finally, a similar computation shows that $C(ab) = \{e, ab, a^2, a^3b\}$ and the conjugacy class of ab is $\{ab, a^3b\}$.   □

Example 7.2.3 (Conjugacy in S_n).

Writing the elements of S_3 in cyclic notation gives a clue to what happens in S_n. Using Example 7.2.1, the conjugacy classes of S_3 are $\{(1)\}$, $\{(1, 2, 3), (1, 3, 2)\}$ and $\{(1, 2), (1, 3), (2, 3)\}$.

We will show that two permutations are conjugate in S_n if and only if they have the same cycle structure. For example, in S_5 the conjugacy class of the identity permutation has only one element, and then in addition there is one conjugacy class for each of the following forms (assuming the permutations are written in cyclic notation): (a, b), (a, b, c), $(a, b)(c, d)$, (a, b, c, d), $(a, b, c)(d, e)$, and (a, b, c, d, e). Thus, in particular, cycles of the same length are always conjugate.

Recall how a permutation $\sigma \in S_n$ is written in cyclic notation. Starting with a number i we construct the cycle $(i, \sigma(i), \sigma^2(i), \ldots)$, and continue with additional disjoint cycles as necessary. If $\tau \in S_n$, then to construct the cyclic representation of the conjugate $\tau\sigma\tau^{-1}$, we can start with the number $\tau(i)$ and then proceed as follows. We have $\tau\sigma\tau^{-1}(\tau(i)) = \tau(\sigma(i))$ as the next entry of the cycle, followed by $\tau\sigma\tau^{-1}(\tau(\sigma(i))) = \tau(\sigma^2(i))$, etc. Thus the cycles of $\tau\sigma\tau^{-1}$ are found by simply applying τ to the entries of the cycles of σ, resulting in precisely the same cycle structure.

On the other hand, if σ and ρ have the same cycle structure, then a simple substitution can be made in which the entries of σ are replaced by the corresponding entries of ρ. For the permutation τ that is defined by this substitution, we have $\tau\sigma\tau^{-1} = \rho$, showing that ρ is conjugate to σ.   □

The equation in the following theorem is called the *conjugacy class equation* of the group G. Recall that the number of left cosets of a subgroup H of a group G is called the index of H in G, and is denoted by $[G:H]$. We will see that a great deal of information can be obtained simply by counting the elements in G according to its conjugacy classes.

7.2.6 Theorem (Class Equation).   Let G be a finite group. Then

$$|G| = |Z(G)| + \sum [G:C(x)]$$

where the sum ranges over one element x from each nontrivial conjugacy class.

Proof. Since conjugacy defines an equivalence relation, the conjugacy classes partition G. The conjugacy classes containing only one element can be grouped together, since they form the center $Z(G)$, and then by Proposition 7.2.5 the number of elements in the conjugacy class of the element x is the index $[G:C(x)]$ of the centralizer of x.   □

7.2.7 Definition.   A group of order p^n, with p a prime number and $n \geq 1$, is called a *p-group*.

7.2.8 Theorem (Burnside).   Let p be a prime number. The center of any p-group is nontrivial.

Proof. Let G be a p-group. In the conjugacy class equation of G, the order of G is by definition divisible by p, and the terms $[G:C(x)]$ are all divisible by p since $x \notin Z(G)$ implies $[G:C(x)] > 1$. Remember that $[G:C(x)]$ is a divisor of $|G|$ and hence is a power of p. This implies that $|Z(G)|$ is divisible by p.   □

7.2.9 Corollary.   Any group of order p^2 (where p is prime) is abelian.

Proof. If $|G| = p^2$ and $Z(G) \neq G$, then let $a \in G - Z(G)$. Then $C(a)$ is a subgroup containing both a and $Z(G)$, with $|Z(G)| \geq p$ by the previous theorem. This shows that $C(a) = G$, a contradiction. Thus $Z(G) = G$, and so G is abelian.   □

The following proof of Cauchy's theorem makes use only of the conjugacy class equation. The next section contains a stronger statement of the theorem, together with a shorter proof, but the motivation for that proof is not as transparent.

7.2.10 Theorem (Cauchy).   If G is a finite group and p is a prime divisor of the order of G, then G contains an element of order p.

Proof. Let $|G| = n$. The proof proceeds by induction on n. We start the induction with the observation that we certainly know that the theorem holds for values of n up to 5. We may assume that the theorem holds for all groups of order less than n. Consider the class equation

$$n = |Z(G)| + \sum [G:C(x)].$$

Case 1. For each $x \notin Z(G)$, p is a divisor of $[G:C(x)]$.
As in the proof of Theorem 7.2.8, in this case $p|Z(G)|$. Now if $Z(G) \neq G$, then the induction hypothesis shows that $Z(G)$ contains an element of order p and we are done. Thus we may assume that $Z(G) = G$, and so G is abelian. Let $a \in G$, with $a \neq e$, and consider $H = \langle a \rangle$. If $H = G$, then G is cyclic and the theorem holds, so we may assume that $|H| < |G|$. If p is a divisor of k, where $k = |H|$, then H has an element of order p by the induction hypothesis and we are done.

Thus we may assume that p is not a divisor of k, and hence must be a divisor of $|G/H|$. But then, since $|G/H| < n$, the group G/H must contain a coset of order p, say bH. Therefore $(bH)^p = H$, or, equivalently, $b^p \in H$. If $c = b^k$, then $c^p = (b^k)^p = (b^p)^k$, and this must give the identity element since $b^p \in H$ and $k = |H|$. If $c = e$, then $b^k = e$, which in turn implies $(bH)^k = H$ in G/H. Since bH has order p, we then have $p|k$, a contradiction. This shows that c is an element of order p.

Case 2. For some $x \notin Z(G)$, p is not a divisor of $[G:C(x)]$.
In this case, for the given element x, it follows that p is a divisor of $|C(x)|$ since p is a divisor of $|G| = |C(x)| \cdot [G:C(x)]$. Then we are done since $|C(x)| < n$ and the induction hypothesis implies that $C(x)$ contains an element of order p.   □

EXERCISES: SECTION 7.2

1. Let H be a subgroup of the group G. Prove that $N(H)$ is a subgroup of G.
2. Let G be a group with subgroups H and K such that $H \subseteq K$. Show that H is a normal subgroup of K if and only if $K \subseteq N(H)$.
3. Find the conjugacy classes of D_5.
4. Describe the conjugacy classes of S_4 and S_5 by listing the types of elements and the number of each type in each class.
5. Find the conjugacy classes of A_4.
   Note: Two elements may be conjugate in S_4 but not in A_4.
6. Find the conjugacy classes of the quaternion group Q defined in Example 3.3.7.
7. Write out the conjugacy class equations for S_4, A_4, and S_5.
8. Let the dihedral group D_n be given by elements a of order n and b of order 2, subject to the identity $ba = a^{-1}b$. Show that a^m is conjugate to only a^{-m}, and that $a^m b$ is conjugate to $a^{m+2k}b$, for any integer k.
9. Show that if a group G has an element a that has precisely two conjugates, then G has a nontrivial proper normal subgroup.
10. Let G be a nonabelian group of order p^3, for a prime number p. Show that $Z(G)$ must have order p.

7.3 GROUPS ACTING ON SETS

Recall that if S is a set and G is a subgroup of the group Sym(S) of all permutations of S, then G is called a group of permutations. Historically, at first the theory of groups meant only the study of groups of permutations. The concept of an abstract group was introduced in studying properties that do not depend on the underlying set. Cayley's theorem states that every abstract group is isomorphic to a group of permutations, so the abstraction really just provides a different point of view. Still, when studying abstract groups, it is often important to be able to relate them to groups in which direct computations can actually be done. Since the appropriate way to relate the algebraic structures of two groups is via a homomorphism, we should study homomorphisms into groups of permutations.

If $\phi: G \rightarrow$ Sym(S) is a group homomorphism, suppose that $g \in G$ and $\phi(g) = \sigma$. Then for any element $x \in S$, $\sigma(x)$ is another element of S, say y, and it makes sense to think of g as "acting" on x to produce y. In many cases it is easier to think of G as "acting" on the set S instead of thinking in terms of a homomorphism. For any group acting on a set, we will be able to obtain a very useful formula that generalizes the conjugacy class equation.

7.3.1 Definition. Let G be a group and let S be a set. A multiplication of elements of S by elements of G (defined by a function from $G \times S \rightarrow S$) is called a *group action* of G on S provided for each $x \in S$:

  (i) $a(bx) = (ab)x$ for all $a, b \in G$;

  (ii) $ex = x$ for the identity element e of G.

Example 7.3.1

If G is a group of permutations of the set S, then the action of permutations on the set S clearly satisfies the two conditions of the previous definition. □

Example 7.3.2

If H is a subgroup of the group G, then H acts on the set G by using the group multiplication defined on G. □

Example 7.3.3

Let G be the multiplicative group $F^\times$ of nonzero elements of a field F. If V is any vector space over F, then scalar multiplication defines an action of G on V. The two conditions that must be satisfied are the only two vector space axioms that deal exclusively with scalar multiplication. □

Example 7.3.4

Let V be an n-dimensional vector space over the field F, and let G be any subgroup of the general linear group $GL_n(F)$ of all invertible $n \times n$ matrices over F. The standard multiplication of (column) vectors by matrices defines a group action of G on V. $\square$

The point of view of the next proposition will be useful in giving some more interesting examples. It explains the introductory statements hinting at the relationship between group actions and representations of abstract groups via homomorphisms into groups of permutations.

7.3.2 Proposition.   Let G be a group and let S be a set. Any group homomorphism from G into the group $\mathrm{Sym}(S)$ of all permutations of S defines an action of G on S. Conversely, every action of G on S arises in this way.

Proof. Let $\phi : G \to \mathrm{Sym}(S)$ be a homomorphism. For $a \in G$, it is convenient to let the permutation $\phi(a)$ be denoted by λ_a. Since ϕ is a homomorphism, we have the formula

$$\lambda_a \lambda_b = \phi(a)\phi(b) = \phi(ab) = \lambda_{ab}$$

for all $a, b \in G$. For $x \in S$ and $a \in G$, we define $ax = \lambda_a(x)$. Since λ_e is the identity permutation, and

$$a(bx) = \lambda_a(\lambda_b(x)) = \lambda_{ab}(x) = (ab)x$$

we have defined a group action.

Conversely, suppose that G acts on S. For each $a \in G$, define a function $\lambda_a : S \to S$ by setting $\lambda_a(x) = ax$, for all $x \in S$. Then λ_a is one-to-one since $\lambda_a(x_1) = \lambda_a(x_2)$ implies that $ax_1 = ax_2$, so multiplying by a^{-1} and using the defining properties of the group action we obtain $x_1 = x_2$. Given $y \in S$, the element $a^{-1}y$ is a solution to the equation $\lambda_a(x) = y$, and thus λ_a is onto. It is not hard to show that $\lambda_a \lambda_b = \lambda_{ab}$, and this in turn can be used to show that the function $\phi : G \to \mathrm{Sym}(S)$ defined by $\phi(a) = \lambda_a$ for each $a \in G$ is a group homomorphism. $\square$

7.3.3 Definition.   Let G be a group acting on the set S. For each element $x \in S$, the set

$$Gx = \{s \in S \mid s = ax \text{ for some } a \in G\}$$

is called the *orbit* of x under G, and the set

$$G_x = \{a \in G \mid ax = x\}$$

is called the *stabilizer* of x in G. The set

$$S^G = \{x \in S \mid ax = x \text{ for all } a \in G\}$$

is called the *subset of S fixed by G*.

   In Example 7.3.2, a subgroup H has a natural action on the entire group G. The orbit of an element $g \in G$ is the right coset Hg. The stabilizer H_g of g is just $\{e\}$. If H is nontrivial, then the fixed subset G^H equals

$$\{g \in G \mid hg = g \text{ for all } h \in H\},$$

and this must be the empty set. The next example is much more interesting.

Example 7.3.5

   For any group G, the homomorphism $\phi : G \to \operatorname{Aut}(G)$ defined by $\phi(a) = i_a$, where i_a is the inner automorphism defined by a, gives a group action of G on itself. The orbit Gg of an element g is just its conjugacy class, and the stabilizer G_g is just the centralizer of g in G. The fixed subset

$$\{g \in G \mid aga^{-1} = g \text{ for all } a \in G\}$$

is the center $Z(G)$. □

Example 7.3.6

   This example is closely related to Example 7.3.5. Let G be a group, and let S be the set of all subgroups of G. If $a \in G$ and H is a subgroup of G, define $a * H = aHa^{-1}$. The fact that $a(bHb^{-1})a^{-1} = abH(ab)^{-1}$ shows that $a * (b * H) = (ab) * H$ for all $H \in S$ and all $a, b \in G$. Since $e * H = H$ for all subgroups H of G, the multiplication $*$ defines a group action of G on S. (We have introduced the $*$ here to avoid confusion with the usual coset notation.)
   The orbit $G * H$ of a subgroup H is the set of all subgroups conjugate to H. The stabilizer of H in G is just the normalizer of H. Finally, the fixed subset S^G is the set of normal subgroups of G.
   To illustrate the many possibilities, in a closely related example we can let H be any subgroup and let S be the set of all subgroups of G that are conjugate to H. If K is any subgroup of G, then for any $k \in K$ and $J \in S$, the subgroup $k * J = kJk^{-1}$ is still a member of S since it is conjugate to H, and so the $*$ operation defines an action of K on S. With this operation the stabilizer of $J \in S$ is $K \cap N(J)$, and the subset of S left fixed by the action is the set of conjugates J of H for which $K \subseteq N(J)$. □

7.3.4 Proposition.   Let G be a group that acts on the set S, and let $x \in S$.
(a) The stabilizer G_x of x in G is a subgroup of G.
(b) There is a one-to-one correspondence between the elements of the orbit Gx of x under G and the left cosets of G_x in G.

Proof. (a) If a, $b \in G_x$, then $(ab)x = a(bx) = ax = x$, and so $ab \in G_x$. Furthermore, $a^{-1}x = a^{-1}(ax) = (a^{-1}a)x$, and then $ex = x$ shows that $a^{-1} \in G_x$, as well as showing that $G_x \neq \emptyset$.

(b) For a, $b \in G$ we have $ax = bx$ if and only if $b^{-1}ax = x$, which occurs if and only if $b^{-1}a \in G_x$. Since this is equivalent to the condition that $aG_x = bG_x$, the function that assigns to the left coset aG_x the element ax in the orbit of x is well-defined and one-to-one. This function is clearly onto, completing the proof. □

Applying the above proposition to Example 7.3.6 shows that the normalizer $N(H)$ of a subgroup H is a subgroup of G. Furthermore, the number of distinct subgroups conjugate to H is equal to $[G:N(H)]$.

7.3.5 Proposition.   Let G be a finite group acting on the set S.
  (a) The orbits of S (under the action of G) partition S.
  (b) For any $x \in S$, $|Gx| = [G:G_x]$.

Proof. (a) For x, $y \in S$ define $x \sim y$ if there exists $a \in G$ such that $x = ay$. This defines an equivalence relation on S, since, to begin with, for all $x \in S$, $x = ex$ implies $x \sim x$. If $x \sim y$, then there exists $a \in G$ with $x = ay$, and then $y = a^{-1}x$ implies that $y \sim x$. If $x \sim y$ and $y \sim z$ for x, y, $z \in S$, then there exist a, $b \in G$ such that $x = ay$ and $y = bz$. Therefore $x = (ab)z$ and $x \sim z$. The equivalence classes of $\sim$ are precisely the orbits Gx.

(b) This follows immediately from Proposition 7.3.4 □

7.3.6 Theorem.   Let G be a finite group acting on the finite set S. Then

$$|S| = |S^G| + \sum_{\Gamma} [G:G_x],$$

where Γ is a set of representatives of the orbits Gx for which $|Gx| > 1$.

Proof. Since the orbits Gx partition S, we have $|S| = \sum |Gx|$. The equation we need to verify simply collects together the orbits with only one element and counts their members as $|S^G|$. □

If we apply Theorem 7.3.6 to Example 7.3.5, we obtain the class equation of Theorem 7.2.6.

7.3.7 Lemma.   Let G be a finite p-group acting on the finite set S. Then

$$|S| \equiv |S^G| \pmod{p}.$$

Proof. Assume that $|G| = p^n$ for some integer n. We simply reduce the equation in Theorem 7.3.6 modulo p. Each term $[G:G_x] > 1$ in the sum $\sum [G:G_x]$ must be a divisor of $|G| = p^n$, and so for some $\alpha \geq 1$ we have $[G:G_x] = p^\alpha \equiv 0 \pmod{p}$. □

7.3.8 Theorem (Cauchy).   If G is a finite group and p is a prime divisor of $|G|$, then the number of solutions of the equation $x^p = e$ is a multiple of p. In particular, G has an element of order p.

Proof. Let $|G| = n$ and let S be the set of all p-tuples $(x_1, x_2, \ldots, x_p)$ such that for each i, $x_i \in G$ and $x_1 x_2 \cdots x_p = e$. The entry x_p is determined by the first $p - 1$ entries, since $x_p = (x_1 x_2 \cdots x_{p-1})^{-1}$, and so $|S| = n^{p-1} \equiv 0 \pmod{p}$, since p is a divisor of n. The motivation for considering this set is that the elements $x \in G$ satisfying $x^p = e$ are precisely the elements such that $(x_1, x_2, \ldots, x_p) \in S$ for $x_1 = x_2 = \cdots = x_p = x$.

Let C be the cyclic subgroup of the permutation group S_p generated by the cycle $\sigma = (1, 2, \ldots, p)$. Then C acts on S by simply permuting indices. That is,

$$\sigma(x_1, x_2, \ldots, x_p) = (x_2, x_3, \ldots, x_1)$$

The product extends to powers of σ in the obvious way. Note that the action produces elements of S, since if $yz = e$, then $zy = e$. We have already observed that the fixed subset S^C consists of p-tuples $(x_1, x_2, \ldots, x_p) \in S$ such that $x_1 = x_2 = \cdots = x_p = x$ and $x^p = e$. Note that S^C is nonempty since it contains $(e, e, \ldots, e)$. The result follows from the previous lemma since

$$|S^C| \equiv |S| \equiv 0 \pmod{p}$$

and $|S^C| \neq 0$. $\square$

If G is a p-group, then by Lagrange's theorem, the order of each element of G is a power of p. The standard definition of a p-group, allowing its usage for infinite groups, is that G is a p-group if each element of G has an order that is some power of p.

EXERCISES: SECTION 7.3

1. Let H be a subgroup of G, and let S denote the set of left cosets of H. For $a, x \in G$, define $a(xH) = axH$.
   (a) Show that the multiplication defined above yields a group action of G on S.
   (b) Let $\phi : G \to \text{Sym}(S)$ be the homomorphism that corresponds to the group action defined above. Show that $\ker(\phi)$ is the largest normal subgroup of G that is contained in H.
   (c) Assume that G is finite and let $[G : H] = n$. Show that if $n!$ is not divisible by $|G|$, then H must contain a nontrivial normal subgroup of G.

2. Let G be a group of order 28. Use Exercise 1 to show that G has a normal subgroup of order 7. Show that if G also has a normal subgroup of order 4, then it must be an abelian group.

3. Let G be any nonabelian group of order 6. By Cauchy's theorem, G has an element, say a, of order 2. Let $H = \langle a \rangle$, and let S be the set of left cosets of H.

(a) Show H is not normal in G.

Hint: If H is normal, then $H \subseteq Z(G)$, and it can then be shown that G is abelian.

(b) Use Exercise 1 and part (a) to show that G must be isomorphic to Sym(S). Thus any nonabelian group of order 6 is isomorphic to S_3.

4. Let G be a p-group with proper subgroup H. Show that there exists an element $a \in G - H$ such that $a^{-1}Ha = H$.

5. Let G be a p-group, say $|G| = p^n$. Show that any subgroup of order p^{n-1} must be normal in G.

6. Let G be a group acting on a set S. Prove that $S^G = \{x \in S \mid G_x = G\}$ and that $S^G = \{x \in S \mid Gx = \{x\}\}$.

7. Prove that if G is a finite p-group acting on a finite set S with $p \nmid |S|$, then G has at least one orbit that contains only one element.

8. If G is a finite group of order n and p is the least prime such that $p \mid n$, show that any subgroup of index p is normal in G.

7.4 THE SYLOW THEOREMS

Lagrange's theorem shows that for any finite group the order of a subgroup is a divisor of the order of the group. The converse is not true. For example, the alternating group A_4 has order 12, but has no subgroup of order 6. Cauchy's theorem gives a weak version of the converse (for prime divisors). The major results in this direction are due to Sylow.

7.4.1 Theorem (First Sylow Theorem). Let G be a finite group. If p is a prime such that p^α is a divisor of $|G|$ for some $\alpha \geq 0$, then G contains a subgroup of order p^α.

Proof. We will use induction on $n = |G|$. The theorem is certainly true for $n = 1$, and so we assume that it holds for all groups of order less than n. Consider the class equation

$$|G| = |Z(G)| + \sum [G : C(x)].$$

where the sum ranges over one entry from each nontrivial conjugacy class. We will consider two cases, depending on whether or not each term in the summation $\Sigma[G : C(x)]$ is divisible by p.

Case 1. For each $x \notin Z(G)$, p is a divisor of $[G : C(x)]$.
In this case the class equation shows that p must be a divisor of $|Z(G)|$, and so $Z(G)$ contains an element a of order p by Cauchy's theorem. Then $\langle a \rangle$ is a normal subgroup of G since $a \in Z(G)$, and so by the induction hypothesis, $G/\langle a \rangle$ contains a subgroup of order $p^{\alpha-1}$, since $p^{\alpha-1}$ is a divisor of $|G/\langle a \rangle|$. The inverse image in G of this subgroup has order p^α since each coset of $\langle a \rangle$ contains p elements.

Case 2. For some $x \notin Z(G)$, p is not a divisor of $[G : C(x)]$.
Since p^α is a divisor of $n = |C(x)| \cdot [G : C(x)]$, it follows that p^α is a divisor of

$|C(x)|$. But then the induction hypothesis can be applied to $C(x)$, since $x \notin Z(G)$ implies $|C(x)| < |G|$, and so $C(x)$ contains a subgroup of order p^α. $\square$

7.4.2 Definition.   Let G be a finite group, and let p be a prime number. A subgroup P of G is called a *Sylow p-subgroup* of G if $|P| = p^\alpha$ for some integer $\alpha \geq 1$ such that p^α is a divisor of $|G|$ but $p^{\alpha+1}$ is not.

The cyclic group $\mathbf{Z}_6$ has unique subgroups of order 2 and order 3, and these are the Sylow p-subgroups for $p = 2$ and $p = 3$. In the symmetric group S_3 there are three subgroups of order 2, so we do not have uniqueness. Nevertheless, at least they are conjugate. The unique subgroup of order 3 is normal, and it turns out to be true in general that there is a unique Sylow p-subgroup if and only if there is a normal Sylow p-subgroup.

7.4.3 Lemma.   Let G be a finite group with $|G| = mp^\alpha$, where $\alpha \geq 1$ and m is not divisible by p. If P is a normal Sylow p-subgroup, then P contains every p-subgroup of G.

Proof. Suppose that $a \in H$ for a p-subgroup H of G. Since P is a normal subgroup of G, we may consider the coset aP as an element of the factor group G/P. On the one hand, the order of a is a power of p since it belongs to a subgroup whose order is a power of p. The order of the coset aP must be a divisor of the order of a, so it is also a power of p. On the other hand, the order of aP must be a divisor of $[G : P]$, but by assumption $[G : P]$ is not divisible by p. This is a contradiction unless $aP = P$, so $a \in P$, and we have shown that $H \subseteq P$. $\square$

7.4.4 Theorem (Second and Third Sylow Theorems).   Let G be a finite group of order n, and let p be a prime number.
 (a) All Sylow p-subgroups of G are conjugate, and any p-subgroup of G is contained in a Sylow p-subgroup.
 (b) Let $n = mp^\alpha$, with $\gcd(m, p) = 1$, and let k be the number of Sylow p-subgroups of G. Then $k|m$ and $k \equiv 1 \pmod{p}$.

Proof. Let P be a Sylow p-subgroup of G with $|P| = p^\alpha$, let S be the set of all conjugates of P, and let P act on S by conjugation. If $Q \in S$ is left fixed by the action of P, then $P \subseteq N(Q)$. Since $|Q| = p^\alpha$, p is not a divisor of $[G : Q]$ and hence p is not a divisor of $[N(Q) : Q]$. Thus the hypothesis of Lemma 7.4.3 is satisfied by Q in $N(Q)$, since Q is normal in $N(Q)$. It follows that $P \subseteq Q$, so $P = Q$ since $|P| = |Q|$. Therefore the only member of S left fixed by the action of P is P itself, so $|S^P| = 1$, and then Lemma 7.3.7 shows that $|S| \equiv 1 \pmod{p}$.
   Next let Q be any maximal p-subgroup. (That is, let Q be any p-subgroup that is not contained in any larger p-subgroup of G.) Let Q act on S by conjugation. Now $|S| \equiv 1 \pmod{p}$ implies by Lemma 7.3.7 that $|S^Q| \equiv 1 \pmod{p}$. In particular, some conjugate K of P must be left fixed by Q. Then $Q \subseteq N(K)$, and as

before it follows from Lemma 7.4.3 that $Q \subseteq K$. But then since Q is a maximal p-subgroup, we must have $Q = K$. This shows that Q is conjugate to P. This implies not only that all Sylow p-subgroups are conjugate, but that any maximal p-subgroup is a Sylow p-subgroup. It is clear that any p-subgroup is contained in a maximal p-subgroup, so we have proved part (a).

Since we now know that S is the set of all Sylow p-subgroups of G, we have $k \equiv 1 \pmod{p}$. Finally, $k = [G : N(P)]$, since this is the number of conjugates of P. Since $P \subseteq N(P)$, we see that $k \mid m$ because $[G : N(P)]$ is a divisor of $[G : P]$. $\square$

Example 7.4.1

To give a simple application of the Sylow theorems, we will show that any group of order 100 has a normal subgroup of order 25. We simply note that the number of Sylow 5-subgroups must be congruent to 1 modulo 5 and also a divisor of 4. The only possibility is that there is just one such subgroup (of order 25), which must then be normal. $\square$

Example 7.4.2

As a slightly less straightforward example, we will show that any group of order 30 must have a nontrivial normal subgroup. The number of Sylow 3-subgroups must be congruent to 1 modulo 3 and a divisor of 10, so it must be either 1 or 10. The number of Sylow 5-subgroups must be congruent to 1 modulo 5 and a divisor of 6, so it must be either 1 or 6. Any Sylow 3-subgroup must have order 3, so the intersection of two distinct such subgroups must be trivial. Therefore ten Sylow 3-subgroups would yield twenty elements of order 3. Similarly, six Sylow 5-subgroups would yield twenty-four elements of order 5. Together, this would simply give too many elements for the group, so we conclude that there must be either one Sylow 3-subgroup or one Sylow 5-subgroup, showing the existence of a nontrivial normal subgroup. $\square$

As further applications of the structure theorems we have proved, we can obtain the following information about the structure of groups of certain types. The amount of work it takes to get even such limited results should make the student appreciate the difficulty of determining the structure of groups.

7.4.5 Proposition.   Let $p > 2$ be a prime, and let G be a group of order $2p$. Then G is either cyclic or isomorphic to the dihedral group D_p of order $2p$.

Proof. By Cauchy's theorem, G contains an element a of order p and an element b of order 2. The cyclic subgroup $\langle a \rangle$ has index 2 in G, and so it must be a normal subgroup. Thus conjugating a by b gives $bab = a^n$ for some n. Then $a = b(bab)b = ba^n b = a^{n^2}$, and so $n^2 \equiv 1 \pmod{p}$. It follows that $n \equiv \pm 1 \pmod{p}$, and thus $bab = a$ or else $bab = a^{-1}$. In the first case, a and b commute, and so ab has order lcm$(2, p) = 2p$ and G is cyclic. In the second case, we obtain $ba = a^{-1}b$ (or $ba = a^{p-1}b$), the familiar equation that defines D_p. $\square$

7.4.6 Proposition.   Let G be a group of order pq, where $p > q$ are primes.
   (a) If q is not a divisor of $p - 1$, then G is cyclic.
   (b) If q is a divisor of $p - 1$, then either G is cyclic or else G is generated by two elements a and b satisfying the following equations:

$$a^p = e, \qquad b^q = e, \qquad ba = a^n b$$

where $n \not\equiv 1 \pmod{p}$ but $n^q \equiv 1 \pmod{p}$.

Proof. The number of Sylow p-subgroups is a divisor of q, so it must be either 1 or q. In the latter case it could not also be congruent to 1 modulo p, since $p > q$. Thus the Sylow p-subgroup is cyclic and normal, say $\langle a \rangle$. There exists an element b of order q, and since $\langle b \rangle$ is a Sylow q-subgroup, there are two cases. If the number of Sylow q-subgroups is 1, then $\langle b \rangle$ is a normal subgroup, and ab has order pq, showing that G is cyclic. (The intersection of Sylow subgroups for different primes is always trivial. Since both are normal subgroups, the element $aba^{-1}b^{-1}$ belongs to both subgroups and hence must be equal to e, showing that $ab = ba$.) In the second case, since $\langle a \rangle$ is normal, we have $bab^{-1} \in \langle a \rangle$, and so $ba = a^n b$ for some n. We can assume n is not congruent to 1 modulo p, since that would imply $ba = ab$, covered in the previous case. Conjugating repeatedly gives $b^q a b^{-q} = a^{n^q}$, or simply $a = a^{n^q}$, which shows that $n^q \equiv 1 \pmod{p}$ since a has order p. $\square$

EXERCISES: SECTION 7.4

   1. Show that A_4 has no subgroup of order 6.
   2. In S_4 find a Sylow 2-subgroup and a Sylow 3-subgroup.
   3. Find all Sylow 3-subgroups of S_4 and show explicitly how they are conjugate.
   4. Show that there is no simple group of order 148.
   5. Show that there is no simple group of order 56.
   6. Let G be a group of order $p^2 q$, where p and q are primes. Show that G must contain a proper nontrivial normal subgroup.
   7. Let G be a finite group, and let H, K be subgroups of G. Prove that

$$|H\,K| = \frac{|H||K|}{|H \cap K|}.$$

   8. Show that there is no simple group of order 48.
   9. Show that there is no simple group of order 36.
   10. Let G be a finite group in which each Sylow subgroup is normal. Prove that G is isomorphic to the direct product of its Sylow subgroups.

7.5 FINITE ABELIAN GROUPS

If $m > 1$ and $n > 1$ are relatively prime numbers, and $\phi : \mathbf{Z}_{mn} \to \mathbf{Z}_m \times \mathbf{Z}_n$ is defined by $\phi([x]_{mn}) = ([x]_m, [x]_n)$, then ϕ is an isomorphism. The statement that ϕ is onto is precisely the statement of the Chinese remainder theorem. Another proof is to observe that the second group is cyclic since the order of the element $([1], [1])$ is $\mathrm{lcm}[m, n] = mn$. Applying this result repeatedly, we can show that for any $n > 1$, the cyclic group $\mathbf{Z}_n$ is isomorphic to a direct product of cyclic groups of prime power order (where the prime powers are those in the prime factorization of n). The goal of this section is to prove a much more general result: Any finite abelian group is isomorphic to a direct product of cyclic groups of prime power order.

7.5.1 Theorem.   A finite abelian group can be expressed as a direct product of its Sylow p-subgroups.

Proof. Let G be a finite abelian group, with $|G| = np^\alpha$, where $p \nmid n$. Let $H_1 = \{a \in G \mid a^{p^\alpha} = e\}$ and let $K_1 = \{a \in G \mid a^n = e\}$. Since G is abelian, both are subgroups, and H_1 is the Sylow p-subgroup of G.

We will show that (i) $H_1 \cap K_1 = \{e\}$ and (ii) $H_1 K_1 = G$. This shows that G is isomorphic to the direct product of H_1 and K_1, by Theorem 7.1.3. Then we can decompose K_1 in a similar fashion, etc., to get $G \cong H_1 \times H_2 \times \cdots \times H_k$, where each subgroup H_i is a Sylow p-subgroup for some prime p.

To prove (i), we simply observe that if $a \in H_1 \cap K_1$, then the order of a is a common divisor of p^α and n, which implies that $a = e$. To prove (ii), let $a \in G$. Then the order k of a is a divisor of $p^\alpha n$, and so $k = p^\beta m$, where $m \mid n$, $\beta \leq \alpha$, and $p \nmid m$. Since $\gcd(p^\beta, m) = 1$, there exist $r, s \in \mathbf{Z}$ with $ms + p^\beta r = 1$. Then $a = (a^m)^s \cdot (a^{p^\beta})^r$, and $a \in H_1 K_1$ since $a^m \in H_1$ and $a^{p^\beta} \in K_1$. The last statement follows from the fact that $(a^m)^{p^\alpha} = e$ and $(a^{p^\beta})^n = e$ since mp^α and np^β are multiples of the order of a. $\square$

7.5.2 Lemma.   Let G be a finite abelian p-group, let $a \in G$ be an element of maximal order, and let $b\langle a \rangle$ be any coset of $G/\langle a \rangle$. Then there exists $d \in G$ such that $d\langle a \rangle = b\langle a \rangle$ and $\langle a \rangle \cap \langle d \rangle = \{e\}$.

Proof. The outline of the proof is to let s be the smallest positive integer such that $b^s \in \langle a \rangle$. Then we solve the equation $b^s = x^s$ for elements $x \in \langle a \rangle$ and let $d = bx^{-1}$.

For convenience, we use $o(x)$ to denote the order of an element x. Let s be the order of $b\langle a \rangle$ in the factor group $G/\langle a \rangle$. Then $b^s \in \langle a \rangle$, and we can write $b^s = a^{qt}$ for some exponent qt such that $t = p^\beta$ for some β and $p \nmid q$. Then a^q is a generator for $\langle a \rangle$, since q is relatively prime to $o(a)$. Since s is a divisor of the order of b, we have $o(b)/s = o(b^s) = o(a^{qt}) = o(a)/t$, or simply, $o(b) \cdot t = o(a) \cdot s$. All of these are powers of p, and so $o(b) \leq o(a)$ implies that $s \mid t$, say $t = ms$. Then $x = a^{qm}$ is a solution of the equation $b^s = x^s$. If $d = bx^{-1}$, then $d\langle a \rangle = b\langle a \rangle$ and so

$d^s = b^s x^{-s} = b^s b^{-s} = e$. Therefore $\langle d \rangle \cap \langle a \rangle = \{e\}$, since $d^n \in \langle a \rangle$ implies $(bx^{-1})^n = b^n x^{-n} \in \langle a \rangle$. Thus $b^n \in \langle a \rangle$ implies $(b\langle a \rangle)^n = \langle a \rangle$ in $G/\langle a \rangle$, so $s|n$ and $d^n = e$.   □

7.5.3 Lemma.   Let G be a finite abelian p-group. If $\langle a \rangle$ is a maximal cyclic subgroup of G, then there exists a subgroup H with $G \cong \langle a \rangle \times H$.

Proof. The outline of the proof is to factor out $\langle a \rangle$ and use induction to decompose $G/\langle a \rangle$ into a direct product of cyclic groups. Then Lemma 7.5.2 can be used to choose the right preimages of the generators of $G/\langle a \rangle$ to generate the complement H of $\langle a \rangle$.

We use induction on the order of G. If $|G|$ is prime, then G is cyclic and there is nothing to prove. Consequently, we may assume that the statement of the theorem holds for all groups of order less than $|G| = p^\alpha$. If G is cyclic, then we are done. If not, let $\langle a \rangle$ be a maximal cyclic subgroup, and use the induction hypothesis repeatedly to write $G/\langle a \rangle$ as a direct product $H_1 \times H_2 \times \cdots \times H_n$ of cyclic subgroups.

We next use Lemma 7.5.2 to choose, for each i, a coset $a_i\langle a \rangle$ that corresponds to a generator of H_i such that $\langle a_i \rangle \cap \langle a \rangle = \{e\}$. We claim that $G \cong \langle a \rangle \times H$ for the smallest subgroup $H = \langle a_1, a_2, \ldots, a_n \rangle$ that contains $a_1, a_2, \ldots, a_n$.

First, if $g \in \langle a \rangle \cap \langle a_1, \ldots, a_n \rangle$, then $g = a_1^{m_1} \cdots a_n^{m_n} \in \langle a \rangle$ for some powers $m_1, \ldots, m_n$. Thus $g\langle a \rangle = a_1^{m_1} \cdots a_n^{m_n} \langle a \rangle = \langle a \rangle$, and since $G/\langle a \rangle$ is a direct product, this implies that $a_i^{m_i}\langle a \rangle = \langle a \rangle$ for each i. But then $a_i^{m_i} \in \langle a \rangle$, and so $a_i^{m_i} = e$ since $\langle a_i \rangle \cap \langle a \rangle = \{e\}$. Thus $g = e$.

Next, given $g \in G$, express the coset $g\langle a \rangle$ as $a_1^{m_1} \cdots a_n^{m_n}\langle a \rangle$ for exponents $m_1, \ldots, m_n$. Then $g \in g\langle a \rangle$, and so $g = a^m a_1^{m_1} \cdots a_n^{m_n}$ for some exponent m.

Thus we have shown that $\langle a \rangle \cap H = \{e\}$ and $G = \langle a \rangle H$, so $G \cong \langle a \rangle \times H$.   □

7.5.4 Theorem (Fundamental Theorem of Finite Abelian Groups).   Any finite abelian group is isomorphic to a direct product of cyclic groups of prime power order. Any two such decompositions have the same number of factors of each order.

Proof. We can use Theorem 7.5.1 to decompose the group G into a direct product of p-groups, and then we can use Lemma 7.5.3 to write each of these groups as a direct product of cyclic subgroups.

Uniqueness is shown by induction on $|G|$. It is enough to prove the uniqueness for a given p-group. We use additive notation and suppose that

$$\mathbf{Z}_{p^{\alpha_1}} \times \mathbf{Z}_{p^{\alpha_2}} \times \cdots \times \mathbf{Z}_{p^{\alpha_n}} = \mathbf{Z}_{p^{\beta_1}} \times \mathbf{Z}_{p^{\beta_2}} \times \cdots \times \mathbf{Z}_{p^{\beta_m}}$$

where $\alpha_1 \geq \alpha_2 \geq \cdots \geq \alpha_n$ and $\beta_1 \geq \beta_2 \geq \cdots \geq \beta_m$. Consider the subgroups in which each element has been multiplied by p. (In multiplicative notation, raising a to the pth power gives a homomorphism of G into G.) By induction, $\alpha_1 - 1 = \beta_1 - 1, \ldots$, which gives $\alpha_1 = \beta_1, \ldots$, with the possible exception of the α_i's and β_j's that equal 1. But the groups have the same order, and this determines that each has the same number of factors isomorphic to $\mathbf{Z}_p$.   □

Example 7.5.1

We will find all finite abelian groups of order 72. The first step is to find the prime factorization; $72 = 2^3 3^2$. There are three possible groups of order 8: $\mathbf{Z}_8$, $\mathbf{Z}_4 \times \mathbf{Z}_2$, and $\mathbf{Z}_2 \times \mathbf{Z}_2 \times \mathbf{Z}_2$. There are two possible groups of order 9: $\mathbf{Z}_9$ and $\mathbf{Z}_3 \times \mathbf{Z}_3$. This gives us the following possible groups:

$$\mathbf{Z}_8 \times \mathbf{Z}_9 \qquad \mathbf{Z}_4 \times \mathbf{Z}_2 \times \mathbf{Z}_9 \qquad \mathbf{Z}_2 \times \mathbf{Z}_2 \times \mathbf{Z}_2 \times \mathbf{Z}_9$$
$$\mathbf{Z}_8 \times \mathbf{Z}_3 \times \mathbf{Z}_3 \qquad \mathbf{Z}_4 \times \mathbf{Z}_2 \times \mathbf{Z}_3 \times \mathbf{Z}_3 \qquad \mathbf{Z}_2 \times \mathbf{Z}_2 \times \mathbf{Z}_2 \times \mathbf{Z}_3 \times \mathbf{Z}_3 . \quad \square$$

Example 7.5.2

There is another way to describe the possible abelian groups of order 72. We can combine the highest powers of each prime by using the fact that $\mathbf{Z}_m \times \mathbf{Z}_n \cong \mathbf{Z}_{mn}$ if $(m, n) = 1$. Then we have the groups in the following form:

$$\mathbf{Z}_{72} \qquad \mathbf{Z}_{36} \times \mathbf{Z}_2 \qquad \mathbf{Z}_{18} \times \mathbf{Z}_2 \times \mathbf{Z}_2$$
$$\mathbf{Z}_{24} \times \mathbf{Z}_3 \qquad \mathbf{Z}_{12} \times \mathbf{Z}_6 \qquad \mathbf{Z}_6 \times \mathbf{Z}_6 \times \mathbf{Z}_2 .$$

Note we have arranged the cyclic factors in such a way that the order of each factor is a divisor of the order of the preceding one. $\square$

7.5.5 Proposition.   Let G be a finite abelian group. Then G is isomorphic to a direct product of cyclic groups $\mathbf{Z}_{n_1} \times \mathbf{Z}_{n_2} \times \cdots \times \mathbf{Z}_{n_k}$ such that $n_i \mid n_{i-1}$ for $i = 2, 3, \ldots, k$.

Proof. We will use induction on the number of prime divisors of $|G|$. If G is a p-group, then we only need to arrange its factors in decreasing order of size, since the divisibility condition automatically follows. If $|G|$ has more than one prime factor, let H_p denote its p-Sylow subgroup, and let K_p be a subgroup with $G \cong H_p \times K_p$. Then the induction hypothesis may be applied to K_p to give $K_p \cong \mathbf{Z}_{n_1} \times \mathbf{Z}_{n_2} \times \cdots$. Furthermore, $H_p \cong \mathbf{Z}_{p^{\alpha_1}} \times \mathbf{Z}_{p^{\alpha_2}} \times \cdots$. Since n_1 and p^{α_1} are relatively prime, the subgroup $\mathbf{Z}_{p^{\alpha_1}} \times \mathbf{Z}_{n_1}$ is cyclic. Similarly, we may combine successive factors of H_p with factors of K_p, and in doing so we maintain the necessary divisibility relations. $\square$

We can use the fundamental theorem of finite abelian groups to give another proof of Proposition 3.5.8. This is the proof that is usually given to show that the multiplicative group of a finite field is cyclic.

7.5.6 Corollary.   Let G be a finite abelian group. If $a \in G$ is an element of maximal order in G, then the order of every element of G is a divisor of the order of a.

Proof. Let G be isomorphic to a direct product of cyclic groups $\mathbf{Z}_{n_1} \times \mathbf{Z}_{n_2} \times \cdots \times \mathbf{Z}_{n_k}$ such that $n_i \mid n_{i-1}$ for $i = 2, 3, \ldots, k$. Recall that the order of an element in a direct product is the least common multiple of the orders of its components.

Thus the largest possible order of an element of G is n_1. Furthermore, it is clear that the order of any element must be a divisor of n_1.   □

We now have enough information to complete a classification of groups of order less than 12. This can be done by using the fundamental theorem of finite abelian groups and Propositions 7.4.5 and 7.4.6. The case of order 8 requires some additional analysis, and hints are given in the exercises.

EXERCISES: SECTION 7.5

1. Give a representative of each isomorphism class of abelian groups of order 64.

2. Using both the form of Theorem 7.5.4 and that of Proposition 7.5.5, list all nonisomorphic abelian groups of the following orders:
   (a) order 108   **(b)** order 200   **(c)** order 900

3. Write each of the following groups as a direct product of cyclic groups of prime power order:
   (a) $\mathbf{Z}_{20}^{\times}$ **(b)** $\mathbf{Z}_{54}^{\times}$ **(c)** $\mathbf{Z}_{180}^{\times}$

4. Prove that if p is a prime and $\mathbf{Z}_{p^\alpha} \cong G_1 \times G_2$, then either $G_1 \cong \mathbf{Z}_{p^\alpha}$ or $G_2 \cong \mathbf{Z}_{p^\alpha}$.

5. Let G be a nonabelian group of order 8.
   (a) Prove that G must have an element of order 4, but none of order 8.
   (b) Let a be an element of order 4, and let $N = \langle a \rangle$. Show that there exists an element b such that $G = N \cup bN$.
   (c) Show that either $b^2 = e$ or $b^2 = a^2$. (Since N is normal, consider the order of bN in G/N.)
   (d) Show that bab^{-1} has order 4 and must be equal to a^3.
   (e) Conclude that either $G \cong D_4$ or else G is determined by the equations $a^4 = e$, $ba = a^3 b$, $b^2 = a^2$. Review Example 3.3.7 (the quaternion group) to verify that the second case can occur.

6. Determine (up to isomorphism) all groups of order less than 12.

7.6 SOLVABLE GROUPS

We are now ready to study groups arising from equations that are solvable by radicals. (We will not be able to give a proof of this correspondence until after we develop some new ideas in the next chapter.) Groups in the class are simply said to be solvable. It is obvious from the following definition that any abelian group is solvable.

   7.6.1 Definition.   The group G is said to be *solvable* if there exists a finite chain of subgroups $G = N_0 \supseteq N_1 \supseteq \ldots \supseteq N_n$ such that (i) N_i is a normal subgroup in N_{i-1} for $i = 1, 2, \ldots, n$, (ii) N_{i-1}/N_i is abelian for $i = 1, 2, \ldots, n$, and (iii) $N_n = \{e\}$.

Example 7.6.1

Let $G = S_3$, the group of all permutations on three elements (the smallest nonabelian group). For the descending chain of subgroups $N_0 = G$, $N_1 = A_3$, and $N_2 = \{e\}$ we have $N_0/N_1 \cong \mathbf{Z}_2$ and $N_1/N_2 \cong \mathbf{Z}_3$. Recall that A_3 is the set $\{(1), (1, 2, 3), (1, 3, 2)\}$ of all even permutations of S_3. This shows that S_3 is a solvable group. □

Example 7.6.2

Let $G = S_4$, and let $N_0 = G$, $N_1 = A_4$. Since A_4 has index 2 in S_4, we must have $N_0/N_1 \cong \mathbf{Z}_2$. Let N_3 be the trivial subgroup $\{e\}$, and let

$$N_2 = \{(1), (1, 2)(3, 4), (1, 3)(2, 4), (1, 4)(2, 3)\}.$$

Then N_2 is a subgroup of G since it is closed under multiplication. Moreover, it is a normal subgroup of both G and N_1, since conjugating an element of N_2 by any element of G must yield an element that has the same cycle structure, and the elements of N_2 are the only permutations in S_4 that can be expressed as products of disjoint transpositions. Since $[N_1 : N_2] = 3$, we have $N_1/N_2 \cong \mathbf{Z}_3$. For all $\sigma \in N_2$, we have $\sigma^2 = (1)$, and so $N_2/N_3 \cong N_2$ is isomorphic to the Klein four-group $\mathbf{Z}_2 \times \mathbf{Z}_2$. Thus each factor group in the given descending chain of subgroups is abelian, and this shows that S_4 is a solvable group. □

In Example 7.6.2 we could have added another term at the bottom of the descending chain of subgroups, by letting $N_3 = \{(1), (1, 2)(3, 4)\}$ and $N_4 = \{e\}$. Then each factor group N_i/N_{i+1} would have been isomorphic to a cyclic group. This can always be done, as the next proposition shows (for finite groups).

7.6.2 Proposition.   Let G be a finite group. Then G is solvable if and only if there exists a finite chain of subgroups $G = N_0 \supseteq N_1 \supseteq \ldots \supseteq N_n$ such that
   (i) N_i is a normal subgroup in N_{i-1} for $i = 1, 2, \ldots, n$,
   (ii) N_{i-1}/N_i is cyclic of prime order for $i = 1, 2, \ldots, n$, and
(iii) $N_n = \{e\}$.

Proof. Assume that G is solvable, with a chain of subgroups $N_0 \supseteq N_1 \supseteq \ldots \supseteq N_n$ satisfying the conditions of Definition 7.6.1. If some factor group N_i/N_{i+1} is not cyclic of prime order, then let p be a prime number such that p divides $|N_i/N_{i+1}|$. By Cauchy's theorem there exists an element aN_{i+1} of N_i/N_{i+1} of order p. Recalling that subgroups of N_i/N_{i+1} correspond to subgroups of N_i that contain N_{i+1}, we let H be the inverse image in N_i of $\langle aN_{i+1}\rangle$. Thus we have $N_i \supseteq H \supseteq N_{i+1}$, and H is a normal subgroup of N_i since $\langle aN_{i+1}\rangle$ is a normal subgroup of the abelian group N_i/N_{i+1}. Furthermore, N_{i+1} is normal in H since N_{i+1} is normal in $N_i \supseteq H$. Since N_i/H is a homomorphic image of the abelian group N_i/N_{i+1}, it is abelian, and $H/N_{i+1} \cong \langle aN_{i+1}\rangle$ is a cyclic group of order p. We next consider the descending chain of subgroups constructed from the original chain by adding K_i. We can apply the same procedure repeatedly, ultimately arriving at a descending chain of subgroups, each normal in the previous one, such that all factors are cyclic of prime order.

The converse is obvious.   □

7.6.3 Theorem.   Let p be a prime. Any finite p-group is solvable.

Proof. Let G be any group of order p^m. First let C_0 be the trivial subgroup. The center $Z(G) = \{g \in G \mid gx = xg$ for all $x \in G\}$ of G is nontrivial, so we let $C_1 = Z(G)$. It follows from the definition of $Z(G)$ that C_1 is abelian, and also that it is normal in G. Since the factor group G/C_1 is defined, it also has nontrivial center $Z(G/C_1)$ since its order is again a power of p. Let C_2 be the subgroup of G that contains C_1 and corresponds to $Z(G/C_1)$. Since normal subgroups correspond to normal subgroups, we see that C_2 is normal in G. Furthermore, $C_2/C_1 \cong Z(G/C_1)$, and so this factor is abelian. We can continue this procedure until we obtain $C_n = G$ for some n. Then we have constructed a descending chain $G = C_n \supseteq \ldots \supseteq C_1 \supseteq C_0$ satisfying the conditions of Definition 7.6.1, and so G is solvable.   □

7.6.4 Definition.   Let G be a group. An element $g \in G$ is called a *commutator* if $g = aba^{-1}b^{-1}$ for elements $a, b \in G$. The smallest subgroup that contains all commutators of G is called the *commutator subgroup* or *derived subgroup* of G, and is denoted by G'.

We note that the commutators themselves do not necessarily form a subgroup.

7.6.5 Proposition.   Let G be a group with commutator subgroup G'.
  (a) The subgroup G' is normal in G, and the factor group G/G' is abelian.
  (b) If N is any normal subgroup of G, then the factor group G/N is abelian if and only if $G' \subseteq N$.

Proof. (a) Let $x \in G'$ and let $g \in G$. Then $gxg^{-1}x^{-1} \in G'$, and since $x \in G'$, we must have $gxg^{-1} = gxg^{-1}x^{-1}x \in G'$.

The factor group G/G' must be abelian since for any cosets aG', bG' we have

$$aG'bG'a^{-1}G'b^{-1}G' = aba^{-1}b^{-1}G' = G'.$$

Thus $aG'bG' = bG'aG'$.

(b) Let N be a normal subgroup of G. If $N \supseteq G'$, then G/N is a homomorphic image of G/G' and must be abelian. Conversely, suppose that G/N is abelian. Then $aNbN = bNaN$ for all $a, b \in G$, or simply $aba^{-1}b^{-1}N = N$, showing that every commutator of G belongs to N. This implies that $G' \subseteq N$.   □

7.6.6 Definition.   Let G be a group. The subgroup $(G')'$ is called the *second derived subgroup* of G. We define $G^{(k)}$ inductively as $(G^{(k-1)})'$, and call it the *kth derived subgroup* of G.

As in the proof of part (a) of Proposition 7.6.5, we note that the kth derived subgroup is always normal. In fact, it can be shown to be invariant under all automorphisms of G. Our reason for considering the commutator subgroups is to develop the following criterion for solvability.

7.6.7 Theorem.   A group G is solvable if and only if $G^{(n)} = \{e\}$ for some positive integer n.

Proof. First assume that G is solvable and that $G = N_0 \supseteq N_1 \supseteq \ldots \supseteq N_n = \{e\}$ is a chain of subgroups such that N_i/N_{i+1} is abelian. Since G/N_1 is abelian, we have $G' \subseteq N_1$. Then we must have $G^{(2)} = (G')' \subseteq (N_1)' \subset N_2$ since N_1/N_2 is abelian. In general, $G^{(k)} \subseteq N_k$, and so $G^{(n)} = \{e\}$.

Conversely, if $G^{(n)} = \{e\}$, then in the descending chain $G \supseteq G' \supseteq \ldots \supseteq G^{(n)} = \{e\}$, each subgroup is normal in G and each factor $G^{(i)}/G^{(i+1)}$ is abelian, showing that G is solvable. $\square$

7.6.8 Corollary.   Let G be a group.
(a) If G is solvable, then so is any subgroup or homomorphic image of G.
(b) If N is a normal subgroup of G such that both N and G/N are solvable, then G is solvable.

Proof. (a) Assume that G is solvable, with $G^{(n)} = \{e\}$, and let H be a subgroup of G. Since $H' \subseteq G'$, it follows inductively that $H^{(n)} \subseteq G^{(n)} = \{e\}$.

To show that any homomorphic image of G is solvable, it suffices to show that G/N is solvable for any normal subgroup N of G. Commutators $aba^{-1}b^{-1}$ of G correspond directly to commutators $aNbNa^{-1}Nb^{-1}N = aba^{-1}b^{-1}N$ of G/N, and so the kth derived subgroup of G/N is the projection of the kth derived subgroup of G onto G/N. It is then obvious that G/N is solvable.

(b) Assume that N is a normal subgroup of G such that N and G/N are solvable. Then $(G/N)^{(n)} = \{N\}$ for some positive integer n, and the correspondence between commutators of G/N and G that we observed in the proof of part (a) shows that $G^{(n)} \subseteq N$. But then $N^{(k)} = \{e\}$ for some positive integer k, and so $G^{(n+k)} = (G^{(n)})^{(k)} \subseteq N^{(k)} = \{e\}$. Thus G is solvable. $\square$

We have seen several methods of determining whether or not a given group is solvable. The methods involved sequences of subgroups, determined in various ways. This raises the question of uniqueness of such sequences. Theorem 7.6.10 shows that if we have a sequence that cannot be lengthened, then there is a certain amount of uniqueness, which we can illustrate with the following example. In $\mathbf{Z}_6$ we have the following two descending chains of subgroups:

$$\mathbf{Z}_6 \supset 3\mathbf{Z}_6 \supset \{0\} \qquad \text{and} \qquad \mathbf{Z}_6 \supset 2\mathbf{Z}_6 \supset \{0\}.$$

In the first chain we have $\mathbf{Z}_6/3\mathbf{Z}_6 \cong \mathbf{Z}_3$ and $3\mathbf{Z}_6 \cong \mathbf{Z}_2$. On the other hand, in the second chain we have $\mathbf{Z}_6/2\mathbf{Z}_6 \cong \mathbf{Z}_2$ and $2\mathbf{Z}_6 \cong \mathbf{Z}_3$. At least we have the same factor groups, even though they occur in a different order.

7.6.9 Definition.   Let G be a finite group. A chain of subgroups $G = N_0 \supseteq N_1 \supseteq \ldots \supseteq N_n$ such that (i) N_i is a normal subgroup in N_{i-1} for $i = 1, 2, \ldots, n$, (ii) N_{i-1}/N_i is simple for $i = 1, 2, \ldots, n$, and (iii) $N_n = \{e\}$ is called a *composition series* for G. The factor groups N_{i-1}/N_i are called the *composition factors* determined by the series.

Note that any finite group G has at least one composition series. Let $N_0 = G$ and then let N_1 be a maximal normal subgroup of G. To continue the series, let N_2 be a maximal normal subgroup of N_1, etc. Since G is finite, the sequence must terminate at the trivial subgroup after at most a finite number of steps.

7.6.10 Theorem (Jordan-Hölder).   Any two compositions series for a finite group have the same length. Furthermore, there exists a one-to-one correspondence between composition factors of the two composition series under which corresponding composition factors are isomorphic.

Proof.   The proof uses induction on the length of a composition series for the group G. That is, we will show that if G is a finite group with a composition series $G \supseteq N_1 \supseteq \ldots \supseteq N_k = \{e\}$ of length k, then any other composition series $G \supseteq H_1 \supseteq \ldots \supseteq H_m = \{e\}$ for G must have $m = k$ and there must exist a permutation $\sigma \in S_k$ such that $N_{i-1}/N_i \cong H_{\sigma(i)-1}/H_{\sigma(i)}$ for $i = 1, 2, \ldots, k$. If $k = 1$, then G must be simple and so there is only one possible composition series.

Assume that G has a composition series of length k, as above. In addition, assume that the induction hypothesis is satisfied for all groups with a composition series of length less than k, and assume that G has another composition series of length m, as above. If $N_1 = H_1$, then we can apply the induction hypothesis to the composition series $N_1 \supseteq N_2 \supseteq \ldots \supseteq N_k = \{e\}$ and thus obtain the result for G. If $H_1 \neq N_1$, then let $N_1 \cap H_1 \supseteq K_3 \supseteq \ldots \supseteq K_n = \{e\}$ be a composition series. This gives the diagram in Figure 7.6.1.

Since N_1 and H_1 are normal in G, so is their intersection. Furthermore, $N_1 H_1$ is normal in G, and so it must be equal to G since it contains both N_1 and H_1, which are maximal normal subgroups. Applying the second isomorphism theorem (Theorem 7.1.2) gives us the following isomorphisms:

$$N_1/(N_1 \cap H_1) \cong (N_1 H_1)/H_1 = G/H_1$$

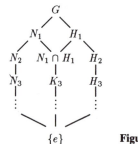

Figure 7.6.1

and

$$H_1/(N_1 \cap H_1) \cong (N_1H_1)/N_1 = G/N_1.$$

This implies that $N_1 \cap H_1$ is a maximal normal subgroup of both N_1 and H_1, so we have the following four composition series for G:

$$G \supseteq N_1 \supseteq N_2 \supseteq \ldots \supseteq N_k = \{e\},$$

$$G \supseteq N_1 \supseteq N_1 \cap H_1 \supseteq \ldots \supseteq K_n = \{e\},$$

$$G \supseteq H_1 \supseteq N_1 \cap H_1 \supseteq \ldots \supseteq K_n = \{e\},$$

and

$$G \supseteq H_1 \supseteq H_2 \supseteq \ldots \supseteq H_m = \{e\}.$$

The first two composition series have N_1 as the first term, so we must have $n = k$, and isomorphic composition factors. The last two composition series have H_1 as the first term, so again we must have $m = n$, and isomorphic composition factors. The isomorphisms given above show that the middle two composition series have isomorphic composition factors, and so by transitivity the first and last composition series have the same length and isomorphic composition factors. $\square$

EXERCISES: SECTION 7.6

1. Let G be a group and let N be a normal subgroup of G. For $a, b \in G$, let $[a, b]$ denote the commutator $aba^{-1}b^{-1}$.
   (a) Show that $g[a, b]g^{-1} = [gag^{-1}, gbg^{-1}]$.
   (b) Show that N' is a normal subgroup of G.

2. Prove that an abelian group has a composition series if and only if it is finite.

3. Give an example of two groups G_1 and G_2 that have the same composition factors, but are not isomorphic.

4. Prove that if $G_1 \times G_2 \times \cdots \times G_s = H_1 \times H_2 \times \cdots \times H_t$, where each of the groups G_i and H_i is simple, then $s = t$ and there exists $\sigma \in S_t$ such that $G_j \cong H_{\sigma(j)}$ for $j = 1, \ldots, t$.

5. Let p and q be primes, not necessarily distinct. A famous theorem of Burnside states that any group of order $p^n q^m$ is solvable for all $n, m \in \mathbf{Z}^+$.
   (a) Show that any group of order pq is solvable.
   (b) Show that any group of order $p^2 q$ is solvable.
   (c) Show that any group of order $p^n q$ is solvable if $p > q$.

6. Let G be a group. A subgroup H of G is called a *characteristic* subgroup if $\phi(H) \subseteq H$ for all $\phi \in \mathrm{Aut}(G)$.
   (a) Prove that any characteristic subgroup is normal.
   (b) Prove that the center of any group is a characteristic subgroup.
   (c) Prove that the commutator subgroup is always a characteristic subgroup.
   (d) Prove that any normal Sylow p-subgroup is a characteristic subgroup.

(e) Prove that if H is a normal subgroup of G, and K is a characteristic subgroup of H, then K is normal in G.

7. Prove that a nontrivial finite solvable group of order ≥ 2 must contain a nontrivial normal abelian subgroup.

8. Prove that if G is a finite group that is not solvable, then G must contain a nontrivial normal subgroup N such that $N' = N$.

7.7 SIMPLE GROUPS

Cyclic groups of prime order form the most elementary class of simple groups. In the introduction to Chapter 7 we presented, without proof, another class of finite simple groups, constructed from the group of invertible $n \times n$ matrices over a finite field. Among the other classes of finite simple groups, the only one we will deal with is the class of alternating groups. We will show that A_n is simple for $n \geq 5$.

7.7.1 Lemma. Every even permutation can be expressed as a product of 3-cycles.

Proof. The product of any two transpositions must have one of the following forms (where different letters represent distinct positive integers):

$$(a, b)(a, b) = (1),$$

$$(a, b)(b, c) = (a, b, c),$$

$$(a, b)(c, d) = (a, b, c)(b, c, d).$$

Since any element of A_n is a product of an even number of transpositions, this shows that any element of A_n can be expressed as a product of 3-cycles. $\square$

7.7.2 Theorem. The symmetric group S_n is not solvable for $n \geq 5$.

Proof. We first show that $(A_n)' = A_n$. Let (a, b, c) be any 3-cycle in A_n. Since $n \geq 5$, we can choose $d, f \in \mathbf{Z}^+$ different from a, b, c. Then

$$(a, b, c) = (a, b, d)(a, c, f)(a, d, b)(a, f, c)$$

and we have shown that any 3-cycle is a commutator. Together with Lemma 7.7.1, this shows that any element of A_n is a product of commutators of elements in A_n.

Finally, we have $(S_n)' \subseteq A_n$ since the factor group S_n/A_n is abelian. In the other direction, we have $A_n = (A_n)' \subseteq (S_n)'$, and so $(S_n)' = A_n$. It follows that $(S_n)^{(k)} = A_n$ for all $k \geq 1$, and thus S_n is not solvable. $\square$

7.7.3 Lemma. If $n \geq 4$, then no proper normal subgroup of A_n contains a 3-cycle.

Proof. Let N be a normal subgroup of A_n that contains a 3-cycle (a, b, c). Note that N must also contain the square (a, c, b) of (a, b, c). Conjugating (a, c, b) by the even permutation $(a, b)(c, x)$, we obtain (a, b, x), which must belong to N since N is normal in A_n. By repeating this argument we can obtain any 3-cycle (x, y, z), and thus N contains all 3-cycles. By Lemma 7.7.1, $N = A_n$.   □

7.7.4 Theorem.   The alternating group A_n is simple if $n \geq 5$.

Proof. Let N be a normal subgroup of A_n. Using the previous lemma, we only need to show that N contains a 3-cycle. Let $\sigma \in N$, and assume that σ is written as a product of disjoint cycles. If σ is itself a 3-cycle, then we are done.

If σ contains a cycle of length ≥ 4, say $\sigma = (a, b, c, d, \ldots) \cdots$, then let $\tau = (b, c, d)$. Since N is a normal subgroup, σ^{-1} and $\tau\sigma\tau^{-1}$ must belong to N, and so $\sigma^{-1}\tau\sigma\tau^{-1} \in N$. A direct computation shows that $\sigma^{-1}\tau\sigma\tau^{-1} = (a, b, d)$, and thus N contains a 3-cycle.

If σ contains a 3-cycle but no longer cycle, then either $\sigma = (a, b, c)(d, f, g) \cdots$ or $\sigma = (a, b, c)(d, f) \cdots$. In the first case, $\sigma^{-1}\tau\sigma\tau^{-1} = (a, b, d, c, g)$, for $\tau = (b, c, d)$, and in the second case, $\sigma^{-1}\tau\sigma\tau^{-1} = (a, b, d, c, f)$. Thus, by the previous argument, N must again contain a 3-cycle.

Finally, if σ consists of only transpositions, then either $\sigma = (a, b)(c, d)$ or $\sigma = (a, b)(c, d) \cdots$. The second case reduces to the first, since $\sigma^{-1}\tau\sigma\tau^{-1} = (a, d)(b, c)$, again using $\tau = (b, c, d)$. In the first case, since $n \geq 5$, there must be a fifth element, say f. For the permutation $\rho = (c, d, f)$, we have $\sigma^{-1}\rho\sigma\rho^{-1} = (c, d, f)$. Thus N must contain a 3-cycle, completing the proof.   □

EXERCISES: SECTION 7.7

1. Let G be a group of order $2m$, where m is odd and $m \geq 1$. Show that G is not simple.

2. **(a)** Let G be a group with a subgroup H of index n. Show that there is a homomorphism $\phi : G \to S_n$ for which $ker(\phi) = \cap_{g \in G} gHg^{-1}$.

   Hint: Let G act on the set of left cosets of H by defining $a(gH) = agH$, for all $a, g \in G$.

   (b) Prove that if an infinite group contains a subgroup of finite index, then it is not simple.

   (c) Prove that if G is a simple group that contains a subgroup of index n, then G can be embedded in S_n.

3. **(a)** Let S be the set $\{1, 2, \ldots\}$ of positive integers, and let $G = \{\sigma \in \mathrm{Sym}(S) \mid \sigma(j) = j$ for all but finitely many $j\}$. Show that G is a subgroup of $\mathrm{Sym}(S)$.

   (b) Let A_∞ be the subgroup of G generated by all 3-cycles (a, b, c). Show that A_∞ is a simple group.

   (c) Show that every finite group can be embedded in A_∞.

4. Let G be a finite group containing a subgroup H of index p, where p is the smallest prime divisor of $|G|$. Prove that H is normal in G, and hence G is not simple if $|G| > p$.

5. Let H be a subgroup of G and let S be the set of all subgroups conjugate to H. Define an action of G on S by letting $a(gHg^{-1}) = (ag)H(ag)^{-1}$, for $a, g \in G$. Show that the corresponding homomorphism $\phi : G \to \text{Sym}(S)$ has kernel $\cap_{g \in G} gN(H)g^{-1}$.

6. Let G be an infinite group containing an element (not equal to the identity) that has only finitely many conjugates. Use the previous exercise to prove that G is not simple.

8

Galois Theory

We have seen in Chapter 4 that a polynomial equation with real coefficients can be solved by radicals, provided it has degree less than five. The solutions for cubic and quartic equations were discovered in the sixteenth century, and from then until the beginning of the nineteenth century, some of the best mathematicians of the period (such as Euler and Lagrange) attempted to find a similar solution by radicals for equations of degree five. All attempts ended in failure, and finally a paper was published by Ruffini 1798, in which he asserted that no method of solution could be found. The proof was not well received at the time, and even further elaborations of the ideas, published in 1802 and 1813, were not regarded as constituting a proof. Ruffini's proof is based on the assumption that the radicals necessary in the solution of the quintic can all be expressed as rational functions of the roots of the equation, and this was only proved later by Abel. It is generally agreed that the first fully correct proof of the insolvability of the quintic was published by Abel in 1826.

Abel attacked the general problem of when a polynomial equation could be solved by radicals. His papers inspired Galois to formulate a theory of solvability of equations involving the structures we now know as groups and fields. Galois worked with fields to which all roots of the given equation had been adjoined. He then considered the set of all permutations of these roots that leave the coefficient field unchanged. The permutations form a group, called the Galois group of the equation. From the modern point of view, the permutations of the roots can be extended to automorphisms of the field, and form a group under composition of functions. Then an equation is solvable by radicals if and only if its Galois group is solvable. Thus to show that there exists an equation that is not solvable by radicals, it is enough to find an equation whose Galois group is S_5.

In 1829, at the age of seventeen, Galois presented two papers on the solution of algebraic equations to the Académie des Sciences de Paris. Both were sent to Cauchy, who lost them. In 1830, Galois presented another paper to the Académie, which was this time given to Fourier, who died before reading it. A third, revised version, was submitted in 1831. This manuscript was reviewed carefully, but was not understood. It was only published in 1846 by Liouville, fourteen years after the death of Galois. Galois had been involved in the 1830 revolution and had been imprisoned twice. In May of 1832 he was forced to accept a duel (the circumstances and motives are unclear), and certain that he would be killed, he spent the night before the duel writing a long letter to his friend Chevalier explaining the basic ideas of his research. This work involved other areas in addition to what we now call Galois theory, and included the construction of finite fields, which are now called Galois fields (see Section 6.5).

In our modern terminology, we let F be the splitting field of a polynomial over the field K that has no repeated roots. The fundamental theorem of Galois theory (in Section 8.3) shows that there is a correspondence between the normal subgroups of the Galois group and the intermediate fields between K and F that are splitting fields for some polynomial. This gives the necessary connection between the structure of the Galois group and the successive adjunctions that must be made in order to solve the equation by radicals.

8.1 THE GALOIS GROUP OF A POLYNOMIAL

We begin by reviewing some facts about automorphisms. Recall that an automorphism ϕ of a field F is a one-to-one correspondence $\phi : F \rightarrow F$ such that for all $a, b \in F$,

$$\phi(a + b) = \phi(a) + \phi(b) \qquad \text{and} \qquad \phi(ab) = \phi(a)\phi(b).$$

That is, ϕ is an automorphism of the additive group of F, and since $a \neq 0$ implies $\phi(a) \neq 0$, it is also an automorphism when restricted to the multiplicative group $F^\times$. We use the notation Aut(F) for the group of all automorphisms of F. The identity element 1 must be left fixed by ϕ, since by elementary results on group homomorphisms we have $\phi(1) = 1$.

Recall that the smallest subfield containing the identity element 1 is called the *prime subfield* of F. If F has characteristic zero, then the prime subfield is isomorphic to $\mathbf{Q}$, and consists of all elements of the form

$$\{(n \cdot 1)(m \cdot 1)^{-1} \mid n, m \in \mathbf{Z}, m \neq 0\}.$$

If F has characteristic p (p a prime), then the prime subfield is isomorphic to $\mathbf{Z}_p$, and consists of all elements of the form

$$\{n \cdot 1 \mid n = 0, 1, \ldots, p - 1\}.$$

For any automorphism ϕ of F, we have

$$\phi(n \cdot 1) = n \cdot \phi(1) = n \cdot 1$$

for any $n \in \mathbf{Z}.$ Furthermore, if $n \cdot 1 \neq 0$, then

$$\phi((n \cdot 1)^{-1}) = (\phi(n \cdot 1))^{-1} = (n \cdot 1)^{-1}.$$

This shows that any automorphism of F must leave the prime subfield of F fixed.

To study solvability by radicals of a polynomial equation $f(x) = 0$, we let K be the field generated by the coefficients of $f(x)$, and let F be a splitting field for $f(x)$ over K. (We know, by Kronecker's theorem, that we can always find roots. The question is whether or not the roots have a particular form.) Galois considered permutations of the roots that leave the coefficient field fixed. The modern approach is to consider the automorphisms determined by these permutations.

8.1.1 Proposition.   Let F be an extension field of K. The set of all automorphisms $\phi : F \to F$ such that $\phi(a) = a$ for all $a \in K$ is a group under composition of functions.

Proof. We only need to show that the given set is a subgroup of $\mathrm{Aut}(F)$. It certainly contains the identity function. If $\phi, \theta \in \mathrm{Aut}(F)$ and $\phi(a) = a$, $\theta(a) = a$ for all $a \in K$, then $\phi\theta(a) = \phi(\theta(a)) = \phi(a) = a$ and $\phi^{-1}(a) = \phi^{-1}\phi(a) = a$ for all $a \in K$. $\square$

8.1.2 Definition.   Let F be an extension field of K. The set

$$\{\theta \in \mathrm{Aut}(F) \mid \theta(a) = a \text{ for all } a \in K\}$$

is called the *Galois group* of F over K, denoted by $\mathrm{Gal}(F/K)$.

8.1.3 Definition.   Let K be a field, let $f(x) \in K[x]$, and let F be a splitting field for $f(x)$ over K. Then $\mathrm{Gal}(F/K)$ is called the *Galois group of* $f(x)$ *over* K, or the *Galois group of the equation* $f(x) = 0$ *over* K.

8.1.4 Proposition.   Let F be an extension field of K, and let $f(x) \in K[x]$. Then any element of $\mathrm{Gal}(F/K)$ defines a permutation of the roots of $f(x)$ that lie in F.

Proof. Let $f(x) = a_0 + a_1 x + \ldots + a_n x^n$, where $a_i \in K$ for $i = 0, 1, \ldots, n$. If $u \in F$ with $f(u) = 0$ and $\theta \in \mathrm{Gal}(F/K)$, then we have

$$\begin{aligned}
\theta(f(u)) &= \theta(a_0 + a_1 u + \ldots + a_n u^n) \\
&= \theta(a_0) + \theta(a_1 u) + \ldots + \theta(a_n u^n) \\
&= \theta(a_0) + \theta(a_1)\theta(u) + \ldots + \theta(a_n)(\theta(u))^n
\end{aligned}$$

since θ preserves sums and products. Finally, since $\theta(a_i) = a_i$ for $i = 0, 1, \ldots, n$, we have

$$\theta(f(u)) = a_0 + a_1\theta(u) + \ldots + a_n(\theta(u))^n.$$

Since $f(u) = 0$, we must have $\theta(f(u)) = 0$, and thus

$$a_0 + a_1\theta(u) + \ldots + a_n(\theta(u))^n = 0,$$

showing that $f(\theta(u)) = 0$.

Thus θ maps roots of $f(x)$ to roots of $f(x)$. Since there are only finitely many roots and θ is one-to-one, θ must define a permutation of those roots of $f(x)$ that lie in F. $\square$

Example 8.1.1

In this example we compute $\text{Gal}(\mathbf{Q}(\sqrt[3]{2})/\mathbf{Q})$. In general, if $F = K(u_1, \ldots, u_n)$, then any automorphism $\theta \in \text{Gal}(F/K)$ is completely determined by the values $\theta(u_1), \ldots, \theta(u_n)$. Thus if we let $\theta \in \text{Gal}(\mathbf{Q}(\sqrt[3]{2})/\mathbf{Q})$, then θ is completely determined by $\theta(\sqrt[3]{2})$. Now $\sqrt[3]{2}$ is a root of the polynomial $x^3 - 2 \in \mathbf{Q}[x]$, so by Proposition 8.1.4, θ must map $\sqrt[3]{2}$ into a root of $x^3 - 2$. The other two roots are $\omega\sqrt[3]{2}$ and $\omega^2\sqrt[3]{2}$, where $\omega = (-1 + \sqrt{2}i)/2$ is a complex cube root of unity. Since these values are not real numbers, they cannot belong to $\mathbf{Q}(\sqrt[3]{2})$, and so we must have $\theta(\sqrt[3]{2}) = \sqrt[3]{2}$. This shows that $\text{Gal}(\mathbf{Q}(\sqrt[3]{2})/\mathbf{Q})$ is the trivial group consisting only of the identity automorphism. Note also that any automorphism of $\mathbf{Q}(\sqrt[3]{2})$ fixes $\mathbf{Q}$, so in fact $\text{Aut}(\mathbf{Q}(\sqrt[3]{2}))$ is also trivial. $\square$

Example 8.1.2

In this example we will compute $\text{Gal}(\mathbf{Q}(\sqrt{2} + \sqrt{3})/\mathbf{Q})$. The field extension $\mathbf{Q}(\sqrt{2} + \sqrt{3})$ is easier to work with when expressed as $\mathbf{Q}(\sqrt{2}, \sqrt{3})$. (We showed in Example 6.2.1 that these two fields are the same.)

Let θ be any automorphism in $\text{Gal}(\mathbf{Q}(\sqrt{2}, \sqrt{3})/\mathbf{Q})$. The roots of $x^2 - 2$ are $\pm\sqrt{2}$ and the roots of $x^2 - 3$ are $\pm\sqrt{3}$, and so we must have $\theta(\sqrt{2}) = \pm\sqrt{2}$ and $\theta(\sqrt{3}) = \pm\sqrt{3}$. Since $\{1, \sqrt{3}\}$ is a basis for $\mathbf{Q}(\sqrt{3})$ over $\mathbf{Q}$ and $\{1, \sqrt{2}\}$ is a basis for $\mathbf{Q}(\sqrt{2}, \sqrt{3})$ over $\mathbf{Q}(\sqrt{3})$, recall that $\{1, \sqrt{2}, \sqrt{3}, \sqrt{2}\sqrt{3}\}$ is a basis for $\mathbf{Q}(\sqrt{2}, \sqrt{3})$ over $\mathbf{Q}$. As soon as we know the action of θ on $\sqrt{2}$ and $\sqrt{3}$, its action on $\sqrt{6} = \sqrt{2}\sqrt{3}$ is determined. This gives a total of four possibilities for θ, which we have labeled with subscripts:

$$\theta_1(a + b\sqrt{2} + c\sqrt{3} + d\sqrt{6}) = a + b\sqrt{2} + c\sqrt{3} + d\sqrt{6}$$

$$\theta_2(a + b\sqrt{2} + c\sqrt{3} + d\sqrt{6}) = a - b\sqrt{2} + c\sqrt{3} - d\sqrt{6}$$

$$\theta_3(a + b\sqrt{2} + c\sqrt{3} + d\sqrt{6}) = a + b\sqrt{2} - c\sqrt{3} - d\sqrt{6}$$

$$\theta_4(a + b\sqrt{2} + c\sqrt{3} + d\sqrt{6}) = a - b\sqrt{2} - c\sqrt{3} + d\sqrt{6}.$$

It is left as an exercise to show that each of these functions defines an automorphism. In each case, repeating the automorphism gives the identity mapping. Thus $\theta^2 = e$ for all $\theta \in \text{Gal}(\mathbf{Q}(\sqrt{2}, \sqrt{3})/\mathbf{Q})$, which shows that this Galois group must be isomorphic to the Klein four-group $\mathbf{Z}_2 \times \mathbf{Z}_2$.

We note that $\mathbf{Q}(\sqrt{2}, \sqrt{3})$ is the splitting field of $(x^2 - 2)(x^2 - 3)$ over $\mathbf{Q}$. The next theorem shows that the number of elements in the Galois group must be equal to $[\mathbf{Q}(\sqrt{2}, \sqrt{3}) : \mathbf{Q}]$. This degree is 4, as was shown in Example 6.1.2. Since we found at most four distinct mappings that carry roots to roots, they must in fact all be automorphisms. Thus if we utilize the theorem, we do not actually have to go through the details of showing that the mappings preserve addition and multiplication. □

8.1.5 Lemma.   Let $f(x) \in K[x]$ be a polynomial with no repeated roots and let F be a splitting field for $f(x)$ over K. If $\phi : K \to L$ is a field isomorphism that maps $f(x)$ to $g(x) \in L[x]$ and E is a splitting field for $g(x)$ over L, then there exist exactly $[F : K]$ isomorphisms $\theta : F \to E$ such that $\theta(a) = \phi(a)$ for all $a \in K$.

Proof. The proof uses induction on the degree of $f(x)$. We follow the proof of Lemma 6.4.4 exactly, except that now we keep careful track of the number of possible isomorphisms. If $f(x)$ has degree 0 or 1, then $F = K$ and $E = L$, so there is nothing to prove. We now assume that the result holds for all polynomials of degree less than n and for all fields K. Let $p(x)$ be an irreducible factor of $f(x)$ of degree d, which maps to the irreducible factor $q(x)$ of $g(x)$. All roots of $p(x)$ belong to F, so we may choose one, say u, which gives $K \subseteq K(u) \subseteq F$. Since $f(x)$ has no repeated roots, the same is true of $q(x)$, so we may choose any one, say v, of the d roots of $q(x)$ in E, which gives $L \subseteq L(v) \subseteq E$. By Lemma 6.4.3 there exist d isomorphisms $\phi' : K(u) \to L(v)$ (one for each root v) such that $\phi'(u) = v$ and $\phi'(a) = \phi(a)$ for all $a \in K$. If we write $f(x) = (x - u)s(x)$ and $g(x) = (x - v)t(x)$, then the polynomial $s(x)$ has degree less than n, the extension F is a splitting field for $s(x)$ over $K(u)$, and the extension E is a splitting field for $t(x)$ over $L(v)$. Thus the induction hypothesis can be applied, and so there exist $[F : K(u)]$ isomorphisms $\theta : F \to E$ such that $\theta(x) = \phi'(x)$ for all $x \in K(u)$. In particular, $\theta(a) = \phi'(a) = \phi(a)$ for all $a \in K$. Thus we have precisely $[F : K] = [F : K(u)][K(u): K]$ extensions of the original isomorphism ϕ, and the proof is complete. □

8.1.6 Theorem.   Let K be a field, let $f(x) \in K[x]$, and let F be a splitting field for $f(x)$ over K. If $f(x)$ has no repeated roots, then

$$|\mathrm{Gal}(F/K)| = [F : K].$$

Proof. This is an immediate consequence of the preceding lemma. □

8.1.7 Corollary.   Let K be a finite field and let F be an extension of K with $[F : K] = m$. Then $\mathrm{Gal}(F/K)$ is a cyclic group of order m.

Proof. Since F is a finite extension of K, it is also a finite field, and so if $\mathrm{char}(K) = p$, then F has p^n elements for some positive integer n. We can apply results in Section 6.5 to show that K has p^r elements, for $r = n/m$, and that F is the splitting field of the polynomial $f(x) = x^{p^n} - x$ over its prime subfield, and hence over K. Since $f(x)$ has no repeated roots, we may apply Theorem 8.1.6 to conclude that $|\text{Gal}(F/K)| = m$.

Define $\theta : F \to F$ by $\theta(x) = x^{p^r}$. As in Lemma 6.5.4, it is easy to show that θ preserves sums and products, so it is a ring homomorphism. Since it is nonzero and F is a field, it must be one-to-one and hence onto since F is finite. We have defined an element of $\text{Gal}(F/K)$ since K is the splitting field of $x^{p^r} - x$ over its prime subfield, and thus θ leaves the elements of K fixed. To compute the order of θ in $\text{Gal}(F/K)$, we first note that θ^m is the identity automorphism since $\theta^m(x) = x^{p^{rm}} = x^{p^n} = x$ for all $s \in F$. Furthermore, θ^s does not equal the identity for any $1 \leq s \leq m$, since this would imply that $x^{p^{rs}} = x$ for all $x \in F$, and this equation cannot have p^n roots. Thus θ is a generator for $\text{Gal}(F/K)$. $\square$

Example 8.1.3

> In Chapter 4 we considered $GF(4)$, which we can work with as the set $\{0, 1, \alpha, 1 + \alpha\}$, where elements are multiplied by using the identity $\alpha^2 = 1 + \alpha$. The previous corollary shows that $\text{Gal}(GF(4)/\mathbf{Z}_2)$ is cyclic of order 2, generated by the automorphism θ defined by $\theta(x) = x^2$ for all $x \in GF(4)$. $\square$

Our immediate goal is a deeper study of splitting fields. In Section 8.3 we will be able to give the following characterization: The following conditions are equivalent for an extension field F of a field K.

(1) F is the splitting field over K of a polynomial with no repeated roots;
(2) There is a finite group G of automorphisms of F such that $a \in K$ if and only if $\theta(a) = a$ for all $\theta \in G$;
(3) F is a finite extension of K; the minimal polynomial over K of any element in F has no repeated roots; and if $p(x) \in K[x]$ is irreducible and has a root in F, then it splits over F.

EXERCISES: SECTION 8.1

1. Prove by a direct computation that the function θ_2 in Example 8.1.2 preserves products.
2. In Example 8.1.2, find $\{x \in \mathbf{Q}(\sqrt{2} + \sqrt{3}) \mid \theta_2(x) = x\}$ and show that it is a subfield of $\mathbf{Q}(\sqrt{2} + \sqrt{3})$.
3. Find a basis for $GF(8)$, and then write out an explicit formula for each of the elements of $\text{Gal}(GF(8)/\mathbf{Z}_2)$.
4. Show that the Galois group of $x^3 - 1$ over $\mathbf{Q}$ is cyclic of order 2.
5. Show that the Galois group of $(x^2 - 2)(x^2 + 2)$ over $\mathbf{Q}$ is isomorphic to $\mathbf{Z}_2 \times \mathbf{Z}_2$.
6. Let E and F be two splitting fields of a polynomial over the field K. We already know that $E \cong F$. Prove that $\text{Gal}(E/K) \cong \text{Gal}(F/K)$.

8.2 MULTIPLICITY OF ROOTS

In the previous section, we showed that the order of the Galois group of a polynomial with no repeated roots is equal to the degree of its splitting field over the base field. If we want to compute the Galois group of a polynomial $f(x)$ over the field K, we can factor $f(x)$ into a product

$$f(x) = p_1(x)^{\alpha_1} p_2(x)^{\alpha_2} \cdots p_n(x)^{\alpha_n}$$

of distinct irreducible factors. The splitting field of F of $f(x)$ over K is the same as the splitting field of

$$g(x) = p_1(x)p_2(x) \cdots p_n(x)$$

over K. Note that distinct irreducible polynomials cannot have roots in common. (If $\gcd(p(x), q(x)) = 1$, then there exist $a(x)$, $b(x)$ such that $a(x)p(x) + b(x)q(x) = 1$, showing that there can be no common roots). Thus $g(x)$ has no repeated roots if and only if each of its irreducible factors has no repeated roots. We will show in this section that over fields of characteristic zero, and over finite fields, irreducible polynomials have no repeated roots. It follows that in these two situations, when computing a Galois group we can always reduce to the case of a polynomial with no repeated roots. The first thing that we need to do in this section is to develop methods to determine whether or not a polynomial has repeated roots.

8.2.1 Definition.   Let $f(x)$ be a polynomial in $K[x]$, and let F be a splitting field for $f(x)$ over K. If $f(x)$ has the factorization

$$f(x) = (x - r_1)^{m_1}(x - r_2)^{m_2} \cdots (x - r_t)^{m_t}$$

over F, then we say that the root r_i has *multiplicity* m_i. If $m_i = 1$, then r_i is called a *simple* root.

In Proposition 4.2.11 we showed that a polynomial $f(x)$ over **R** has no repeated factors if and only if it has no factors in common with its derivative $f'(x)$. We can extend this result to polynomials over any field and use it to check for multiple roots.

8.2.2 Definition.   Let $f(x) \in K[x]$, with $f(x) = \Sigma_{k=0}^t a_k x^k$. The *formal derivative* $f'(x)$ of $f(x)$ is defined by the formula $f'(x) = \Sigma_{k=0}^t k a_k x^{k-1}$, where $k a_k$ denotes the sum of a_k added to itself k times.

It is not difficult to show from this definition that the standard differentiation formulas hold. The next proposition gives a test for multiple roots, which can be carried out over K without actually finding the roots.

8.2.3 Proposition.   The polynomial $f(x) \in K[x]$ has no multiple roots if and only if $\gcd(f(x), f'(x)) = 1$.

Proof. Let F be a splitting field for $f(x)$ over K. In using the Euclidean algorithm to find the greatest common divisor of $f(x)$ and $f'(x)$, we make use of the division algorithm. We can do the necessary computations in both $K[x]$ and $F[x]$. But the quotients and remainders that occur over $K[x]$ also serve as the appropriate quotients and remainders over $F[x]$. Using the fact that these answers must be unique over $F[x]$, it follows that it makes no difference if we do the computations in the Euclidean algorithm over the splitting field F.

If $f(x)$ has a root r of multiplicity $m > 1$ in F, then we can write $f(x) = (x - r)^m g(x)$ for some polynomial $g(x) \in F[x]$. Thus

$$f'(x) = m(x - r)^{m-1} g(x) + (x - r)^m g'(x),$$

and so $(x - r)^{m-1}$ is a common divisor of $f(x)$ and $f'(x)$, which shows that $\gcd(f(x), f'(x)) \neq 1$.

On the other hand, if $f(x)$ has no multiple roots, then we can write $f(x) = (x - r_1)(x - r_2) \cdots (x - r_t)$. Applying the product rule, we find that $f'(x)$ is a sum of terms, each of which contains all but one of the linear factors of $f(x)$. It follows that each of the linear factors of $f(x)$ is a divisor of all but one of the terms in $f'(x)$, and so $\gcd(f(x), f'(x)) = 1$. $\square$

8.2.4 Proposition. Let $f(x)$ be an irreducible polynomial over the field K. Then $f(x)$ has no multiple roots unless $\operatorname{char}(K) = p \neq 0$ and $f(x)$ has the form $f(x) = a_0 + a_1 x^p + a_2 x^{2p} + \ldots + a_n x^{np}$.

Proof. Using the previous proposition, the only case in which $f(x)$ has a multiple root is if $\gcd(f(x), f'(x)) \neq 1$. Since $f(x)$ is irreducible and $\deg(f'(x)) < \deg(f(x))$, the only way this can happen is if $f'(x)$ is the zero polynomial. This is impossible over a field of characteristic zero. However, over a field of characteristic $p > 0$, it is possible if every coefficient of $f'(x)$ is a multiple of p. Thus $f(x)$ has no multiple roots unless it has the form specified in the statement of the proposition. $\square$

8.2.5 Definition. A polynomial $f(x)$ over the field K is called *separable* if its irreducible factors have only simple roots. An algebraic extension field F of K is called *separable over* K if the minimal polynomial of each element of F is separable. The field F is called *perfect* if every polynomial over F is separable.

8.2.6 Theorem. Any field of characteristic zero is perfect. A field of characteristic $p > 0$ is perfect if and only if each of its elements has a pth root.

Proof. Proposition 8.2.4 implies that any field of characteristic zero is perfect, and so we consider the case of a field F of characteristic $p > 0$ and a polynomial $f(x)$ irreducible over F.

If $f(x)$ has a multiple root, then it has the form $f(x) = \sum_{i=0}^{n} a_i(x^p)^i$. If each element of F has a pth root, then for each i we may let b_i be the pth root of a_i.

Thus

$$f(x) = \sum_{i=0}^{n} (b_i)^p (x^p)^i = \sum_{i=0}^{n} (b_i x^i)^p = \left(\sum_{i=0}^{n} b_i x^i \right)^p$$

and so $f(x)$ is reducible, a contradiction.

To prove the converse, suppose that there exists an element $a \in F$ that has no pth root. Consider the polynomial $f(x) = x^p - a$ and let $g(x)$ be an irreducible factor of $f(x)$. Next, consider the extension field $F(b)$, where b is a root of $g(x)$. Then $b^p = a$, and so over $F(b)$ the polynomial $f(x)$ has the form $f(x) = x^p - b^p = (x - b)^p$. But then $g(x)$ must have the form $g(x) = (x - b)^k$ with $1 < k \le p$ since $b \notin F$. Thus we have produced an irreducible polynomial with repeated roots.   □

8.2.7 Corollary.   Any finite field is perfect.

Proof. Let F be a finite field of characteristic p. Consider $\phi : F \to F$ defined by $\phi(x) = x^p$. It is clear that ϕ preserves products. Furthermore, ϕ preserves sums since in the binomial expansion of $(a + b)^p$ each coefficient except those of a^p and b^p is zero. The coefficient $(p!)/(k!(p - k)!)$ contains p in the numerator but not the denominator since p is prime, and so it must be equal to zero in a field of characteristic p. Thus ϕ is a ring homomorphism, and since it is not the zero mapping ($\phi(1) = 1$), it must be one-to-one. Since F is finite, ϕ must be onto, showing that every element of F has a pth root.   □

It can be shown that in the field $\mathbf{Z}_p(x)$ of rational functions over $\mathbf{Z}_p$, the element x has no pth root. Thus this rational function field is not perfect.

In the final result of this section, we will use some of the ideas we have developed to investigate the structure of finite extensions. We began by studying field extensions of the form $F = K(u)$. If F is a finite field, then we showed in Theorem 6.5.10 that the multiplicative group $F^\times$ is cyclic. If the generator of this group is a, then it is easy to see that $F = K(a)$ for any subfield K. We now show that any finite separable extension has this form. Recall Definition 6.1.5: The extension field F of K is called a simple extension if there exists an element $u \in F$ such that $F = K(u)$. In this case, u is called a *primitive* element.

8.2.8 Theorem.   Let F be a finite extension of the field K. If F is separable over K, then it is a simple extension of K.

Proof. Let F be a finite separable extension of K. If K is a finite field, then F is also a finite field. As remarked above, we must have $F = K(u)$ for any generator u of the simple group $F^\times$.

If we can prove the result in case $F = K(u_1, u_2)$, then it is clear how to extend it by induction to the case $F = K(u_1, u_2, \ldots, u_n)$. Thus we may assume that K is infinite and $F = K(u, v)$ for elements $u, v \in F$.

Let $f(x)$ and $g(x)$ be the minimal polynomials of u and v, and assume their degrees to be m and n, respectively. Let E be an extension of F over which both $f(x)$ and $g(x)$ split. Since F is separable, the roots $u = u_1, u_2, \ldots, u_m$ and $v = v_1, v_2, \ldots, v_n$ of $f(x)$ and $g(x)$ are distinct. If $j \neq 1$, then the equation $u_i + v_j x = u + vx$ has a unique solution $x = (u - u_i)/(v_j - v)$ in E. Therefore, since K is infinite, there must exist an element $a \in K$ such that $u + av \neq u_i + av_j$ for all i and all $j \neq 1$.

We will show that $F = K(t)$ for $t = u + av$. It is clear that $K(t) \subseteq K(u, v) = F$. If we can show that $v \in K(t)$, then it will follow easily that $u \in K(t)$, giving us the desired equality. Our strategy is to show that the minimal polynomial $p(x)$ of v over $K(t)$ has degree 1, which will force v to belong to $K(t)$.

Let $h(x) = f(t - ax)$. This polynomial has coefficients in $K(t)$ and has v as a root since $h(v) = f(t - av) = f(u) = 0$. Since we also have $g(v) = 0$, it follows that $p(x)$ is a common divisor of $h(x)$ and $g(x)$. Now we consider all three polynomials $p(x)$, $h(x)$, and $g(x)$ over the extension field E. Since $t \neq u_i + av_j$, it follows that $t - av_j \neq u_i$ for all i and all $j \neq 1$. Thus v_j is not a root of $h(x)$, for $j = 2, 3, \ldots, n$. Since $g(x)$ splits over E and $x - v_j$ is not a divisor of $h(x)$ for $j = 2, 3, \ldots, n$, we can conclude that over E we have $\gcd(h(x), g(x)) = x - v$. But $p(x)$ is a common divisor of $h(x)$ and $g(x)$ over E as well as over $K(t)$, and so $p(x)$ must be linear. $\square$

EXERCISES: SECTION 8.2

1. Show that any algebraic extension of a perfect field is perfect.
2. Show that if $F \supseteq E \supseteq K$ are fields and F is separable over K, then F is separable over E.
3. Let E be a separable algebraic extension of F and let F be a separable algebraic extension of K. Show that E is a separable extension of K.
4. Show that the product rule holds for the derivative defined in Definition 8.2.2.
5. Let $\omega = (-1 + \sqrt{3}i)/2$ (a primitive cube root of unity). Find a primitive element for the extension $\mathbf{Q}(\omega, \sqrt[3]{2})$ of $\mathbf{Q}$.
6. Show that in the field $\mathbf{Z}_p(x)$ of rational functions over $\mathbf{Z}_p$, the element x has no pth root.
7. Let F be a field of characteristic $p \neq 0$, and let $a \in F$. Show that in $F[x]$, the polynomial $x^p - a$ is either irreducible or a pth power.

8.3 THE FUNDAMENTAL THEOREM OF GALOIS THEORY

In this section we study the connection between subgroups of $\mathrm{Gal}(F/K)$ and fields between K and F. This is a critical step in proving that a polynomial is solvable by radicals if and only if its Galois group is solvable.

8.3.1 Proposition. Let F be a field, and let G be a subgroup of $\mathrm{Aut}(F)$. Then $\{a \in F \mid \theta(a) = a$ for all $\theta \in G\}$ is a subfield of F.

Proof. If a and b are elements of F that are left fixed by all automorphisms in G, then for any $\theta \in G$ we have

$$\theta(a \pm b) = \theta(a) \pm \theta(b) = a \pm b \qquad \text{and} \qquad \theta(ab) = \theta(a)\theta(b) = ab.$$

For $a \neq 0$ we have $\theta(a^{-1}) = (\theta(a))^{-1} = a^{-1}$, and thus we have shown that the given set is a subfield. $\square$

8.3.2 Definition.   Let F be a field, and let G be a subgroup of Aut(F). Then $\{a \in F \mid \theta(a) = a \text{ for all } \theta \in G\}$ is called the *G-fixed subfield* of F, or the *G-invariant subfield* of F, and is denoted by F^G.

8.3.3 Proposition.   If F is the splitting field over K of a separable polynomial and $G = \text{Gal}(F/K)$, then $F^G = K$.

Proof. Let $E = F^G$, so that we have $K \subseteq E \subseteq F$. It is clear that F is a splitting field over E as well as over K, and that $G = \text{Gal}(F/E)$. By Theorem 8.1.6 we have both $|G| = [F : E]$ and $|G| = [F : K]$. This implies that $[E : K] = 1$, and so $E = K$. $\square$

8.3.4 Lemma (Artin).   Let G be a finite group of automorphisms of the field F, and let $K = F^G$. Then $[F : K] \leq |G|$.

Proof. Let $G = \{\theta_1, \theta_2, \ldots, \theta_n\}$, with the identity element of G denoted by θ_1. Suppose that there exist $n + 1$ elements $\{u_1, u_2, \ldots, u_{n+1}\}$ of F which are linearly independent over K. We next consider the following system of equations:

$$\theta_1(u_1)x_1 + \theta_1(u_2)x_2 + \ldots + \theta_1(u_{n+1})x_{n+1} = 0$$

$$\theta_2(u_1)x_1 + \theta_2(u_2)x_2 + \ldots + \theta_2(u_{n+1})x_{n+1} = 0$$

$$\cdot \quad \cdot \quad \cdot \quad \cdot \quad \cdot$$

$$\theta_n(u_1)x_1 + \theta_n(u_2)x_2 + \ldots + \theta_n(u_{n+1})x_{n+1} = 0.$$

There are n equations corresponding to the n elements of G and $n + 1$ unknowns corresponding to the $n + 1$ linearly independent elements of F, and so it follows from the elementary theory of systems of linear equations that there exists a nontrivial solution $(a_1, a_2, \ldots, a_{n+1})$ in F. Among all such solutions, choose one with the smallest number of nonzero terms. By rearranging the terms in the solution set, we may assume that $a_1 \neq 0$. Dividing each of the solutions by a_1 still gives us a set of solutions, and so we may assume that $a_1 = 1$.

Since θ_1 is the identity, the first equation is $u_1x_1 + u_2x_2 + \ldots + u_{n+1}x_{n+1} = 0$. The solutions we have chosen cannot all belong to K, since this would contradict the fact that $u_1, u_2, \ldots, u_{n+1}$ are assumed to be linearly independent over K. We can rearrange the solutions to guarantee $a_2 \notin K$. Since $K = F^G$, this means that a_2 is not invariant under all automorphisms in G, say $\theta_i(a_2) \neq a_2$. If we apply

θ_i to each of the equations in the system, we do not change the system, since G is a group and multiplying each element by θ_i merely permutes the elements. On the other hand, since θ_i is an automorphism, this leads to a second solution $(1, \theta_i(a_2), \ldots, \theta_i(a_{n+1}))$. Subtracting the second solution from the first gives a nontrivial solution (since $a_2 - \theta_i(a_2) \neq 0$) that has fewer nonzero entries. This is a contradiction, completing the proof. $\square$

The next theorem will show that the following property holds for the splitting field of a separable polynomial. Let F be the splitting field of $f(x)$ over K, and assume that $f(x)$ has no repeated roots. Then if $p(x) \in K[x]$ is the minimal polynomial of any element of F, it follows that $p(x)$ splits into linear factors in $F[x]$. This is equivalent to saying that F contains a splitting field for any polynomial in $K[x]$ that has a root in F. It is convenient to make the following definition.

8.3.5 Definition.   Let F be an algebraic extension of the field K. Then F is said to be a *normal* extension of K if every irreducible polynomial in $K[x]$ that contains a root in F is a product of linear factors in $F[x]$.

The choice of the term normal is justified by part (c) of Theorem 8.3.8.

8.3.6 Theorem.   The following conditions are equivalent for an extension field F of K:
  (1) F is the splitting field over K of a separable polynomial;
  (2) $K = F^G$ for some finite group G of automorphisms of F;
  (3) F is a finite, normal, separable extension of K.

Proof. (1) implies (2): If F is the splitting field over K of a separable polynomial, then by Proposition 8.3.3 we must have $K = F^G$ for $G = \mathrm{Gal}(F/K)$. The Galois group is finite by Theorem 8.1.6.

(2) implies (3): Assume that $K = F^G$ for some finite group G of automorphisms of F. Then Artin's lemma shows that $[F : K] \leq |G|$, and so F is a finite extension of K. Let $f(x)$ be a monic irreducible polynomial in $K[x]$ that has a root r in F. Since G is a finite group, the set $\{\theta(r) \mid \theta \in G\}$ is finite, say with distinct elements $r_1 = r, r_2, \ldots, r_m$. These elements are roots of $f(x)$, since G is a group of automorphisms of F that fixes the coefficients of $f(x)$, so $\deg(f(x)) \geq m$. Let

$$h(x) = (x - r_1)(x - r_2) \cdots (x - r_m).$$

It follows from our choice of the elements r_i that applying any automorphism $\theta \in G$ yields $\theta(r_i) = r_j$, for some j. Therefore applying $\theta \in G$ to the polynomial $h(x)$ simply permutes the factors, showing that $\theta(h(x)) = h(x)$. Using the fact that θ is an automorphism and equating corresponding coefficients shows that every coefficient of $h(x)$ must be left fixed by G, so by assumption the coefficients of $h(x)$ belong to $K = F^G$. Now $f(x)$ is the minimal polynomial of r over K, and $h(r) = 0$, so $f(x) \mid h(x)$. Since $f(x)$ and $h(x)$ are monic and $\deg(f(x)) \geq \deg(h(x))$, we must have $f(x) = h(x)$. In particular, we have shown that $f(x)$ has distinct roots, all belonging to F. This implies that F is both normal and separable over K.

(3) implies (1): If F is a finite separable extension of K, then by Theorem 8.2.8 it is a simple extension, say $F = K(u)$. If F is a normal extension of K and u has the minimal polynomial $f(x)$ over K, then F contains all of the roots of $f(x)$, and so F is a splitting field for $f(x)$. $\square$

8.3.7 Corollary.   If F is an extension field of K such that $K = F^G$ for some finite group G of automorphisms of F, then $G = \mathrm{Gal}(F/K)$.

Proof. By assumption $K = F^G$, so G is a subgroup of $\mathrm{Gal}(F/K)$, and the result follows since

$$[F : K] \leq |G| \leq |\mathrm{Gal}(F/K)| = [F : K].$$

The first inequality comes from Artin's lemma, and the last equality is implied by Theorems 8.1.6 and 8.3.6. $\square$

Example 8.3.1

In Corollary 8.1.7 we showed that $\mathrm{Gal}(GF(p^n)/\mathbf{Z}_p)$ is cyclic of order n, generated by the automorphism ϕ defined by $\phi(x) = x^p$, for all $x \in GF(p^n)$. (This automorphism is usually known as the *Frobenius automorphism* of $GF(p^n)$.) In Section 6.5 we showed that the subfields of $GF(p^n)$ are of the form $GF(p^r)$, for all divisors r of n, and then Corollary 8.1.7 shows that $\mathrm{Gal}(GF(p^n)/GF(p^r))$ has order m, where $n = mr$, and is generated by ϕ^r. Note that as the value of r is increased, to give a larger subfield, the power of ϕ is increased, and so the corresponding subgroup it defines is smaller. It happens very generally that there is an order-reversing correspondence between subfields and subgroups of the Galois group, and this is proved in the following fundamental theorem of Galois theory.

The Galois group of $GF(p^r)$ is also generated by the Frobenius automorphism, when restricted to the smaller field. Is there a connection between $\mathrm{Gal}(GF(p^n)/\mathbf{Z}_p)$ and $\mathrm{Gal}(GF(p^r)/\mathbf{Z}_p)$? Since any power of ϕ^r leaves elements of $GF(p^r)$ fixed, for any integer we have $\phi^t(x) = \phi^{t+rs}(x)$ for all $x \in GF(p^r)$. This suggests that the cosets of $\langle \phi^r \rangle$ in $\langle \phi \rangle$ might correspond to nontrivial automorphisms in $\mathrm{Gal}(GF(p^r)/\mathbf{Z}_p)$, and indeed this is the case, as will be shown in Theorem 8.3.8. $\square$

8.3.8 Theorem (Fundamental Theorem of Galois Theory).   Let F be the splitting field of a separable polynomial over the field K, and let $G = \mathrm{Gal}(F/K)$.
  (a) There is a one-to-one order-reversing correspondence between subgroups of G and subfields of F that contain K:
    (i) If H is a subgroup of G, then the corresponding subfield is F^H, and $H = \mathrm{Gal}(F/F^H)$;

   (ii) If E is a subfield of F that contains K, then the corresponding subgroup
   of G is $\text{Gal}(F/E)$, and $E = F^{\text{Gal}(F/E)}$.

 (b) For any subgroup H of G, we have $[F : F^H] = |H|$ and $[F^H : K] = [G : H]$.

 (c) Under the above correspondence, the subgroup H is normal if and only if the
   subfield $E = F^H$ is a normal extension of K. In this case, $\text{Gal}(E/K) \cong$
   $\text{Gal}(F/K)/\text{Gal}(F/E)$.

   Proof. (a) If we verify that (i) and (ii) hold, then it is clear that the mapping
that assigns to each subgroup H of G the subfield F^H has an inverse mapping, and
so it defines a one-to-one correspondence. We will first show that the one-to-one
correspondence reverses the natural ordering in the lattices of subgroups and
subfields. If $H_1 \subseteq H_2$ are subgroups of G, then it is clear that the subfield left fixed
by H_2 is contained in the subfield left fixed by H_1. On the other hand, if $E_1 \subseteq E_2$,
then it is clear that $\text{Gal}(F/E_2) \subseteq \text{Gal}(F/E_1)$.
   Given the subgroup H, Corollary 8.3.7 implies that $H = \text{Gal}(F/F^H)$. On the
other hand, given a subfield E with $K \subseteq E \subseteq F$, it is clear from the initial assumption that F is also a splitting field over E, and then $E = F^{\text{Gal}(F/E)}$ by Proposition
8.3.3.
   (b) Given the subgroup H of G, since F is a splitting field over F^H and $H =$
$\text{Gal}(F/F^H)$, it follows from Theorem 8.1.6 that $[F : F^H] = |H|$. Since $[F : K] = |G|$
by Theorem 8.1.6, the desired equality $[F^H : K] = [G : H]$ follows from the two
equalities $[F : K] = [F : F^H][F^H : K]$ and $|G| = |H| \cdot [G : H]$.
   (c) Let E be a normal extension of K in F, and let ϕ be any element of G. If
$u \in E$ with minimal polynomial $p(x)$, then $\phi(u)$ is also a root of $p(x)$, and so we
must have $\phi(u) \in E$ since E is a normal extension of K. Thus for any $\theta \in$
$\text{Gal}(F/E)$, we have $\theta\phi(u) = \phi(u)$, and so $\phi^{-1}\theta\phi(u) = \phi^{-1}\phi(u) = u$. This implies
that $\phi^{-1}\theta\phi \in \text{Gal}(F/E)$, and therefore $\text{Gal}(F/E)$ is a normal subgroup of G.
   Conversely, let H be a normal subgroup of G, and let $E = F^H$. If $\phi \in G$ and
$\theta \in H$, then by the normality of H we must have $\phi^{-1}\theta\phi = \theta'$ for some $\theta' \in H$, and
then $\theta\phi = \phi\theta'$. For any $u \in E$ we have by definition $\theta'(u) = u$, and so $\theta(\phi(u)) =$
$\phi(\theta'(u)) = \phi(u)$. This shows that $\phi(u) \in F^H$, and so ϕ maps E into E. Because ϕ
leaves K fixed, it is a K-linear transformation and is one-to-one when restricted to
E, so a dimension argument shows that it maps E onto E.
   The above argument shows that the restriction mapping defines a function
(which is easily seen to be a group homomorphism) from $G = \text{Gal}(F/K)$ into
$\text{Gal}(E/K)$. By Theorem 8.2.8, F is a simple extension of E, and so Lemma 6.4.4
implies that any element of $\text{Gal}(E/K)$ can be extended to an element of $\text{Gal}(F/K)$.
Thus the restriction mapping is onto, and since the kernel of this mapping is
clearly $\text{Gal}(F/E) = H$, by the fundamental homomorphism theorem we must have
$\text{Gal}(E/K) \cong \text{Gal}(F/K)/\text{Gal}(F/E)$.
   By part (b) the index of $\text{Gal}(F/E)$ in G equals $[E : K]$, and so $|\text{Gal}(E/K)| =$
$[E : K]$. Since $|\text{Gal}(E/K)|$ is finite, Theorem 8.3.6 implies that E is a normal
separable extension of $E^{\text{Gal}(E/K)}$. But then $[E : E^{\text{Gal}(E/K)}] = [E : K]$, and so $E^{\text{Gal}(E/K)} =$
K, which finally shows that E is a normal extension of K. $\quad\square$

In the statement of the fundamental theorem we could have simply said that normal subgroups correspond to normal extensions. In the proof we noted that if E is a normal extension of K, then $\phi(E) \subseteq E$ for all $\phi \in \text{Gal}(F/K)$. In the context of the fundamental theorem, we say that two intermediate subfields E_1 and E_2 are *conjugate* if there exists $\phi \in \text{Gal}(F/K)$ such that $\phi(E_1) = E_2$. We now show that the subfields conjugate to an intermediate subfield E correspond to the subgroups conjugate to $\text{Gal}(F/E)$. Thus E is a normal extension if and only if it is conjugate only to itself.

8.3.9 Proposition.   Let F be the splitting field of a separable polynomial over the field K, and let E be a subfield such that $K \subseteq E \subseteq F$, with $H = \text{Gal}(F/E)$. If $\phi \in \text{Gal}(F/K)$, then $\text{Gal}(F/\phi(E)) = \phi H \phi^{-1}$.

Proof. Since $\phi \in \text{Gal}(F/K)$, for any $\theta \in \text{Gal}(F/E)$ we have

$\theta \in \text{Gal}(F/\phi(E))$	if and only if	$\theta\phi(x) = \phi(x)$ for all $x \in E$
	if and only if	$\phi^{-1}\theta\phi(x) = x$ for all $x \in E$
	if and only if	$\phi^{-1}\theta\phi \in \text{Gal}(F/E) = H$
	if and only if	$\theta \in \phi H \phi^{-1}$.

This shows that $\text{Gal}(F/\phi(E)) = \phi H \phi^{-1}$ and completes the proof. $\square$

Example 8.3.2

In this example we will compute the Galois group G of the polynomial $p(x) = x^3 - 2$ over $\mathbf{Q}$. We will also illustrate the fundamental theorem by investigating the subfields of its splitting field. The roots of the polynomial are $\sqrt[3]{2}$, $\omega\sqrt[3]{2}$, and $\omega^2\sqrt[3]{2}$, where $\omega = (-1 + \sqrt{3}i)/2$ is a primitive cube root of unity. Thus the splitting field of $p(x)$ over $\mathbf{Q}$ can be constructed as $\mathbf{Q}(\omega, \sqrt[3]{2})$. Since $x^3 - 2$ is irreducible over $\mathbf{Q}$, adjoining $\sqrt[3]{2}$ gives an extension of degree 3, which is contained in $\mathbf{R}$. The minimal polynomial of ω over $\mathbf{Q}$ is $x^2 + x + 1$, and it is irreducible since its roots are ω and ω^2, which are not real. This implies that $[\mathbf{Q}(\omega, \sqrt[3]{2}) : \mathbf{Q}] = 6$, and so $|G| = 6$. The set $\{1, \sqrt[3]{2}, \sqrt[3]{4}, \omega, \omega\sqrt[3]{2}, \omega\sqrt[3]{2}\}$ is a basis for $\mathbf{Q}(\omega, \sqrt[3]{2})$ over $\mathbf{Q}$.

Note that any automorphism $\phi \in G$ is completely determined by its values on $\sqrt[3]{2}$ and ω. Since ϕ must preserve roots of polynomials, the only possibilities are $\phi(\sqrt[3]{2}) = \sqrt[3]{2}, \omega\sqrt[3]{2}, \omega^2\sqrt[3]{2}$, and $\phi(\omega) = \omega, \omega^2$. This gives six possible functions, and so they must determine the six possible elements of G. Let α be defined by $\alpha(\sqrt[3]{2}) = \omega\sqrt[3]{2}$ and $\alpha(\omega) = \omega$, and let β be defined by $\beta(\sqrt[3]{2}) = \sqrt[3]{2}$ and $\beta(\omega) = \omega^2$. Since ω^2 is the complex conjugate of ω, complex conjugation defines an automorphism of $\mathbf{Q}(\sqrt[3]{2}, \omega)$ that has the properties required of β, so evidently we have defined $\beta(x) = \bar{x}$, for all $x \in \mathbf{Q}(\sqrt[3]{2}, \omega)$. A direct computation shows that $\alpha^2(\sqrt[3]{2}) = \omega^2\sqrt[3]{2}$ and then $\alpha^3(\sqrt[3]{2}) = \sqrt[3]{2}$, so α has order 3. Similarly, β has order 2. Furthermore, $\beta\alpha(\sqrt[3]{2}) = \beta(\omega\sqrt[3]{2}) = \omega^2\sqrt[3]{2}$ and $\beta\alpha(\omega) = \beta(\omega) = \omega^2$. On the other hand,

$\alpha^2\beta(\sqrt[3]{2}) = \alpha^2(\sqrt[3]{2}) = \omega^2\sqrt[3]{2}$ and $\alpha^2\beta(\omega) = \alpha^2(\omega^2) = \omega^2$. This shows that $\beta\alpha = \alpha^2\beta$, and it follows that G is isomorphic to S_3.

The subfield $\mathbf{Q}(\omega)$ is the splitting field of $x^2 + x + 1$ over $\mathbf{Q}$, and so it must be the fixed subfield of the only proper nontrivial normal subgroup $\{1, \alpha, \alpha^2\}$ of G. The subfield $\mathbf{Q}(\sqrt[3]{2})$ is the fixed subfield of the subgroup $H = \{1, \beta\}$, since $\beta(\sqrt[3]{2}) = \sqrt[3]{2}$. By Proposition 8.3.9, the subfield $\mathbf{Q}(\omega\sqrt[3]{2}) = \alpha(\mathbf{Q}(\sqrt[3]{2}))$ must be the fixed subfield of the subgroup $\alpha H\alpha^{-1} = \{1, \alpha^2\beta\}$. (A direct computation shows that $\alpha^2\beta(\omega\sqrt[3]{2}) = \omega\sqrt[3]{2}$.) The remaining subfield is $\mathbf{Q}(\omega^2\sqrt[3]{2}) = \alpha^2(\mathbf{Q}(\sqrt[3]{2}))$, and this suffices to determine the correspondence between subgroups and subfields. We give the respective lattices in Figures 8.3.1 and 8.3.2   □

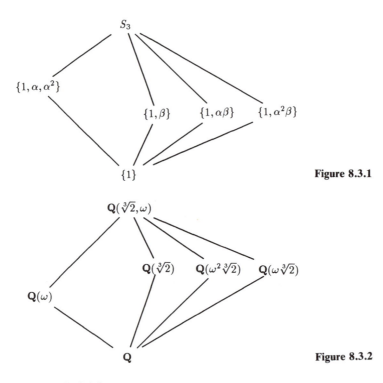

Figure 8.3.1

Figure 8.3.2

Example 8.3.3

In the final example of this section we compute the Galois group G of $x^4 - 2$ over $\mathbf{Q}$ and investigate the subfields of its splitting field. We have $x^4 - 2 = (x^2 + \sqrt{2})(x^2 - \sqrt{2})$, and this factorization leads to the roots $\pm\sqrt[4]{2}$, $\pm i\sqrt[4]{2}$ of $x^4 - 2$. We can obtain the splitting field as $\mathbf{Q}(\sqrt[4]{2}, i)$ by first adjoining the root $\sqrt[4]{2}$ of $x^4 - 2$ to obtain $\mathbf{Q}(\sqrt[4]{2})$ and then adjoining the root i of the irreducible polynomial $x^2 + 1$. Since any element of the Galois group must simply permute the roots of $x^4 - 2$ and $x^2 + 1$, the eight possible permutations determine the Galois group, which has order 8.

Let α be the automorphism defined by $\alpha(\sqrt[4]{2}) = i\sqrt[4]{2}$ and $\alpha(i) = i$, and let β be defined by $\beta(\sqrt[4]{2}) = \sqrt[4]{2}$ and $\beta(i) = -i$. Then α represents a cyclic permutation of the roots of $x^4 - 2$ and has order 4, while β represents complex conjugation and has order 2. A direct computation shows that $\beta\alpha = \alpha^3\beta$, and so $G \cong D_4$.

To compute the fixed subfields of the various subgroups of D_4, we first note that all powers of α leave i fixed. Furthermore, α^2 also leaves $\sqrt{2}$ fixed since $\alpha(\sqrt{2}) = \alpha(\sqrt[4]{2}\sqrt[4]{2}) = (i\sqrt[4]{2})(i\sqrt[4]{2}) = -\sqrt{2}$. There are various ways in which to obtain the splitting field of $x^4 - 2$ by adjoining, at each stage, a root of a quadratic. These are reflected in the various chains of subfields. For example, we can first adjoin one of $\sqrt{2}$, i or $i\sqrt{2}$ to obtain the splitting field of $x^2 - 2$, $x^2 + 1$, or $x^2 + 2$, respectively. Continuing, for example, given $\mathbf{Q}(i\sqrt{2})$, adjoining $\sqrt{i\sqrt{2}} = ((\sqrt{2} + \sqrt{2}i)\sqrt[4]{2})/2$ and then multiplying this by $i\sqrt{2}$ gives $(i - 1)\sqrt[4]{2}$. The extension $\mathbf{Q}(\sqrt{i\sqrt{2}}) = \mathbf{Q}((i - 1)\sqrt[4]{2})$ belongs to $\mathbf{Q}(\sqrt[4]{2}, i)$, and so it must define an intermediate subfield, which turns out to be the fixed subfield of the subgroup $\{1, \alpha^3\beta\}$. For the remaining subgroups it is not difficult to determine the corresponding fixed subfields. The appropriate lattice diagrams are given in Figures 8.3.3 and 8.3.4. $\square$

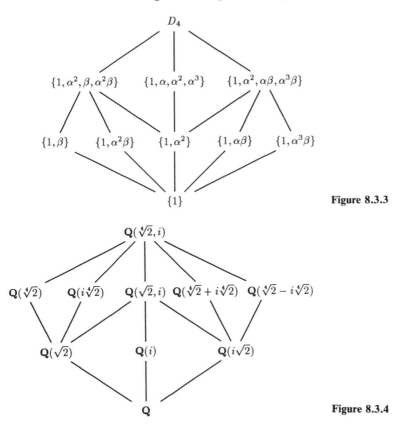

Figure 8.3.3

Figure 8.3.4

We end this section by proving the fundamental theorem of algebra. The proof uses algebraic techniques, although we do need to know that any polynomial over **R** of odd degree must have a root. If $f(x) \in \mathbf{R}[x]$ had odd degree, then $f(n)$ and $f(-n)$ have opposite signs for sufficiently large n, so $f(x)$ must have a root since it is a continuous function.

8.3.10 Theorem (Fundamental Theorem of Algebra).   Any polynomial in $\mathbf{C}[x]$ has a root in **C**.

Proof. Let $f(x) \in \mathbf{C}[x]$ and let L be a splitting field for $f(x)$ over **C**. Let F be a splitting field for $(x^2 + 1)f(x)\overline{f(x)}$ over **R**, where $\overline{f(x)}$ is the polynomial whose coefficients are obtained by taking the complex conjugates of the coefficients of $f(x)$. Since char(**R**) $= 0$, we are in the situation of the fundamental theorem of Galois theory. We will use $G = \mathrm{Gal}(F/\mathbf{R})$ to show that $F = \mathbf{C}$. Then $L = \mathbf{C}$ since $F \supseteq L \supseteq \mathbf{C}$.

Let H be a Sylow 2-subgroup of G, and let $E = F^H$. Since H is a Sylow 2-subgroup, $[G : H]$ is odd, which implies by Theorem 8.3.8 that $[E : \mathbf{R}]$ is odd. By Theorem 8.2.8 there exists $u \in E$ such that $E = \mathbf{R}(u)$, and so the minimal polynomial of u over **R** has odd degree. The minimal polynomial is irreducible, so it can only have degree 1 (by our comment preceding the theorem that any polynomial of odd degree over **R** has a root in **R**), and this implies that $E = \mathbf{R}$. Now, since $F^H = \mathbf{R}$, we must have $H = G$, and so G is a 2-group.

The subgroup $G_1 = \mathrm{Gal}(F/\mathbf{C})$ of G is also a 2-group, and is normal since it has index 2. Thus F is a normal extension of **C**, and so we can again apply Theorem 8.2.8. If G_1 is not the trivial group, then the first Sylow theorem implies that it has a normal subgroup N of index 2. If $K = F^N$, then $[K : \mathbf{C}] = [G : N] = 2$, so $K = \mathbf{C}(v)$ for some $v \in K$ with a minimal polynomial that is a quadratic. We know that square roots exist in **C**, so the quadratic formula is valid, which implies that no polynomial in $\mathbf{C}[x]$ of degree 2 is irreducible. Therefore G_1 must be the trivial group, showing that $F = \mathbf{C}$.   □

EXERCISES: SECTION 8.3

1. In Example 8.3.3 determine which subfields are conjugate, and in each case find an automorphism under which the subfields are conjugate.
2. Find the Galois group of $x^4 + 1$ over **Q**.
3. Find the Galois group of $x^4 - x^2 - 6$ over **Q**.
4. Find the Galois group of $x^8 - 1$ over **Q**.
5. Let F be the splitting field of a separable polynomial over K, and let E be a subfield between K and F. Show that if $[E : K] = 2$, then E is the splitting field of some polynomial over K.

8.4 SOLVABILITY BY RADICALS

In most results in this section we will assume that the fields have characteristic zero, in order to guarantee that no irreducible polynomial has multiple roots. When we say that a polynomial equation is solvable by radicals, we mean that the solutions can be obtained from the coefficients in a finite sequence of steps, each of which may involve addition, subtraction, multiplication, division, or taking nth roots. Only the extraction of an nth root leads to a larger field, and so our formal definition is phrased in terms of subfields and adjunction of roots of $x^n - a$ for suitable elements a. The definition is reminiscent of the condition for constructibility given in Section 6.3.

8.4.1 Definition.   An extension field F of K is called a *radical extension* of K if there exist elements $u_1, u_2, \ldots, u_m \in F$ such that (i) $F = K(u_1, u_2, \ldots, u_m)$ and (ii) $u_1^{n_1} \in K$ and $u_i^{n_i} \in K(u_1, \ldots, u_{i-1})$ for $i = 2, \ldots, n$ and integers $n_1, n_2, \ldots, n_m$.

For $f(x) \in K[x]$, the polynomial equation $f(x) = 0$ is said to be *solvable by radicals* if there exists a radical extension F of K that contains all roots of $f(x)$.

We must first determine the structure of the Galois group of a polynomial of the form $x^n - a$. Then we will make use of the fundamental theorem of Galois theory to see what happens when we successively adjoin roots of such polynomials.

We first need to recall a result from group theory. We showed in Example 7.1.2 that $\mathrm{Aut}(\mathbf{Z}_n) \cong \mathbf{Z}_n^\times$. This follows from the fact that any automorphism of $\mathbf{Z}_n$ has the form $\phi_a([x]) = [ax]$ for all $[x] \in \mathbf{Z}_n$, where $\gcd(a, n) = 1$. It is easy to see that $\phi_a \phi_b = \phi_{ab}$, and it then follows immediately that $\mathrm{Aut}(\mathbf{Z}_n) \cong \mathbf{Z}_n^\times$.

8.4.2 Proposition.   Let F be the splitting field of $x^n - 1$ over a field K of characteristic zero. Then $\mathrm{Gal}(F/K)$ is an abelian group.

Proof. Since $\mathrm{char}(K) = 0$, the polynomial $x^n - 1$ has n distinct roots, and it is easy to check that they form a subgroup C of $F^\times$. Furthermore, if the group C has exponent m, then there are n solutions of the equation $x^m - 1 = 0$, which forces $n \leq m$. It follows from Proposition 3.5.8 that C must be cyclic. Every element of $\mathrm{Gal}(F/K)$ defines an automorphism of C, and using this observation it is apparent that $\mathrm{Gal}(F/K)$ is isomorphic to a subgroup of $\mathrm{Aut}(C)$. Since C is cyclic of order n, the remarks preceding the proposition show that $\mathrm{Gal}(F/K)$ is isomorphic to a subgroup of $\mathbf{Z}_n^\times$, and so it must be abelian. $\square$

The roots of $x^n - 1$ are called the *nth roots of unity*. Any generator of the group of all nth roots of unity is called a *primitive nth root of unity*.

8.4.3 Theorem.   Let K be a field of characteristic zero that contains all nth roots of unity, let $a \in K$, and let F be the splitting field of $x^n - a$ over K. Then $\mathrm{Gal}(F/K)$ is a cyclic group whose order is a divisor of n.

Proof. If u is any root of $x^n - a$ and ζ is a primitive nth root of unity, then all other roots of $x^n - a$ have the form $\zeta^i u$, for $i = 2, 3, \ldots, n - 1$. Thus $F = K(u)$, and so any element $\phi \in \text{Gal}(F/K)$ is completely determined by its value on u, which must be another root, say $\phi(u) = \zeta^i u$. If $\theta \in \text{Gal}(F/K)$ with $\theta(u) = \zeta^j u$, then we have $\phi\theta(u) = \phi(\zeta^j u) = \zeta^i\zeta^j u = \zeta^{i+j} u$. Assigning to $\phi \in \text{Gal}(F/K)$ the exponent i of ζ in $\phi(u) = \zeta^i u$ defines a one-to-one homomorphism from $\text{Gal}(F/K)$ into $\mathbf{Z}_n$. $\square$

8.4.4 Theorem. Let p be a prime, let K be a field that contains all pth roots of unity, and let F be an extension of K. If $|\text{Gal}(F/K)| = p$, then $F = K(u)$ for some $u \in F$ such that $u^p \in K$.

Proof. Let $G = \text{Gal}(F/K)$. If $w \in F$ is not in K, then $F = K(w)$ since G is simple. Let $C = \{\zeta_1, \zeta_2, \ldots, \zeta_p\} \subseteq K$ be the pth roots of unity, and let θ be a generator of G. Let $w_1 = w$ and $w_i = \theta(w_{i-1})$ for $1 < i \le p$. Equivalently, $w_i = \theta^{i-1}(w)$ for $1 < i \le p$. Since θ^p is the identity, we have $w_1 = \theta(w_p)$. For each i we let

$$w_1 + \zeta_i w_2 + \zeta_i^2 w_3 \ldots + \zeta_i^{p-1} w_p = u_i.$$

Applying θ leaves the roots of unity fixed, and since $\zeta_i^p = 1$ we obtain

$$\theta(u_i) = \theta(w_1) + \zeta_i\theta(w_2) + \ldots + \zeta_i^{p-1}\theta(w_p)$$
$$= w_2 + \zeta_i w_3 + \ldots + \zeta_i^{p-2} w_p + \zeta_i^{p-1} w_1$$
$$= \zeta_i^{-1}(\zeta_i w_2 + \zeta_i^2 w_3 + \ldots + \zeta_i^{p-1} w_p + w_1)$$
$$= \zeta_i^{-1} u_i$$

and therefore we have

$$\theta(u_i^p) = (\theta(u_i))^p = (\zeta_i^{-1} u_i)^p = \zeta_i^{-p} u_i^p = u_i^p.$$

This shows that u_i^p is left fixed by the Galois group G, since θ is a generator of G, and it follows that $u_i^p \in K$.

Writing the definition of the u_i's in matrix form

$$\begin{bmatrix} 1 & \zeta_1 & \zeta_1^2 & \cdots & \zeta_1^{p-1} \\ 1 & \zeta_2 & \zeta_2^2 & \cdots & \zeta_2^{p-1} \\ \vdots & \vdots & \vdots & & \vdots \\ 1 & \zeta_p & \zeta_p^2 & \cdots & \zeta_p^{p-1} \end{bmatrix} \begin{bmatrix} w_1 \\ w_2 \\ \vdots \\ w_p \end{bmatrix} = \begin{bmatrix} u_1 \\ u_2 \\ \vdots \\ u_p \end{bmatrix}$$

we see that the coefficient matrix has the form of a Vandermonde matrix, which is invertible since the ζ_i's are distinct. One way to see that the matrix must be invertible is to observe that it is the matrix used to solve for the coefficients of a polynomial $f(x) = a_0 + a_1 x + \ldots + a_{p-1} x^{p-1}$ over K such that $f(\zeta_i) = 0$ for all i. The only solution is the zero polynomial, since no other polynomial of degree $p - 1$ can have p roots. This means that it is possible to solve for $w = w_1$ as a linear combination (with coefficients in K) of the elements u_i. Thus if $u_i \in K$ for all i, then $w \in K$, a contradiction. We conclude that $u_j \notin K$ for some j, and so $F = K(u)$, for $u = u_j$, and $u^p \in K$, as we had previously shown. $\square$

8.4.5 Lemma.   Let K be a field of characteristic zero, and let E be a radical extension of K. Then there exists an extension F of E that is a normal radical extension of K.

Proof. Let E be a radical extension of K with elements $u_1, u_2, \ldots, u_m \in E$ such that (i) $E = K(u_1, u_2, \ldots, u_m)$ and (ii) $u_1^{n_1} \in K$ and $u_i^{n_i} \in K(u_1, \ldots, u_{i-1})$ for $i = 2, \ldots, n$ and integers $n_1, n_2, \ldots, n_m$. Let F be the splitting field of the product $f(x)$ of the minimal polynomials of u_i over K, for $i = 1, 2, \ldots, n$. The proof that condition (2) implies condition (3) in Theorem 8.3.6 shows that in F each root of $f(x)$ has the form $\theta(u_i)$ for some integer i and some automorphism $\theta \in \mathrm{Gal}(F/K)$. For any $\theta \in \mathrm{Gal}(F/K)$, we have $\theta(u_1)^{n_1} \in K$ and $\theta(u_i)^{n_i} \in K(\theta(u_1), \ldots, \theta(u_{i-1}))$ for $i = 2, \ldots, m$. Thus if $\mathrm{Gal}(F/K) = \{\theta_1, \theta_2, \ldots, \theta_k\}$, then the elements $\{\theta_j(u_i)\}$ for $i = 1, \ldots, m$ and $j = 1, \ldots, k$ satisfy the conditions of Definition 8.4.1, showing that F is a radical extension of K. $\square$

8.4.6 Theorem.   Let $f(x)$ be a polynomial over a field K of characteristic zero. The equation $f(x) = 0$ is solvable by radicals if and only if the Galois group of $f(x)$ over K is solvable.

Proof. We first assume that the equation $f(x) = 0$ is solvable by radicals. Let F be a radical extension of K that contains a splitting field E of $f(x)$ over K. By the previous lemma we may assume that F is a splitting field over K, with elements $u_1, u_2, \ldots, u_m \in F$ such that (i) $F = K(u_1, u_2, \ldots, u_m)$ and (ii) $u_1^{n_1} \in K$ and $u_i^{n_i} \in K(u_1, \ldots, u_{i-1})$ for $i = 2, \ldots, m$ and integers $n_1, n_2, \ldots, n_m$. Let n be the least common multiple of the exponents n_i. By adjoining a primitive nth root of unity ζ, we obtain a normal radical extension $F(\zeta)$ of $K(\zeta)$. Note that $K(\zeta)$ contains all of the n_ith roots of unity for $i = 1, \ldots, m$. It follows from the fundamental theorem of Galois theory that $\mathrm{Gal}(E/K)$ is a factor group of $\mathrm{Gal}(F(\zeta)/K)$. Since any factor of a solvable group is again solvable, it suffices to show that $\mathrm{Gal}(F(\zeta)/K)$ is solvable.

Let $K(\zeta, u_1, \ldots, u_i) = F_i$ and let $\mathrm{Gal}(F(\zeta)/K) = G$, $\mathrm{Gal}(F(\zeta)/K(\zeta)) = N$, and $\mathrm{Gal}(F(\zeta)/F_i) = N_i$, for $i = 1, 2, \ldots, m - 1$. Since F_{i-1} contains all n_ith roots of unity, F_i is the splitting field of $x^{n_i} - a_i$ over F_{i-1}, for some $a_i \in F_{i-1}$. Therefore N_i is a normal subgroup of N_{i-1} and $\mathrm{Gal}(F_i/F_{i-1}) \cong N_{i-1}/N_i$ by the fundamental theorem. Furthermore, by Theorem 8.4.3, $\mathrm{Gal}(F_i/F_{i-1})$ is cyclic. Finally, N is normal in $\mathrm{Gal}(F(\zeta)/K)$, and $G/N \cong \mathrm{Gal}(K(\zeta)/K)$ is abelian by Proposition 8.4.2. The descending chain of subgroups

$$G \supseteq N \supseteq N_1 \supseteq \ldots \supseteq N_m = \{e\}$$

shows that $G = \mathrm{Gal}(F(\zeta)/K)$ is a solvable group.

To prove the converse, assume that the Galois group G of $f(x)$ over K is solvable, and let E be a splitting field for $f(x)$ over K. If $|G| = n$, let ζ be a primitive nth root of unity, and let $F = E(\zeta)$. Any element ϕ of $\mathrm{Gal}(E(\zeta)/K(\zeta))$ leaves ζ fixed, so $\phi(E) = E$ and the restriction of ϕ to E defines an element of $\mathrm{Gal}(E/K) = G$. Thus $\mathrm{Gal}(F/K(\zeta))$ is isomorphic to a subgroup of G and hence is solvable. By Proposition 7.6.2 there exists a finite chain of subgroups

$$\mathrm{Gal}(F/K(\zeta)) \supset N_1 \supset \ldots \supset \{1\}$$

such that each subgroup is normal in the one above it and the factor groups N_i/N_{i+1} are cyclic of prime order p_i. By the fundamental theorem of Galois theory there is a corresponding ascending chain of subfields

$$K(\zeta) \subset F_1 \subset \ldots \subset F$$

with $N_i = \mathrm{Gal}(F/F_i)$ and $\mathrm{Gal}(F_{i+1}/F_i) \cong N_i/N_{i+1}$. Since $p_i | n$ and F_i contains a primitive nth root of unity, it contains all p_ith roots of unity, and we can apply Theorem 8.4.4 to show that $f(x)$ is solvable by radicals over $K(\zeta)$, and hence over K. $\square$

This seems an appropriate point at which to remind the student that this fundamental result has motivated almost all of our work on groups and extension fields.

Theorem 7.7.2 shows that S_n is not solvable for $n \geq 5$, and so to give an example of a polynomial equation of degree n that is not solvable by radicals, we only need to find a polynomial of degree n whose Galois group over $\mathbf{Q}$ is S_n. We will give such an example of degree 5 over $\mathbf{Q}$, a special case of a more general construction.

Let m be a positive even integer, and let $n_1 < n_2 < \ldots < n_{k-2}$ be even integers, where k is odd and $k > 3$. Let

$$g(x) = (x^2 + m)(x - n_1)(x - n_2) \cdots (x - n_{k-2}).$$

It can be shown that the polynomial $f(x) = g(x) - 2$ has exactly two nonreal roots in $\mathbf{C}$, and if k is prime, then the Galois group of $f(x)$ over $\mathbf{Q}$ is S_k. For degree 5, one of the simplest cases is

$$f(x) = (x^2 + 2)(x + 2)(x)(x - 2) - 2 = x^5 - 2x^3 - 8x - 2.$$

To complete the proof that $f(x)$ has Galois group S_5, we need the following group theoretic lemma.

8.4.7 Lemma.   Any subgroup of S_5 that contains both a transposition and a cycle of length 5 must be equal to S_5 itself.

Proof. By renaming the elements of S_5 we may assume that the given transposition is $(1,2)$. We can then replace the cycle of length 5 with one of its powers to obtain $(1,2,a,b,c)$, and then we can again rename the elements so that we may assume without loss of generality that we are given $(1,2)$ and $(1,2,3,4,5)$. We have $(1,2)(1,2,3,4,5) = (2,3,4,5)$, and conjugating $(1,2)$ by powers of

(2,3,4,5) gives (1,3), (1,4), and (1,5). Then it follows from the formula $(1,n)(1,m)(1,n) = (m,n)$ that any subgroup of S_5 that contains the two given elements must contain every transposition.   □

8.4.8 Theorem.   There exists a polynomial of degree 5 with rational coefficients that is not solvable by radicals.

Proof. Let $f(x) = x^5 - 2x^3 - 8x - 2$. Then $f'(x) = 5x^4 - 6x^2 - 8$, and the quadratic formula can be used to show that the solutions of $f'(x) = 0$ are $x^2 = 2$, $-4/5$, yielding two real roots. Then $f(x)$ has one relative maximum and one relative minimum, and since the values of $f(x)$ change sign between -2 and -1, between -1 and 0, and between 2 and 3, it must have precisely three real roots.

By Theorem 8.3.10 (the fundamental theorem of algebra) there exists a splitting field F for $f(x)$ with $F \subseteq \mathbf{C}$. The polynomial $f(x)$ is irreducible by Eisenstein's criterion, and so adjoining a root of $f(x)$ gives an extension of degree 5. By the fundamental theorem of Galois theory, the Galois group of $f(x)$ over $\mathbf{Q}$ must contain a subgroup of index 5, so since its order is divisible by 5, it follows from Cauchy's theorem that it must contain an element of order 5. By Proposition 8.1.4, every element of the Galois group of $f(x)$ gives a permutation of the roots, and so the Galois group is easily seen to be isomorphic to a subgroup of S_5. This subgroup must contain an element of order 5, and it must also contain the transposition that corresponds to the element of the Galois group defined by complex conjugation. Therefore, by the previous lemma, the Galois group must be isomorphic to S_5. Applying Theorem 8.4.6 completes the proof, since S_5 is not a solvable group.   □

EXERCISES: SECTION 8.4

1. Show that $2x^5 - 10x + 5$ is irreducible over $\mathbf{Q}$ and is not solvable by radicals.

2. The determinant given below is called a Vandermonde determinant. Show that

$$\begin{vmatrix} 1 & \zeta_1 & \zeta_1^2 & \cdots & \zeta_1^{p-1} \\ 1 & \zeta_2 & \zeta_2^2 & \cdots & \zeta_2^{p-1} \\ \vdots & \vdots & \vdots & & \vdots \\ 1 & \zeta_p & \zeta_p^2 & \cdots & \zeta_p^{p-1} \end{vmatrix} = \prod_{1 \leq i < j \leq p} (\zeta_j - \zeta_i).$$

Hint: Use induction. To make the inductive step, start from the right and subtract from each column ζ_p times the column to its left.

3. Let H be a subgroup of S_p, where p is prime. Show that if H contains a transposition and a cycle of length p, then $H = S_p$.

4. Prove that if $f(x) \in \mathbf{Q}[x]$ is irreducible of prime degree and has exactly two nonreal roots in $\mathbf{C}$, then the Galois group of $f(x)$ over $\mathbf{Q}$ is S_p.

Hint: Show that complex conjugation gives a transposition in the Galois group, and apply the previous exercise.

Bibliography

Abstract Algebra

BIRKHOFF, G., AND S. MACLANE, *A Survey of Modern Algebra* (4th ed.). New York: Macmillan Publishing Co., Inc., 1977.

CLARK, A., *Elements of Abstract Algebra.* New York: Dover Publications, Inc., 1984.

FRALEIGH, J., *A First Course in Abstract Algebra* (4th ed). Reading, Mass.: Addison-Wesley Publishing Co., 1989.

GOLDSTEIN, L. J., *Abstract Algebra: A First Course.* Englewood Cliffs, N. J.: Prentice-Hall, 1973.

HERSTEIN, I. N., *Abstract Algebra.* New York: Macmillan Publishing Co., Inc., 1986.

———, *Topics in Algebra* (2nd ed.). New York: John Wiley & Sons, Inc., 1973.

HUNGERFORD, T., *Algebra.* New York: Springer-Verlag New York, Inc., 1974.

JACOBSON, N. *Basic Algebra I* (2nd ed.). San Francisco: W. H. Freeman & Company Publishers, 1985.

LANG, S., *Algebra* (2nd ed.). Reading, Mass.: Addison-Wesley Publishing Co., Inc., 1984.

SHAPIRO, L. W., *Introduction to Abstract Algebra.* New York: McGraw-Hill Book Company, 1975.

VAN DER WAERDEN, B. L., *Algebra* (7th ed.). vol. 1. New York: Frederick Unger Publishing Co., Inc., 1970.

Linear Algebra

HERSTEIN, I. N. AND D. J. WINTER, *Matrix Theory and Linear Algebra.* New York: Macmillan Publishing Co., Inc., 1988.

HOFFMAN, K. AND R. KUNZE, *Linear Algebra* (2nd ed.). Englewood Cliffs, N. J.: Prentice-Hall, 1971.

Number Theory

ADAMS, W. W. AND L. J. GOLDSTEIN, *Introduction to Number Theory*. Englewood Cliffs, N. J.: Prentice-Hall, 1976.

HARDY, G. H., AND E. M. WRIGHT, *The Theory of Numbers* (5th ed.). Oxford, England: Oxford University Press, 1979.

NIVEN, I. *Irrational Numbers* (Carus Mathematical Monograph No. 11). Washington, D. C.: The Mathematical Association of America, 1956.

Foundations

COHEN, L. AND G. EHRLICH, *The Structure of the Real Number System*. Princeton, N.J.: D. Van Nostrand Company, 1963.

LANDAU, E., *Foundations of Analysis* (2nd ed.). New York: Chelsea, 1960.

Theory of Equations

DICKSON, L. E., *New First Course in the Theory of Equations*. New York: John Wiley & Sons, Inc., 1939.

USPENSKY, J. V., *Theory of Equations*. New York: McGraw-Hill Book Company, 1948.

Group Theory

HALL, M., *The Theory of Groups*. New York: Macmillan Publishing Co., Inc., 1959.

ROTMAN, J. J., *An Introduction to the Theory of Groups* (3rd ed.). Boston, Mass.: Allyn & Bacon, Inc., 1984.

Ring Theory

HERSTEIN, I. N., *Noncommutative Rings* (Carus Mathematical Monographs No. 15). Washington, D. C.: The Mathematical Association of America, 1968.

LAMBEK, J., *Lectures on Rings and Modules*. New York: Chelsea, 1976.

Field Theory

ADAMSON, I. T., *Introduction to Field Theory*. Edinburgh: Oliver & Boyd, 1964.

ARTIN, E., *Galois theory* (2nd ed.). Notre Dame Mathematical Lectures No. 2. Notre Dame, Ind., 1959.

LIDL, R. AND H. NIEDERREITER, *Finite Fields*. Reading, Mass.: Addison-Wesley Publishing Co., Inc., 1983.

History

VAN DER WAERDEN, B. L., *A History of Algebra*. New York: Springer-Verlag New York, Inc., 1985.

Index